KT-592-684

Social Behaviour in Farm Animals

Edited by

L.J. Keeling
Department of Animal Environment and Health
Swedish University of Agricultural Sciences
Skara
Sweden

and

H.W. Gonyou
Prairie Swine Centre
Saskatoon
Canada

CABI *Publishing*

CABI *Publishing* **is a division of CAB** *International*

CABI Publishing
CAB International
Wallingford
Oxon OX10 8DE
UK

Tel: +44 (0)1491 832111
Fax: +44 (0)1491 833508
Email: cabi@cabi.org
Web site: http://www.cabi.org

CABI Publishing
10 E 40th Street
Suite 3203
New York, NY 10016
USA

Tel: +1 212 481 7018
Fax: +1 212 686 7993
Email: cabi-nao@cabi.org

© CAB *International* 2001. All rights reserved. No part of this publication may
be reproduced in any form or by any means, electronically, mechanically, by
photocopying, recording or otherwise, without the prior permission of the
copyright owners.

A catalogue record for this book is available from the British Library, London, UK.

Library of Congress Cataloging-in-Publication Data
Social behavior in farm animals / edited by L.K. Keeling and H.W. Gonyou.
    p.   cm.
  Includes bibliographical references (p.).
  ISBN 0-85199-397-4 (alk. paper)
    1. Domestic animals--Behaviour. 2. Social behavior in animals. I. Keeling, L.K.
(Linda K.) II. Gonyou, H.W. (Harold W.)

SF756.7.S58 2001
636′.001′5915--dc21   00-056479

ISBN 0 85199 397 4

Typeset by AMA DataSet Ltd, UK.
Printed and bound in the UK by Biddles Ltd, Guildford and King's Lynn.

Contents

LIVERPOOL
JOHN MOORES UNIVERSITY
AVRIL ROBARTS LRC
TEL. 0151 231 4022

Contributors

Anders Alanärä, *Department of Aquaculture, Swedish University of Agricultural Sciences, 90183 Umea, Sweden. Anders.Alanara@ vabr.slu.se*

Alain Boissy, *Unité de Recherches sur les Herbivores, INRA, 63122 Saint-Genès-Champanelle, France. aboissy@clermont.inra.fr*

Marie-France Bouissou, *Laboratoire d'Etude du Comportement Animal, INRA, 37380 Nouzilly, France. bouissou@tours.inra.fr*

Eva Brännäs, *Department of Aquaculture, Swedish University of Agricultural Sciences, 90183 Umea, Sweden. Eva.Brannas@vabr. slu.se*

Anne Marie de Passillé, *Dairy and Swine Research and Development Centre, Agriculture and Agri-Food Canada, Lennoxville, Quebec J1M 1Z3, Canada. depassilleam@em.agr.ca*

Ian J.H. Duncan, *Department of Animal and Poultry Science, University of Guelph, Guelph, Ontario N1G 2W1, Canada. Iduncan@APS.UoGuelph.ca*

Hans Erhard, *Macaulay Land Use Research Institute, Aberdeen AB15 8QH, UK. h.erhard@mluri. sari.ac.uk*

Andrew Fisher, *AgResearch, Ruakura Agricultural Research Centre, PB 3123, Hamilton, New Zealand. fishera@agresearch.cri.nz*

Harold W. Gonyou, *Prairie Swine Centre, POB 21057, Saskatoon, Saskatchewan S7H 5N9, Canada. gonyou@usask.ca*

Suzanne Held, *Department of Clinical Veterinary Science, University of Bristol, Langford House, Langford, Bristol BS40 5DU, UK. suzanne.held@bristol.ac.uk*

Per Jensen, *Department of Animal Environment and Health, Section of Ethology, Swedish University of Agricultural Sciences, POB 234, 532 23 Skara, Sweden. Per.Jensen@hmh.slu.se*

Linda Keeling, *Department of Animal Environment and Health, Section of Ethology, Swedish University of Agricultural Sciences, POB 234, 532 23 Skara, Sweden. Linda.Keeling@hmh.slu.se*

Pierre Le Neindre, *Unité de Recherches sur les Herbivores, INRA, 63122 Saint-Genès-Champanelle, France. pln@clermont.inra.fr*

A. Cecilia Lindberg, *Department of Clinical Veterinary Science, University of Bristol, Langford House, Langford, Bristol BS40 5DU, UK. a.c.lindberg@bristol.ac.uk*

Carin Magnhagen, *Department of Aquaculture, Swedish University of Agricultural Sciences, 90183 Umea, Sweden. Carin.Magnhagen@vabr.slu.se*

Lindsay Matthews, *AgResearch, Ruakura Agricultural Research Centre, PB 3123, Hamilton, New Zealand. matthewsl@agresearch.cri.nz*

Joy Mench, *Department of Animal Science, University of California, Davis, CA 95616, USA. jamench@ucdavis.edu*

Michael Mendl, *Department of Clinical Veterinary Science, University of Bristol, Langford House, Langford, Bristol BS40 5DU, UK. mike.mendl@bristol.ac.uk*

Suzanne T. Millman, *Humane Society of the United States, 2100 L Street NW, Washington, DC 20037, USA. smillman@hsus.org*

Lene Munksgaard, *Department of Health and Welfare, Danish Institute of Agricultural Science, Research Centre Foulum, PO Box 50, Tjele, Denmark. lene.munksgaard@agrsci.dk*

Ruth Newberry, *Center for the Study of Animal Well-being, Department of Veterinary and Comparative Anatomy, Pharmacology and Physiology and Department of Animal Sciences, Washington State University, Pullman, WA 99164-6520, USA. rnewberry@wsu.edu*

Jeff Rushen, *Dairy and Swine Research and Development Centre, Agriculture and Agri-Food Canada, Lennoxville, Quebec J1M 1Z3, Canada. rushenj@em.agr.ca*

Willem Schouten, *Department of Animal Husbandry, Wageningen University, PO Box 338, 6700 AH Wageningen, The Netherlands. Willem.Schouten@Etho.vh.wau.nl*

W. Ray Stricklin, *Department of Animal and Avian Sciences, University of Maryland, College Park, MD 20742, USA. Ws31@umail.umd.edu*

Janice Swanson, *Department of Animal Sciences and Industry, 134C Weber Hall, Kansas State University, Manhattan, KS 66506, USA. jswanson@oznet.ksu.edu.*

Hajime Tanida, *Department of Animal Science, Faculty of Applied Biological Science, Hiroshima University, 1-4-4 Kagamiyama, Higashi-Hiroshima 724, Japan. htanida@hiroshima-u.ac.jp*

Isabelle Veissier, *Unité de Recherches sur les Herbivores, INRA, 63122 Saint-Genès-Champanelle, France. veissier@clermont.inra.fr*

Natalie K. Waran, *Institute of Ecology and Resource Management, The University of Edinburgh, West Mains Road, Edinburgh EH9 3JG, UK. nwaran@srv0.bio.ed.ac.uk*

Acknowledgements

The editors would like to thank the following people for the help they gave either to the authors directly or by acting as referees for the chapters in this book: Jack Albright, Thierry Boujard, Öje Danell, Nell Davidson, David Fraser, Pete Goddard, Paul Hemsworth, Kathy Houpt, Per Jensen, Anne Larsen, Don Lay, Linda Mars, Mike Mendl, Neil Metcalfe, John Milne, Christine Nicol, Chris Sherwin, Angela Sibbald, Joe Stookey, Bob Wootton and Iain Wright.

Introduction

Why would people want to publish a book on social behaviour in farm animals? That is to say, why a book dealing only with the social behaviour of farm animals and not all aspects of farm animal behaviour? Or why focus only on farm animals and not on the social behaviour of all animal species: insects, reptiles, amphibians, birds and mammals included? The answer is simple: we feel there is a need for it. There are no other current books on this topic and there is already so much work in the area that the difficulty in editing this book has been in choosing what we must exclude, not what we should include. Of course all classic ethology textbooks include a chapter on social behaviour, but, by directing the book at the social behaviour of farm animals, we have given ourselves the opportunity to examine a whole new dimension. In nature animals choose their own social groups, but in agriculture it is we humans who select which individuals are housed together in the same group. How we group them and how the individuals respond have major implications for the management, productivity and welfare of these animals.

So, what do we mean by the words 'social behaviour'? The term can be considered in broad or narrow contexts. *Webster's Dictionary* (1980) defines social as, 'tending to form cooperative and interdependent relationships' and 'living and breeding in more or less organized communities'. Wilson (1980) states that the essential criterion of a society is 'reciprocal communication of a cooperative nature, transcending mere sexual activity'. In attempting to develop topics for various chapters, our thoughts have perhaps more closely reflected those of Banks and

©CAB *International* 2001. *Social Behaviour in Farm Animals*
(eds L.J. Keeling and H.W. Gonyou)

Heisey (1977), who stated that, 'Social behaviour is comprised of those patterns of behaviour that involve two or more members of a species and includes aggression and spacing, reproduction, parental care or aid-related behaviour, and social organization. In almost all cases, social transactions involve communication.' We have chosen to include most of these aspects of social behaviour in this book. As Wilson suggests, we have not included sexual activity as a major class of social behaviour. McFarland (1982) describes a social interaction as, 'where each individual influences the behaviour of the other'. If we were to coin a statement defining social behaviour, it would be that behaviour which is influenced by the presence or absence of another individual.

We said at the start of this introduction that by focusing on farm animals we had the opportunity to combine information of the type usually found in classical ethology textbooks with information usually only found in the applied ethology literature. That is exactly what we have done. Part I of the book deals with concepts in social behaviour and Part II concentrates on species-specific animal behaviour. However, even that combination would have been incomplete, so we created Part III, where we take up what we have called contemporary topics in social behaviour.

No matter how practical or applied the topic, we are of the firm opinion that you cannot really appreciate the behaviour of domesticated farm animals if you do not have at least some basic understanding of the principles underlying all animal behaviour. It would be like trying to run before you could walk. The first three chapters in this book, therefore, take up the essentials in social behaviour and form the basis for the following chapters. Our authors have not assumed any previous knowledge of animal behaviour and have given examples from farm animals whenever possible. The fourth chapter in this section deals with domestication. The study of farm animal behaviour came along somewhat later than studies on wild animals. A contributing factor in this was the assumption that the behaviour of domestic animals was for some reason less interesting. The fact that you are reading this, we hope, indicates that you do not subscribe to that false view of farm animals.

The middle and main part of the book presents the social behaviour of six farm animal species (or groups of species): cattle, pigs, domestic birds, sheep, horses and fish. Here each chapter has the same basic structure. There is a section on the basic social characteristics of the species. Following the same argument as we used to include the section on concepts in social behaviour, we believe that to understand species-specific behaviour you have to know the evolutionary pressures that have shaped it. But it is of course also necessary to know how present-day constraints and influences affect an animal's behaviour. The section on social behaviour under commercial conditions, therefore, takes up common husbandry practices and their effects on production, and leads logically into the final section of each chapter dealing with social effects on management and welfare. If you are surprised that we included horses and fish in our list of farm animals, then a quick glance through those chapters will convince you of their similarities to the other species.

As the title 'Contemporary topics in Social Behaviour' implies, the final part of the book presents relatively new or controversial research areas. The topics we have chosen to include, however, are ones that we felt are sufficiently well accepted and documented to warrant inclusion in a textbook. Breaking bonds is perhaps not a new topic, but it is one that is sorely neglected. Since this book deals with establishing and maintaining relationships between individuals, it is inevitable that we take up the consequences of breaking them up. The next three chapters reflect society's changing attitudes to animals. Accepting that individuals differ and that animals may have what in everyday language is called a personality also affects the way we treat them. Thus chapters on individual differences, the man–animal relationship, and animal cognition and consciousness seem a fitting way to end Part III.

The book is intended to serve as a reference book for many of its readers. We hope that the extensive reference lists at the end of each chapter will be useful if you decide to pursue a topic further. In addition, we have attempted to integrate the different chapters by means of cross-referencing. We are aware that many people already established in their interests will jump directly to 'their' species. We hope that they will follow the references to other chapters in the book. We have also organized the species chapters somewhat similarly, in order that readers can easily find similar topics across species.

While the main theme in this book is social behaviour, there are two other topics that are worth mentioning specifically in this introduction, and these are animal management and animal welfare. The management of animals is becoming increasingly specialized, demanding a greater knowledge and expertise on a wider variety of topics than it did only decades ago. Not only does a farm manager nowadays need to have a grasp of the basics and an appreciation of the potential benefits of nutrition, genetics and veterinary medicine, we argue that he or she also needs the same level of understanding of the discipline of animal behaviour. In times of increasing economic pressure, such as these, farm managers need to use all the resources at their disposal to maximize the potential of their animals.

The other topic that runs through the book is that of animal welfare. Animal behaviour and animal welfare are different disciplines and should not be confused. Research and knowledge on behaviour can help in animal production without necessarily involving welfare and there are many examples of this in the book. On the other hand, ethologists have made major contributions to several topics related to animal welfare, many of which are taken up in this book. We need only point to the number of times the words 'aggression' and 'behavioural problems' are mentioned throughout this book to emphasize this. Even though farm animals choose to be in groups when given a choice, often the way we manage them leads to conflicts and social stress. This need for improved knowledge in the area of social behaviour is going to be even greater if the current trend away from confinement systems and towards loose housing continues. The aim should be to help the animals use their social behaviour to make the groups function successfully for them and for us.

We hope this book will be useful and helpful in a number of ways. For undergraduate students studying animal behaviour, whether it be in biology, agriculture or veterinary science, we hope that this textbook will give structure and information relevant to your course. For postgraduate students and researchers we hope that the book will act as a good introduction to areas related to your subject and act as a valuable resource, leading you quickly to key references in an area. Finally, we hope that all people interested in animals can find something new and relevant to them and that we have managed to awaken an interest in the social behaviour of farm animals.

References

Banks, E.M. and Heisey, J.A. (1977) *Animal Behavior.* Educational Methods, Chicago, pp. 111–112.

McFarland, D. (1982) *The Oxford Companion to Animal Behavior.* Oxford University Press, New York, 517 pp.

Webster (1980) *Webster's New Collegiate Dictionary.* G. & C. Merriam Company, Springfield, Massachusetts, 1094 pp.

Wilson, E.O. (1980) *Sociobiology: the Abridged Edition.* The Belknap Press of Harvard University Press, Cambridge, Massachusetts, 7 pp.

Concepts in Social Behaviour

I

In this section we want to give a general introduction to the basic biological theories and concepts of social behaviour. Although this book deals with farm animals, the principles underlying their behaviour are the same as for all animals. These chapters are therefore intended to form the foundation for the detailed information contained in the species-specific chapters that follow in Part II.

Basic concepts in behaviour are based on the theory of natural selection and the assumption that evolution of adaptive behaviour is no different to the evolution of adaptive physical characteristics. That is to say, behaviour that improves the survival of an individual and its relatives will spread in the population. How this process led to certain species living in groups and how the structure and social organization of these groups came to be as they are today are the basis of the first chapter.

Since all our commonly used farm animals live in groups, then the next important question becomes how do these groups function and what are the relationships between individuals within a group? This is the focus of the second chapter in this section. As previously, this chapter addresses the social dynamics of group life from its basic principles such as formation and maintenance of the dominance hierarchy and communication between individuals.

Social behaviour can be broad, including all types of interactions between two or more individuals, or narrow, including only specific behavioural interactions between individuals belonging to the same social group. For reasons explained in the general introduction, we have not included a chapter on sexual behaviour, but we could not produce a book on social behaviour and fail to include a chapter on the very special social relationship that occurs between parents and their offspring. The third chapter in this section covers all aspects from the formation of the mother–young bond to the difficulties of parent–offspring conflict.

The final chapter in this section on basic concepts deals with domestication. The first three chapters include many examples from wild animal species, because researchers choose the most appropriate species to test out their hypotheses and theories. Although our authors have given references to farm animal species whenever possible, it is inevitable after reading these chapters that the reader questions whether the principles that are presented can really apply to cows, pigs, chickens, etc., that have been domesticated for many thousands of years. This last chapter tackles these issues and tries to synthesize views of how domestication really works and what it has changed in our modern farm animals and what it has not.

It is our opinion that people who have read and digested these basic concepts have the essential ethological framework. One could say these principles are the bones of applied ethology.

Living in Groups: an Evolutionary Perspective

Michael Mendl and Suzanne Held

Department of Clinical Veterinary Science, University of Bristol, Bristol BS40 5DU, UK

(Editors' comments: This chapter deals with why animals live in groups at all and it addresses this question from the evolutionary point of view, by examining the costs and benefits of group life. The concepts of selfish individuals, kin selection and inclusive fitness are all explained in relation to competitive and cooperative behaviour between group members.

The costs and benefits of group life are usually associated with finding food and avoiding predation. For example, many individuals working together may be more likely to find food, which is a benefit, but when they do find it then there are more individuals between whom it must be divided, and that is a cost. More individuals may be more likely to detect an approaching predator and the probability of an individual being the victim is less in a large group. But then again, larger groups are more conspicuous, so attracting more predator attacks. For group life to be favoured, its overall benefits must outweigh its costs, and the resulting fitness pay-off to each group member must be greater than for a solitary individual. It seems that for most of our farm animal species the odds have favoured living in groups. Nevertheless the distribution of resources, food, mates, etc., changes over time, so one can therefore expect the costs and benefits of living in a group to vary. Thus groups are dynamic, changing in size and structure. The optimal group size and structure at one time of year may not be optimal at another. Nevertheless, knowledge gained using this evolutionary approach helps us to understand and predict farm animal social behaviour and the authors conclude this chapter by presenting practical ways such information can be used to guide animal husbandry design.)

©CAB *International 2001. Social Behaviour in Farm Animals*
(eds L.J. Keeling and H.W. Gonyou)

1.1   Introduction

In environments which give animals the opportunity to range freely and to adopt patterns of dispersion and social organization freed from the constraints of captivity, farm animal species will live in groups for at least part of their lives (e.g. chickens, Wood-Gush *et al.*, 1978; sheep, Lawrence and Wood-Gush, 1988; pigs, Stolba and Wood-Gush, 1989; cattle, Howery *et al.*, 1996). The same is true for their ancestral species (e.g. Collias and Collias, 1967; Mauget, 1981; see Clutton-Brock, 1987). Group structures may range from loose aggregations of large numbers of individuals (e.g. herds or flocks) to tight-knit groups in which well developed social relationships are evident (e.g. family groups, harems). Within any one species, different forms of group structure may be evident at different times of the year and in different age/sex classes (e.g. pigs, Mauget, 1981; sheep, Lawrence, 1990), and individuals may also adopt solitary lifestyles at some stages of the annual or life cycles (see Lott, 1991). Group life thus varies both within and between species.

One consequence of life in groups is the development of social interactions and relationships between group members. The nature of this social behaviour and its species-typical characteristics are the subject of the rest of this book. In this chapter, we examine why it is that animals live in groups at all. We address this question from an evolutionary perspective by asking what the benefits and costs of group life are, thus giving us clues as to why it evolved. We also examine how evolutionary theory can help to explain the diversity of group structures between and within species, and the balance of competitive and cooperative behaviour that occurs between group members. Where we describe animals as choosing between different courses of action or assessing their environment, we use this as a convenient short-hand without implying anything about the animals' intentions, thought or awareness.

How relevant is an evolutionary approach to understanding the social organization of domesticated species? The process of domestication is tackled in detail in Chapter 4, and we can only address it briefly here. For the majority of the evolutionary history of domesticated species, the main factor influencing evolutionary adaptive change has been natural selection. For most domesticated species the actions of man and the process of artificial selection have only been evident for the last 10,000–15,000 years (Clutton-Brock, 1987), a brief moment in evolutionary time. Nevertheless, domestication does appear to have had effects on brain structure and various aspects of behaviour. Hemmer (1990) suggests that domesticated species have smaller brains, are generally less active, have weaker alarm reactions and are socially more tolerant of others than related wild species.

Despite these changes, the relatively short period of domestication means that modern farm animal species almost certainly retain a strong evolutionary legacy of their naturally selected past (cf. Newberry, 1993). Price (1984) argues that although domestication may have altered the threshold and frequency at which some behaviour patterns are expressed, the basic social characteristics of domestic animals remain similar to those of their wild conspecifics or ancestral species. Free-living domestic pigs adopt a similar social organization to that observed in the European wild boar (Wood and Brenneman, 1980; Mauget, 1981) (see Section 6.1). The same seems to be true for domestic fowl and the ancestral red jungle fowl (Collias and Collias, 1967; Wood-Gush *et al.*, 1978) (see Section 7.1). It also seems likely that domestic animals retain a general propensity to behave in ways that increase their chances of survival and reproductive success; most farm animals compete for resources such as food and mates and can reproduce successfully. The underlying rules organizing their behaviour are thus likely to be similar to those of related wild species. Therefore, we suggest that an evolutionary approach can provide species-specific information about group structure which may still have relevance to domesticated species. Furthermore, it provides a theoretical framework predicting general principles of behavioural organization and function which we believe can help us understand and interpret the social behaviour of farm animals (see also Fraser *et al.*, 1995; Spinka and Algers, 1995), and can even suggest methods for improving their care and management (e.g. Mendl, 1994; Mendl and Newberry, 1997).

We start the chapter by outlining basic principles of natural selection as they relate to group living. We then consider the benefits and costs of group life. The rest of the chapter addresses the general issue of group structure. Is there an 'optimal group size'? How flexible are group structures, and under what conditions do we expect individuals to join or leave groups? To what extent does natural selection favour cooperative and competitive behaviour within groups? What influence does the species' mating system have on group organization? Much of this chapter introduces theoretical ideas based on studies of wild animals, so we conclude by considering briefly the usefulness and limitations of the evolutionary approach to understanding farm animal social behaviour.

1.2 Basic Principles: Natural Selection and Group Life

1.2.1 Natural selection and behaviour

In wild animal populations, the genotypes which are to contribute to the next generation are selected naturally. Individuals vary in their genetic characteristics, and this typically leads to variation in phenotypical characteristics. Consequently some individuals are better able than others to thrive and reproduce in the environmental conditions prevailing during their lifetimes. More of their genes are passed on to the next generation than those of their competitors. Natural selection, thus, works by favouring certain genotypes over others.

It is easy to see how this leads to the evolution of adaptive morphological and physiological characteristics: genes code for proteins, which interact to shape an animal's form and physiological function. The evolution of adaptive behaviour is no different in principle. As Krebs and Davies (1993) put it:

> natural selection can only work on genetic differences and so for behaviour to evolve (a) there must be, or must have been in the past, behavioural alternatives in the population, (b) the differences must be, or must have been, heritable; in other words a proportion of the variation must be genetic in origin, and (c) some behavioural alternatives must confer greater reproductive success than others.

Points (a) and (c) are easy to accept, but the relationship between genes and behaviour (point b) is more difficult to understand (see Oyama, 1985). We cannot discuss this complex issue here, but the important point for our purposes is that many studies, including studies of farmed animals, have shown that an animal's genetic make-up can influence its behaviour. For example, Dwyer and Lawrence (1998) demonstrated that genotype differences account for the observed differences in the maternal and neonatal behaviour of Suffolk and Blackface sheep. In a cross-fostering experiment, Sinclair *et al.* (1998) showed that maternal genotype appeared to have a strong influence on pig maternal behaviour. It would be wrong to conclude from these and other examples that genes alone determine an animal's behaviour. Behaviour is influenced by the environment as well as by the animal's genetic make-up (for full discussions see Oyama, 1985; McFarland, 1993; Manning and Dawkins, 1998). Nevertheless, it is clear that genes do contribute to differences in behaviour between individuals, although exactly how their contribution is expressed may depend on the development and current environment of each individual.

Behaviour thus can evolve through the process of natural selection acting on heritable behavioural characteristics. Those characteristics which enable individuals to contribute the most reproducing offspring to the next generation will spread through the population. Consequently, natural selection should result in the evolution of animals who employ behaviour patterns and make behavioural decisions which are most likely to maximize their chances of survival and lifetime reproductive success, often referred to as their 'fitness' (see Dunbar, 1982). Indeed, the language of economics, the 'costs', 'benefits' and 'pay-offs' of behavioural decisions, has come to characterize much research in behavioural ecology. Animals are expected to make proximate behavioural decisions which generate the highest fitness benefits for the lowest fitness costs, thus maximizing their fitness pay-offs. There are of course many problems in demonstrating that this is the case (e.g. how do you measure fitness consequences of single behavioural decisions?), and many reasons for not expecting fitness maximizing or 'optimal' behaviour (e.g. there may be limits to the animal's ability to assess fitness consequences of proximate decisions, behavioural changes in the population may lag behind recent changes in selection pressures). For an excellent discussion of the optimality concept, see Dawkins (1995). Nevertheless, the language of costs and benefits provides a useful shorthand with which to convey the logic behind theoretical predictions about animal behaviour and behavioural decisions and, as such, we will use it throughout this chapter.

1.2.2 Kin selection and cooperation between relatives

Having considered the general principles underlying the evolution of adaptive behaviour, we now turn our attention to the particular case of group living. Group living poses a particular problem for the application of evolutionary theory to behaviour, although it is a characteristic of many farmed species and their undomesticated relatives. Group living is often facilitated by cooperation between individuals. We therefore start by looking at the conditions under which cooperative behaviour might evolve before going on to identify some of the costs and benefits associated with group life.

If evolution results in 'selfish' individuals attempting to maximize their own reproductive success, why do animals appear to cooperate with each other at all? The reason is that producing large numbers of offspring is not the only way in which individuals can make a genetic contribution to following generations. They can also do so indirectly through increasing the chances of survival and reproduction of relatives who share a certain proportion of their genes. By helping sisters, brothers or other relatives to survive and raise their offspring, an animal increases the probability of shared copies of its own genes being passed on to the future (Fig. 1.1). The benefit to the reproduction chances of the helped relative, weighted by the probability that the relative shares the helper's genes, has to outweigh the cost to the helper's own reproductive success incurred through helping (formalized in 'Hamilton's rule'; Hamilton, 1964). Only then will helping or 'altruistic' behaviour towards relatives be favoured over 'selfish' behaviour. This type of selection, by which traits are promoted that increase the survival and reproduction of close relatives in

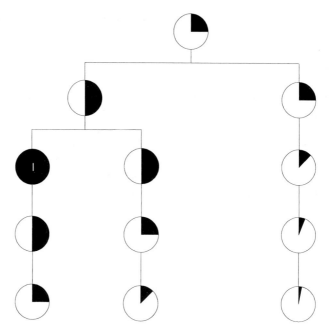

Fig. 1.1. Proportion of genes belonging to individual 'I' which are shared by various relatives. Black slices indicate the proportion, but not the identity, of gene copies which relatives share with 'I'.

addition to those of direct offspring, is referred to as 'kin selection' (as introduced by Maynard Smith, 1964).

Hamilton (1964) also introduced the concept of an individual's 'inclusive fitness' as the currency for assessing the conditions under which kin selection might operate. 'Inclusive fitness' encompasses the individual's own number of reproducing offspring (minus its increase in offspring caused by being helped) as well as the number of offspring of any helped relative (for discussions on the uses and abuses of 'inclusive fitness' see Grafen, 1984, and Dawkins, 1995).

1.2.3   Cooperation between non-relatives

Why would animals help or cooperate with non-relatives? Under what conditions could cooperative behaviour between non-relatives evolve, or have evolved in the past? At first sight it would appear that helping a non-relative would only incur a cost to one's own survival and reproduction chances. However, this may not always be so. In some cases, cooperating, for example in the defence of a resource, directly increases the net survival and reproductive chances of all participants. Mutual help of this type is referred to as mutualism, and is selected for because the fitness of all parties is increased. In other cases, an animal might help another at a cost to itself but without receiving help in return until some time in the future (reciprocal altruism; Trivers, 1971). Helping without receiving any immediate return is easily exploited when helper and helped meet only once or a few times. The helped individual may cheat by defecting. Reciprocal altruism is thus more likely to evolve when animals live in stable social groups and meet repeatedly and over a long time period, that is when cheating now may incur a cost later. Thus, in both these cases and in the case of kin selection, the evolution of cooperative behaviour can be explained by positing that it results in increased inclusive fitness for the individuals involved. And, since cooperation is favoured under these conditions, such conditions also facilitate the evolution of cooperative group living.

1.2.4 Studying past and present benefits and costs of group life

So far, we have briefly introduced basic principles of evolutionary theory which show how natural selection can influence the evolution of behaviour and the conditions under which it favours cooperative behaviour. We now consider how we can investigate the fitness advantages of group living. This may allow us to identify current benefits and putative selection pressures which may have favoured group life in the past.

One approach is to compare the reproductive output or some other fitness measure between members of the same species or different but related species that show alternative behaviours. Differences in fitness pay-offs between solitary and group-living conspecifics, for example, might inform us about the advantages of group living in a given environment.

We can also learn about the selection pressures which may have led to the evolution of group life by comparing conspecifics or related species living in different environments. Jarman (1974) used this comparative approach in his classic study of social organization in wild African bovids. On the basis of correlated variations in body size, habitat type, feeding ecology and social organization of 74 species, he concluded that group size and structure in wild African bovids were determined by body size and habitat type. This approach is powerful, but shares its main weaknesses with all correlational methods; correlation does not necessarily mean causation, cause and effect may be obscured, and there may be underlying confounding variables (cf. Elgar, 1989). It is also possible that behavioural differences between species represent alternative adaptive peaks in that they are equally adaptive, successful responses to the same selection pressures rather than reflecting different pressures.

These problems are overcome to some extent by using an experimental approach in which one factor is tested at a time, either in a unifactorial design or a multifactorial design by controlling statistically for the effects of the other variables. Controlling for the effects of other variables, the fitness costs and benefits of different behaviours can be compared. To investigate, for example, whether grouping would be advantageous for individuals of a given species in a given environment, group sizes can be manipulated and some individual fitness parameters compared at the different group sizes (e.g. Penning *et al.*, 1993). Similarly, one can manipulate the environment and record the behavioural changes taking place over several generations to test whether one particular aspect of the environment exerted a major selection pressure. In one such experiment, Endler (1980) and Magurran *et al.* (1992) investigated the effect of reduced predation pressure on the morphology and behaviour of a population of guppies. Guppies were taken from one stream where they suffered very high predation risks and introduced into a predator-free stream. Removing predation as a selection pressure not only changed the guppies' outward appearance after a few generations, but also their behaviour: among other changes, the guppies became less likely to form schools, thereby indicating that predation had been a major selection pressure favouring schooling (see Section 10.1.3).

Thus, methods exist which allow us to investigate the role of various factors in the evolution of group life. However, we should remember that just because a behaviour appears to have a particular fitness value now, this does not mean that it was selected for this reason in the past. Similarly, a variable which appears to favour group life in the present need not necessarily have been an important selection pressure in evolutionary history.

1.3   Feeding Without Being Fed Upon: Benefits and Costs of Group Life

In the previous section, we saw that natural selection promotes behaviours which will maximize an individual's inclusive fitness. Animals that live in groups must gain fitness advantages which exceed those available from solitary life. In this section we attempt to identify some of the benefits and costs of group living, focusing in particular on its effects on foraging efficiency and predator avoidance to suggest two important selection pressures which may have favoured group living over solitary life during evolution.

1.3.1 Foraging advantages of group living

Increased efficiency in detecting food

An animal's chances of detecting food may increase with group size, for example by learning about the location of a food source by seeing others exploit it (local enhancement) (see Section 14.2). The reduction in search time resulting for each individual is thus an important benefit of group living particularly for animals relying on food that is distributed unpredictably in space or time. Fish, for example, find food faster in larger shoals as demonstrated in experiments on two species of naturally shoaling cyprinids (Pitcher *et al.*, 1982). Ward and Zahavi (1973) suggested that communal roosts and colonies in birds act as 'information centres'. Individuals find out about the location of feeding sites by following successful birds on the latter's next foray. This is less likely to apply to group living in animals such as grazing ungulates which feed on more predictably distributed food (O'Brien, 1988). Alternatively, colonies may act as recruitment centres for group foraging when the benefits to a successful forager of feeding in the company of others outweigh the costs of sharing the food (Richner and Heeb, 1996; Danchin and Wagner, 1997).

Increased efficiency in acquiring food

For an individual of a predator species, associating with conspecifics can be advantageous because it makes available prey items that the animal would be incapable of catching alone. Female lions, for example, have very low success rates when hunting large prey by themselves. They are, therefore, more likely to participate in group hunts of zebra or buffalo than of warthog for which the probability of a successful solitary hunt is one and a half times higher than for larger prey (Scheel and Packer, 1991). So group living may broaden the range of available prey. It may also increase an individual's chances of catching prey too elusive to be caught by a solitary hunter (e.g. group hunting in killer whales; Baird and Dill, 1996). In both cases, the benefit of group living decreases with increasing group size if food resources are limited.

Increased efficiency in defending food

Food is often distributed in such a way that individuals are unable to defend it on their own. In these circumstances, grouping may enable several individuals to successfully defend this resource together. For example, Davies and Houston (1981) found that pied wagtails may defend their territories either singly or allow a non-territory holder to share their patch, depending on the abundance of food. The food was distributed over such a large area that shared territory defence between the two birds doubled the percentage of intruders immediately spotted and chased off. But, given that the sharing non-territory holder consumed some of the food as well as defending it, it paid the territory holder to share only when food was abundant. Similarly, Schaller (1972) describes how groups of lions defend carcasses against hyena clans more successfully than solitary lions. Generally, where two species compete for a food resource, grouping in the species that would lose out in an inter-species contest can increase the chances of successful resource defence against individuals of the superior species (Pulliam and Caraco, 1984).

1.3.2 Avoiding predators: detection, dilution and defence

Group life may also have evolved in response to predation pressure. With more eyes scanning the environment, the chances of detecting a predator increase with group size. Group members can therefore lower their vigilance levels without increasing their predation risk as compared with solitary animals as long as the detecting individual signals the predator's presence to the rest of the group ('many eyes hypothesis'; Pulliam, 1973). Lowering vigilance levels, for example by scanning for predators less frequently, frees time for feeding. In groups of wild boar, for example, time spent on vigilance behaviour was found to decrease with group size (Quenette and Gerard, 1992) and in a study on Scottish Halfbred ewes, Penning *et al.* (1993) found that the proportion of time individuals spent grazing in a 24-h day increased as a function of group size as shown in Fig. 1.2. Since Pulliam's original model (1973) there have been numerous theoretical explorations and empirical studies of this group size effect (e.g. Elgar, 1989; McNamara and Houston, 1992; Roberts, 1996; Bednekoff and Lima, 1998).

Another way in which group life enhances the survival prospects of individuals relative to solitary life is by diluting their chances of being attacked by a predator. Assuming a predator can take one prey item per attack, an individual would halve its predation risk by joining another animal, and its chances of getting preyed on would generally decrease in proportion to group size. By clustering in groups, individual Carmargue horses, for example, reduce the number of flies attacking them: individuals in larger groups have fewer flies (Duncan and Vigne, 1979; cf. cattle: Schmidtmann and Valla, 1982). Aggregating in groups can also reduce predation risk by making it more difficult for the predator to single out and attack an individual, thereby reducing the predator's attack : kill ratio (confusion effect; see Krakauer, 1995).

Some social ungulates and colonial nesting birds provide good illustrations of the advantages of grouping in terms of communal predator defence. Musk oxen and water buffalo (Wilson, 1975), for example, fall into protective group formations around their young

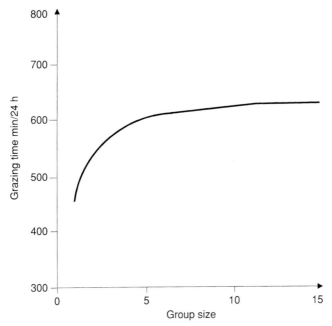

Fig. 1.2. Grazing time increases as a function of group size in sheep. (After Penning *et al.*, 1993.)

when faced by a predator. In this way, each individual benefits by decreasing the chances of losing any offspring in an attack. In eland, several adult females may attack large predators together, which increases their chances of driving them off (Hillman, 1987).

1.3.3 Thermal benefits

Group life can also aid thermal regulation in endotherms. Small homeotherms may conserve energy by huddling together. Male grey and fox squirrels, for example, form sleeping groups in winter although they hardly act cohesively away from the nest. During the warmer periods when the need to conserve thermal energy is lower, they sleep singly and interact agonistically (Koprowski, 1996).

1.3.4 Disadvantages of group living: competition, conspicuousness, contamination and cuckoldry

While an individual may benefit from group living by gaining increased access to resources there is also an obvious associated cost. The resource now has to be shared with the other group members.

Shared resource defence in both pied wagtails and lions illustrates this point as briefly summarized above (Schaller, 1972; Davies and Houston, 1981).

Grouping may benefit individuals through better predator detection, attack dilution and joint predator defence, but it also increases their conspicuousness. In shoaling fish, for example, visual conspicuousness rather than shoal size alone has been shown to be a key factor in determining predation risk (Krause and Godin, 1995). Conspicuousness can also work against groups of predators. In predator species feeding on mobile prey, for example, disturbance of the prey by group members can decrease the number of prey items caught by each individual (e.g. Amat and Rilla, 1994).

Individuals may not only decrease, but also increase their chances of parasite attack by grouping (e.g. prairie dogs: Hoogland, 1979; cliff swallows: Brown and Brown, 1986). Whether grouping is costly or beneficial in terms of probability of infection depends on the mode of parasite transmission: grouping increases an individual's chances of infection with contact-transmitted parasites and decreases its rate of infection with mobile parasites which do not rely on host proximity for transmission (Cote and Poulin, 1995).

Cuckoldry can be yet another cost of social living, especially to males. Colonial birds and fish, for example, may lose a significant number of fertilizations to extra-pair males (e.g. Jennings and Phillip, 1992; Møller and Birkhead, 1993).

1.3.5 Confusing the consequences of group living with selection pressures – and other pitfalls

Group living also facilitates the development of more complex social behaviours and interactions such as alloparenting, social learning and helper systems. This raises the question of which of these have played a part in the evolution of group life and which are a current utility or consequences of group living. Without any evidence, this is not an easy question to answer satisfactorily. But it emphasizes the general point made earlier that just because group life has certain advantages today, this does not necessarily mean that these were the benefits underlying its evolution. Overlooking confounding variables or other alternative explanations may lead to wrong conclusions about the evolutionary function of group life, especially where conclusions are drawn on the basis of correlations.

A comprehensive list of all possible costs and benefits has not been given here. However, it serves to point to some major selection pressures which may have favoured group life over solitary life. We discuss below in more detail how resource distribution acts as a selection pressure on group size and group structure. But we should remind ourselves at this point, that none of these factors work in isolation: group living in any species is likely to have resulted from the interaction of various selection pressures. And, as mentioned above, the evolution of group life is also facilitated by the fitness gains resulting from cooperating with relatives and non-relatives. For group life to be favoured, its overall benefits must outweigh its costs, and the resulting fitness pay-off to each group member must be greater than for a solitary individual.

1.4   Group Size and the Dynamic Nature of Groups

So far, we have considered, in general terms, the costs and benefits of group life. In the next sections, we examine how principles from evolutionary theory may help us to make general predictions about the sorts of group size and structure that animals, including farm animals, are likely to adopt. We start by considering whether group sizes are likely to be stable or dynamic. As we have seen, evolutionary theory predicts that animals will behave in ways which maximize their reproductive success. In the context of group life, we thus expect them to join groups (immigrate) when it benefits them to do so, and to leave groups (emigrate) when the pay-offs of solitary life exceed those of group living. Group size can thus be viewed as the result of the individual animal's decisions (Pulliam and Caraco, 1984). Clearly, the animals' behaviour is constrained by the extent to which they can assess the fitness pay-offs of these decisions, and such mechanisms of assessment have received relatively little investigation in behavioural ecology research (but see e.g. Bateson and Kacelnik, 1997). Furthermore, it is possible that these abilities have been somewhat blunted by domestication. Nevertheless, the general prediction suggests that group-living animals, including farm animals, should be predisposed to live in groups whose size and composition may change across time.

1.4.1   Resource distribution: ideal free distribution theory

What factors influence group size? A fundamental underlying factor is the way in which resources such as food or mates are distributed in time and space. An important theoretical framework which has been developed to predict how animals should distribute themselves around resources is ideal free distribution theory (Fretwell and Lucas, 1970). Its basic prediction is that animals should distribute themselves around resource sites such that each individual is able to acquire resources at the same rate, and cannot profit by moving to another site. It assumes that animals have complete information about the value of resource sites and are free to move between these sites unconstrained by each other.

These are clearly simplistic assumptions. In circumstances where resources arrive at each site at constant rates and animals thus have a good chance of obtaining complete information about the value of sites, there is some evidence that animals show an approximation of the predicted ideal free distribution. For example, if there are two equally productive sites and all animals are equally able competitors, then group sizes at the two sites should be equal (see Milinski and Parker, 1991). However, usually the situation is complicated by the fact that some individuals are more able competitors than others. In this case, ideal free distribution theory makes predictions based on the distribution of competitive ability rather than individual animals (Parker and Sutherland, 1986). Thus, an ideal free distribution could be achieved with a small number of highly competitive individuals and a large number of less competitive individuals grouping around two equally productive resources (see Harper, 1982). More complex scenarios include those in which resources are depleting rather than remaining constant. Despite its limitations, ideal free distribution theory shows how the way in which resources are distributed can affect the distribution of individuals within aggregations or groups.

1.4.2   Optimal group size

The concept of an 'optimal' group size is sometimes mentioned when discussing how farm animals should be housed (e.g. Stricklin *et al.*, 1995). In theory, as group size increases in free-ranging populations, there will come a point, the optimal group size, at which the fitness pay-off to each individual of being in a group (e.g. benefits of locating and defending resources – costs of competing with others) is maximized. In a natural population, it is unlikely that the optimal group size will be stable. This is because individuals in smaller groups will benefit from joining the optimally sized group, making it bigger and hence suboptimal once more. Individual decisions to join groups of different size can be modelled using ideal free distribution theory which shows that groups will indeed be larger than the predicted optimal size, and this is often observed in empirical studies (see Sibly, 1983; Pulliam and Caraco, 1984).

For captive animals, it is theoretically possible to calculate an optimal group size given known constraints such as space and resource availability, and an agreed way of assessing fitness. For example, the fitness pay-offs of being in the group may initially increase with group size due to factors such as the thermal benefits of huddling. They may then remain level ('optimal') until a certain group size is reached at which point within-group competition for space or resources increases and pay-offs decrease. From an animal production and welfare perspective, it is possible to try and assess optimal group size on the basis of welfare rather than, or as well as, fitness parameters (see Mendl, 1991, for discussion of the relationship between measures of fitness and welfare). For example, some studies have attempted to determine the maximum number of individuals that can be housed in a fixed space without obvious detrimental effects on productivity and welfare (e.g. Zayan, 1985; Weng *et al.*, 1998).

1.4.3 Resource distribution and defence

In theory, individuals should become more resistant to immigration into their group as the group reaches its optimal size, in order to preserve this group size. This prediction draws our attention to the fact that groups are only sometimes simple aggregations of individuals coming and going as they please. If the quality and location of resources has long-term stability, for example, food patches of differing productivity, then it may pay individuals to defend these from intruders when the benefits of resource defence (e.g. priority of access) outweigh the costs (e.g. energy and time expenditure, injury risk). The distribution of the resource can then influence how many individuals are required to defend it. If resources are tightly clumped and easily defendable by one individual, then resource defence by single individuals will be favoured. For example, during the breeding season male horses are able to defend groups of females from the attention of other males (Waring *et al.*, 1975). If resources are distributed more widely, individuals may not benefit from any form of resource defence, or may benefit by grouping together to defend them, as in the pied wagtail example given earlier (Davies and Houston, 1981). Kin relationships or familiarity may predispose particular individuals to group together to defend resources. Thus, defendable resources can encourage aggressive territorial behaviour and influence group size. This general principle is relevant when deciding how to distribute resources for captive animals (Fig. 1.3).

1.5 Within-group Dynamics: Effects of Dominance

We have seen how resource distribution can act as an important determinant of group size and how group size can be viewed as the outcome of individual decisions to join or to leave. Here, we consider how these decisions may depend on the individual's status in the group. Within

Fig. 1.3.   This electronic sow feeder system is a potentially defendable resource. Only one sow can access it at any one time. However, because the system dispenses only a limited amount of food to each sow (which the system recognizes according to the sow's individually identifiable transponder collar), there is little point in sows attempting to defend and re-enter the system once they have received their daily ration. (Photo M. Mendl.)

most groups, not all individuals will achieve the same fitness pay-offs. This may occur for a variety of reasons, but is particularly likely in groups which have a clear dominance structure. The concepts of dominance and subordination and their behavioural manifestations are discussed in detail in Chapter 2. Broadly speaking, dominant or high-ranking individuals are able to maintain priority of access to resources over subordinate, low-ranking individuals, and hence are likely to achieve better fitness pay-offs from group life. Consequently, high- and low-ranking animals may maximize their fitness pay-offs by living in groups of quite different size (Pulliam and Caraco, 1984). For example, high rankers may acquire more fitness benefits as group size increases, especially if their reproductive success is enhanced by the presence of other animals (Brown *et al.*, 1982), while low rankers are likely to incur more fitness costs as group size and within-group competition increase.

1.5.1   Competition or cooperation between dominant and subordinate group members

These rank-related differences in 'optimal group size' and fitness pay-offs can affect the amount of within-group competitive or cooperative behaviour that is observed, and hence may have implications for how captive groups should be structured. If high rankers impose too great a cost on low rankers, there will come a point at which the latter benefit from deserting the group. The question thus arises as to how much subordination should be tolerated by low rankers before they decide to leave. Similarly, when should high rankers 'encourage' them to stay, for example by ceding reproductive concessions to them? Theory suggests that under circumstances where the low ranker's fitness pay-offs from remaining in the group are relatively good (e.g. if it is closely related to the dominant), there are few opportunities for it to leave or be successful outside the group, and there is little chance of overthrowing the dominant, it should tolerate subordination and few reproductive opportunities (e.g. Vehrencamp, 1983; Emlen, 1997). In these sorts of group, which may be common in captivity where leaving groups is often not an option, overt exploitation and aggression by dominants is relatively unconstrained by the threat of emigration by subordinates.

Conversely, when the opposite circumstances are present, high rankers are predicted to actively induce subordinates to stay in the group, for example by allowing them reproductive opportunities (see Emlen, 1997; Pusey and Packer, 1997). Cooperative, concessionary behaviour by dominants may thus be observed in circumstances in which, for example, animals have the opportunity to leave the group. In free-living groups, there is limited evidence that dominants actively provide incentives (e.g. breeding opportunities) to subordinates to stay in groups, and it may well be that subordinates are able to stay and breed because their behaviour is not under complete control of the dominants (Clutton-Brock, 1998). Subordinates may also choose to stay because dominance and subordination are usually transient features of an animal's status and not lifetime characteristics. Thus, over a long period of time, the fitness pay-offs of individuals in a group may actually be quite similar if most of them experience periods as both low- and high-ranking animals (Clutton-Brock, 1988). In captivity, however, groups are usually not together long enough for this to outweigh the potential costs of subordination.

It is worth noting briefly here that different 'roles' within groups need not imply dominance–subordination relationships or differing reproductive success. Group members may adopt alternative tactics or strategies which have the same fitness pay-offs (see Dominey, 1984). For example, male coho salmon adopt two strikingly different growth and mating strategies. Small jack males sneak for matings which their fellow large hooknose males fight for, and both ways of behaving seem to have equal fitness pay-offs (Gross, 1985). In captivity it is possible that marked differences in within-group behaviour do, in fact, have similar fitness and welfare pay-offs (see Mendl and Deag, 1995).

1.6 Group Structure: the Influence of Mating Systems

Groups can take many different forms ranging from male/female pairs, to female groups with attendant males, to large mixed-sex or segregated-sex herds or flocks. Indeed, different farm animal species adopt different group structures. Why should this be? The species' ecology, resource distribution and the selection pressures already considered are obviously important determinants of group structure (e.g. Jarman, 1974; Wrangham, 1979). For example, a large group size may be a more effective anti-predator response for a prey species living in an open habitat than it is for one living in a forest habitat. This may be one of the reasons why sheep adopt larger group sizes than do pigs. Superimposed on these basic influences on group structure are the effects of sex differences and the species' mating system. These act to determine group composition and how males and females associate.

1.6.1 Sex differences and parental care as determinants of mating systems

The roles of the two sexes in reproduction have an important effect on mating systems and group structure. Avian and mammalian females usually produce a limited number of eggs which can often be fertilized by one or a few matings. Males, on the other hand, produce much larger numbers of sperm and so have the potential to father offspring at a faster rate than a female can produce them. The reproductive success of a male is thus limited by his access to females, whereas the opposite is not usually true; female reproductive success is limited primarily by access to environmental resources such as food or nest sites. The implication of these sex differences is that, generally speaking, females distribute themselves around these resources, while males organize themselves around and compete for females (Emlen and Oring, 1977; Wrangham, 1979).

However, the extent to which males can compete for females is limited by the amount of parental care they provide for their offspring. If the survival and successful rearing of the offspring depends on male parental care, then males are constrained in the extent to which they can desert females and search for further copulations. The requirement for male parental care depends on a number of factors. These include whether the young require postnatal care at all (e.g. precocial or altricial) and whether males are physiologically capable of providing this care (e.g. lactation in mammals). Environmental factors, such as whether resources are plentiful or easily defendable enough for one parent to successfully rear the young, may also play a role. The degree of male parental care varies from species to species but, in general, paternal care is considerably more common in birds than in mammals.

1.6.2   Resource distribution and types of mating system

The term 'mating system' refers to the ways in which individuals obtain and defend mates, how many mates they have and how they care for offspring (Emlen and Oring, 1977; Davies, 1991). In general, mating systems can be categorized as monogamous when there is evidence of long-term pair bonds and both parents contribute to rearing the young, polygynous where males compete for matings with many females and provide little or no parental care, and polyandrous where one female monopolizes several males.

Most mammalian species, including farm animals and related species, show low levels of male parental care and are thus predisposed to form polygynous mating systems. In these species, female distribution and group structure are influenced primarily by the factors we have discussed earlier (e.g. predation pressure, food distrubution and defendability). Male distribution is also influenced by these factors but, at least during the breeding season, is more strongly influenced by inter-male competition for access to females (female defence), or for resource-rich sites (resource defence) around which females are expected to gather. Mating systems and corresponding group structures are thus strongly influenced by the defendability of the resources which each sex values most highly.

If females occur in small ranges or in small, stable groups, then single males may be able to defend them from others resulting in a single-male polygynous mating system which may take the form of a harem (Clutton-Brock, 1989). This appears to be the case in horses, red deer and also in pigs, where males may compete for control of female groups during the breeding season (e.g. Frädrich, 1974; Wood and Brenneman, 1980; Clutton-Brock *et al.*, 1982; Kaseda and Khalil, 1996). Domestic fowl and other galliforms also show a harem-type mating system, with males defending up to 12 females in some cases (Collias and Collias, 1967; McBride *et al.*, 1969).

When females occur in large or unstable groups, it is likely to be uneconomic for males to defend several females at once and so the mating system is more likely to be monogamous or involve competition for females as they come into oestrus. Feral domestic sheep and related species such as the moufflon seem to adopt this form of mating system, with a clear rutting season during which male and female groups intermingle and males compete for access to receptive females (e.g. Bon *et al.*, 1993; Rowell and Rowell, 1993). Free-ranging domestic cattle living in mixed-sex herds (e.g. Howery *et al.*, 1996, 1998) may adopt a similar system in which male–male dominance relationships are maintained throughout the year and influence priority of access to cows when they come into oestrus (e.g. Chillingham cattle; Hall, 1989).

1.6.3 Mating systems and offspring emigration

The mating system that a species shows is one factor influencing offspring emigration (dispersal) from the natal group. Dispersal may function to prevent inbreeding and also because it benefits the dispersing sex to leave rather than compete with older residents. In mammals, males tend to disperse more than females. This may be partly because, in polygynous mating systems typical of many mammalian species, males defend female groups. Therefore, it pays young males to leave their natal group to avoid mating with related females and to avoid competing with their fathers for mates. Females, on the other hand, may benefit from inheriting or sharing a good home range with their mothers and sisters. In contrast, in birds it is females who tend to disperse. In these species where males often defend breeding territories, it may pay a male to remain near his birth place and perhaps inherit a part of his father's territory (Greenwood, 1980). One consequence of dispersal is that the non-dispersing sex, especially females, form family groups in which kin-selected 'altruistic' behaviours can evolve.

Although these general principles of dispersal apply to many mammal and bird species (Greenwood, 1980), the dispersal of an individual will be influenced by the relative costs and benefits of doing so at the time. In several species, principally birds, dispersal may be delayed when the fitness pay-offs to an individual of staying in the natal group and helping to raise related young are greater than those to be had from leaving and attempting to breed (Emlen, 1997).

1.6.4 Seasonal variation in group structure

Mating systems may themselves vary within a species according to prevailing ecological conditions (see Davies, 1991). In species with distinct breeding seasons, group structure outside these seasons may be quite different from that observed during breeding, and determined more by the factors discussed earlier in this chapter. For example, outside the rutting season sheep adopt sex-segregated groups which often occupy non-overlapping home ranges (e.g. Ruckstuhl, 1998). Furthermore, mating systems may stay constant, but group sizes can fluctuate according to the seasonal availability of resources. In Cantabrian chamois, for example, Perez-Barberia and Nores (1994) showed that herd size varied seasonally according to food availability and the nature of the escape terrain. All these points again emphasize the flexibility of natural social groupings, and that groups adjust according to resource distribution and availability.

Finally, it is worth noting that, although mating systems may appear to be of one type, this does not always reflect the true distribution of reproductive success. For example, Kaseda and Khalil (1996) used blood types to test paternity in 99 feral horse foals and showed that, despite the existence of apparently stable harem structures in the population, 15% of foals born were not sired by their harem stallion.

1.7   Conclusions: What Use is the Evolutionary Approach for Understanding Farm Animal Social Behaviour?

Throughout this chapter we have used an evolutionary framework to answer questions about the existence of group life and the variety of group structures. To conclude, we will now consider in more detail how this approach can help us understand, predict and manage farm animal social behaviour. For the evolutionary approach to be of much use, we need to assume that the social organization and behaviour of farmed species still retain characteristics and 'design features' resulting from natural selection prior to domestication. Clearly, the assumption is most likely to be valid for recently domesticated species and for wild species which humans have only just started to manage and farm (e.g. some species of deer and fish). As outlined in the Introduction, there is also support for the assumption in species which have been domesticated for several thousand years. If the assumption is valid, we suggest that the evolutionary approach is useful in two main ways.

1.7.1   General principles may guide husbandry design and practice

First, it provides general principles about group life which can help us predict the social behaviour of farm animals under certain conditions, and hence design humane and efficient housing and husbandry systems which take these predictions into account. The evolutionary approach emphasizes the inherent flexibility of natural social groups. Individuals are predicted to join and leave groups as the relative benefits that they can realize from group life change. We should thus design group housing with this in mind. Animals may need to be removed from groups or at least be able to escape the attention of other group members. Furthermore, the ability of animals to leave groups may act as a constraint on the expression of despotic behaviour. In natural groups, high-ranking animals risk losing group members if they overexert their dominance. In captive environments where emigration is not possible, this proximate constraint on the behaviour of high-ranking animals is removed and potentially allows them to behave in a much more despotic way than would normally be tolerated by other group members (Mendl and Newberry, 1997).

The defendability of resources, be they food, home ranges or females, has repeatedly emerged as an important factor influencing group structure. Defendable resources encourage territorial behaviour, and the distribution of resources influences the number of individuals required to defend them, hence affecting group size. The existence of defendable resources within a group may also exacerbate the effects of dominance on subordinate individuals (see Hansen and Hagelso, 1980). If resources such as food are widely distributed in space and arrive simultaneously, this decreases their defendability and minimizes within-group competition. Resource distribution of this sort has been used in farming as one way of preventing dominant animals from out-competing their subordinate group members (e.g. fish farming; see Metcalfe, 1990). Recent research has started to examine whether pigs distribute themselves in direct proportion to the availability of food at different sites. If so, this would increase the chances of all pigs achieving an equal food intake rate (the ideal free distribution). Pigs show a propensity to distribute themselves in this way, but it appears that they use trough length rather than actual amount of food as the main cue guiding their dispersal around the troughs (Done *et al.*, 1996) (Fig. 1.4). Research of this sort, building on principles from behavioural ecology, could be used to ensure that resource distribution in farm animal housing is equitable and favours low levels of within-group competition (see Metcalfe, 1990).

1.7.2   Species-specific knowledge and husbandry design

We suggest that the second main way in which the evolutionary approach can be of use in predicting and understanding farm animal behaviour is by providing species-specific knowledge. It seems likely that farm animals have evolved to deal most effectively with the group sizes and structures typical of ancestral or related species (cf. Price, 1984). By understanding the group structures of these species, it may be possible to predict the types of group which lead to harmonious or damaging social interactions in farm animals. At the simplest level, the social behaviour, communication and recognition abilities of individuals are likely to be adapted to the range of group sizes adopted by the species in free-ranging situations. Housing the species in groups of radically different size (e.g. chickens housed in groups of thousands (Nicol and Dawkins, 1990)) may interfere with these abilities and lead to welfare problems. However, the inherent flexibility

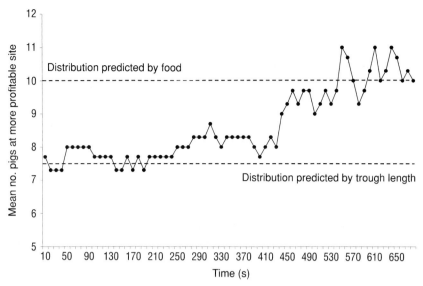

Fig. 1.4. Distribution of a group of 15 pigs around two troughs of the same length, but with twice as much food in one (the 'more profitable trough') as the other. The figure shows the mean number of pigs at the more profitable trough, starting at time 0 when food was delivered to the troughs, across four separate trials. The lower broken line shows the predicted distribution if pigs distribute themselves according to trough length. The upper broken line shows the predicted distribution if they use amount of food available as the main cue. Pigs initially appear to distribute themselves according to trough length. As time passes, the distribution of food becomes more influential. (Adapted from Done *et al.*, 1996.)

of social life may predispose animals to be able to adopt a broad range of social groupings. For example, domestic cats are able to adopt a variety of different social groupings depending on environmental conditions and resource distribution (see Kerby and Macdonald, 1988).

A knowledge of the species' mating system may also provide important predictive information. For example, in species with a polygynous mating system where females and resources are defended vigorously, the housing of adult males together is likely to be problematic (e.g. domestic pig, deer). In seasonal breeders, this is most likely during the breeding season. Outside this season, housing males together may cause few problems. An understanding of the patterns of dispersal of offspring from natal groups can also be useful in guiding the design of husbandry systems for growing animals.

1.7.3   Limitations of the evolutionary approach

Although evolutionary theory provides a basic framework for understanding the phenomena of group life, there are problems with the
approach which must temper any conclusions that we make. One
fundamental problem relates to the ability of animals to behave
optimally. Throughout the chapter, we have spoken about animals
being expected to behave in certain ways in order to maximize their
reproductive success. However, this raises the question of exactly how
animals might evaluate fitness consequences of behavioural decisions.
How does an animal determine whether it is in its best fitness interests
to join a group or remain solitary? There may be constraints on the
accuracy of such assessments, or use of simple 'rules of thumb' which
mean that animals are limited in their ability to do exactly what
evolutionary theory predicts they should do.

Another general problem is that we can end up proposing adaptive
'explanations' for the origins of group life which are based more on the
observer's hunches than on hard evidence (Gould and Lewontin, 1979).
This is exacerbated if we are unable to test hypotheses which derive
from these explanations, and, even if we can test them, we must be
careful not to confuse the apparently adaptive current utility of a
behaviour with its presumed function in the past history of the species.
Similarly, we can never be entirely sure about exactly how a species'
social organization has evolved, so any predictions we make are
necessarily based on assumptions rather than precise facts about the
species' evolutionary history. Finally, the relevance of evolutionary
theory to domestic species remains a matter of debate. For example,
artificial selection may have inadvertently selected for social tolerance
of others (Hemmer, 1990), thus allowing domestic species to be kept in
social conditions which ancestral or related species would not tolerate.
While this is certainly possible, we still feel that the principles and
concepts offered by an evolutionary framework provide a valuable
background against which to consider the social behaviour of farm
animals.

Acknowledgements

We acknowledge the support of the UK Biotechnology and Biological
Sciences Research Council.

References

Amat J.A. and Rilla, F.D. (1994) Foraging behavior of white-faced ibises (*Plegadis chihi*) in relation to habitat, group-size, and sex. *Colonial Waterbirds* 17, 42–49.

Baird, R.W. and Dill, L.M. (1996) Ecological and social determinants of group size in transient killer whales. *Behavioral Ecology* 7, 408–416.

Bateson, M. and Kacelnik, A. (1997) Starlings' preferences for predictable and unpredictable delays to food. *Animal Behaviour* 53, 1129–1142.

Bednekoff, P.A. and Lima, S.L. (1998) Randomness, chaos, confusion in the study of antipredator vigilance. *Trends in Ecology and Evolution* 13, 284–287.

Bon, R., Badia, J., Maublanc, M.L. and Recarte, J.M. (1993) Social dynamics of mouflon (*Ovis ammon*) during rut. *International Journal of Mammalian Biology* 58, 294–301.

Brown, C.R. and Brown, M.B. (1986) Ectoparasitism as a cost of coloniality in cliff swallows (*Hirudino pyrrhonata*). *Ecology* 67, 1206–1218.

Brown, J.L., Brown, E.R., Brown, S.D. and Dow, D.D. (1982) Helpers: effects of experimental removal on reproductive success. *Science* 215, 421–422.

Clutton-Brock, J. (1987) *A Natural History of Domesticated Mammals.* Cambridge University Press, Cambridge.

Clutton-Brock, T.H. (1988) Reproductive success. In: Clutton-Brock, T.H. (ed.) *Reproductive Success.* Chicago University Press, Chicago, pp. 472–485.

Clutton-Brock, T.H. (1989) Mammalian mating systems. *Proceedings of the Royal Society London B* 236, 339–372.

Clutton-Brock, T.H. (1998) Reproductive skew, concessions and limited control. *Trends in Ecology and Evolution* 13, 288–292.

Clutton-Brock, T.H., Albon, S.D. and Guiness, F.E. (1982) *Red Deer: Behavior and Ecology of Two Sexes.* Edinburgh University Press, Edinburgh.

Collias, N.E. and Collias, E.C. (1967) A field study of the Red Jungle Fowl in North-Central India. *Condor* 69, 360–386.

Cote, I.M. and Poulin, R . (1995) Parasitism and group-size in social animals – a metaanalysis. *Behavioral Ecology* 6, 159–165.

Danchin, E. and Wagner, R.H. (1997) The evolution of coloniality: the emergence of new perspectives. *Trends in Ecology and Evolution* 12, 342–347.

Davies, N.B. (1991) Mating systems. In: Krebs, J.R. and Davies, N.B. (eds) *Behavioural Ecology*, 3rd edn. Blackwell Scientific Publications, Oxford, pp. 263–294.

Davies, N.B and Houston, A.I. (1981) Owners and satellites: the economics of territory defence in the pied wagtail, *Motacilla alba. Journal of Animal Ecology* 50, 157–180.

Dawkins, M.S. (1995) *Unravelling Animal Behaviour*, 2nd edn. Longman, Harlow, 183 pp.

Dominey, W.J. (1984) Alternative mating tactics and evolutionarily stable strategies. *American Zoologist* 24, 385–396.

Done, E., Wheatley, S. and Mendl, M. (1996) Feeding pigs in troughs: a preliminary study of the distribution of individuals around depleting resources. *Applied Animal Behaviour Science* 47, 255–262.

Dunbar, R.I.M. (1982) Adaptation, fitness and the evolutionary tautology. In: King's College Sociobiology Group (ed.) *Current Problems in Sociobiology.* Cambridge University Press, Cambridge.

Duncan, P. and Vigne, P. (1979) The effects of group size in horses on the rate of attacks by blood-sucking flies. *Animal Behaviour* 27, 623–625.

Dwyer, C.M. and Lawrence, A.B. (1998) Variability in the expression of maternal behaviour in primiparous sheep: effects of genotype and litter size. *Applied Animal Behaviour Science* 58, 311–330.

Elgar, M.A. (1989) Predator vigilance and group size in mammals and birds: a critical review of the empirical evidence. *Biological Reviews* 64, 13–33.

Emlen, S.T. (1997) Predicting family dynamics in social vertebrates. In: Krebs, J.R. and Davies, N.B. (eds) *Behavioural Ecology*, 4th edn. Blackwell Scientific Publications, Oxford, pp. 228–253.

Emlen, S.T. and Oring, L.W. (1977) Ecology, sexual selection, and the evolution of mating systems. *Science* 197, 215–223.

Endler. J.A. (1980) Natural selection on colour patterns in *Poecilia reticulata. Evolution* 34, 76–91.

Frädrich, H. (1974) A comparison of behaviour in the Suidae. In: Geist, V. and Walther, F. (eds) *The Behaviour of Ungulates and its Relation to Management*, Vol. 1. IUCN, Morges, pp. 133–143.

Fraser, D., Kramer, D.L., Pajor, E.A. and Weary, D.M. (1995) Conflict and cooperation: sociobiological principles and the behaviour of pigs. *Applied Animal Behaviour Science* 44, 139–157.

Fretwell, S.D. and Lucas, H.L. (1970) On territorial and other factors influencing habitat distribution in birds. I. Theoretical development. *Acta Biotheoretica* 19, 16–36.

Gould, S.J. and Lewontin, R.C. (1979) The spandrels of San Marco and the Panglossian paradigm: a critique of the adaptationist programme. *Proceedings of the Royal Society London B* 205, 581–598.

Grafen, A. (1984) Natural selection, kin selection and group selection. In: Krebs, J.R. and Davies, N.B. (eds) *Behavioural Ecology*, 2nd edn. Blackwell Scientific Publications, Oxford, pp. 62–84.

Greenwood, P.J. (1980) Mating systems, philopatry and dispersal in birds and mammals. *Animal Behaviour* 28, 1140–1162.

Gross, M.R. (1985) Disruptive selection for alternative life histories in salmon. *Nature* 313, 47–48.

Hall, S.J.G. (1989) Chillingham cattle: social and maintenance behaviour in an ungulate that breeds all year round. *Animal Behaviour* 38, 215–225.

Hamilton, W.D. (1964) The genetical evolution of social behaviour. *Journal of Theoretical Biology* 7, 1–52.

Hansen, L.L. and Hagelso, A.M. (1980) A general survey of environmental influences on the social hierarchy function in pigs. *Acta Agricultura Scandinavica* 30, 388–392.

Harper, D.G.C. (1982) Competitive foraging in mallards: 'ideal free ducks'. *Animal Behaviour* 30, 575–584.

Hemmer, H. (1990) *Domestication. The Decline of Environmental Appreciation.* Cambridge University Press, Cambridge.

Hillman, J.C. (1987) Group size and association patterns of the common eland (*Tragelaphus oryx*). *Journal of Zoology* 213, 641–663.

Hoogland, J.L. (1979) Aggression, ectoparasitism and other possible costs of prairie dog (Sciuridae: *Cynommys* spp.) coloniality. *Behaviour* 69, 1–35.

Howery, L.D., Provenza, F.D., Banner, R.E. and Scott, C.B. (1996) Differences in distribution patterns among individuals in a cattle herd. *Applied Animal Behaviour Science* 49, 305–320.

Howery, L.D., Provenza, F.D., Banner, R.E. and Scott, C.B. (1998) Social and environmental factors influence cattle distribution on rangeland. *Applied Animal Behaviour Science* 55, 231–244.

Jarman, P.J. (1974) The social organization of antelope in relation to their ecology. *Behaviour* 48, 215–267.

Jennings, M.J. and Phillip, D.P. (1992) Female choice and male competition in longear sunfish. *Behavioural Ecology* 3, 84–94.

Kaseda, Y. and Khalil, A.M. (1996) Harem size and reproductive success of stallions in Misaki feral horses. *Applied Animal Behaviour Science* 47, 163–173.

Kerby, G. and Macdonald, D.W. (1988) Cat society and the consequences of colony size. In: Turner, D.C. and Bateson, P. (eds) *The Domestic Cat.* Cambridge University Press, Cambridge, pp. 67–81.

Koprowski, J.L. (1996) Natal philopatry, communal nesting, and kinship in fox squirrels and gray squirrels. *Journal of Mammalogy* 77, 1006–1016.

Krakauer, D.C. (1995) Groups confuse predators by exploiting conceptual bottlenecks: a connectionist model of the confusion effect. *Behavioural Ecology and Sociobiology* 36, 421–429.

Krause, J. and Godin, J.J. (1995) Predator preferences for attacking particular group sizes: consequences for predator hunting success and prey predation risk. *Animal Behaviour* 50, 465–473.

Krebs, J.R. and Davies, N.B. (1993) *An Introduction to Behavioural Ecology,* 3rd edn. Blackwell Science Ltd, Oxford, 420 pp.

Lawrence, A.B. (1990) Mother–daughter and peer relationships of Scottish hill sheep. *Animal Behaviour* 39, 481–486.

Lawrence, A.B. and Wood-Gush, D.G.M. (1988) Home range behaviour and social organisation of Scottish Blackface sheep. *Journal of Applied Ecology* 25, 25–40.

Lott, D.F. (1991) *Intraspecific Variation in the Social Systems of Wild Vertebrates.* Cambridge University Press, Cambridge.

Magurran, A.E., Seghers, B.H., Carvalho, G.R. and Shaw, P.W. (1992) Behavioural consequences of an artificial introduction of guppies (*Poecilia reticulata*) in N. Trinidad: evidence for the evolution of anti-predator behaviour in the wild. *Proceedings of the Royal Society London B* 248, 117–122.

Manning, A. and Dawkins, M.S. (1998) *An Introduction to Animal Behaviour,* 5th edn. Cambridge University Press, Cambridge, 450 pp.

Mauget, R. (1981) Behavioural and reproductive strategies in wild forms of *Sus scrofa* (European wild boar and feral pigs). In: Sybesma, W. (ed.) *The Welfare of Pigs.* Martinus Nijhoff, The Hague, pp. 3–13.

Maynard Smith, J. (1964) Group selection and kin selection. *Nature* 201, 1145–1147.

McBride, G., Parer, I.P. and Foenander, F. (1969) The social organization and behaviour of the feral domestic fowl. *Animal Behaviour Monographs* 2, 125–181.

McFarland, D. (1993) *Animal Behaviour: Psychobiology, Ethology and Evolution*, 2nd edn. Longman, Harlow, 576 pp.

McNamara, J.M. and Houston, A.I. (1992) Evolutionarily stable levels of vigilance as a function of group-size. *Animal Behaviour* 43, 641–658.

Mendl, M. (1991) Some problems with the concept of a cut-off point for determining when an animal's welfare is at risk. *Applied Animal Behaviour Science* 31, 139–146.

Mendl, M. (1994) The social behaviour of non-lactating sows and its implications for managing sow aggression. *The Pig Journal* 34, 9–20.

Mendl, M, and Deag, J.M. (1995) How useful are the concepts of alternative strategy and coping strategy in applied studies of social behaviour? *Applied Animal Behaviour Science* 44, 119–137.

Mendl, M. and Newberry, R.C. (1997) Social conditions. In: Appleby, M.C. and Hughes, B.O. (eds) *Animal Welfare*. CAB International, Wallingford, UK, pp. 191–203.

Metcalfe, N.B. (1990) Aquaculture. In: Monaghan, P. and Wood-Gush, D. (eds) *Managing the Behaviour of Animals*. Chapman & Hall, London, pp. 125–154.

Milinski, M. and Parker, G.A. (1991) Competition for resources. In: Krebs, J.R. and Davies, N.B. (eds) *Behavioural Ecology*, 3rd edn. Blackwell Scientific, Oxford, pp. 137–168.

Møller, A.P. and Birkhead,T.R. (1993) Cuckoldry and sociality – a comparative study of birds. *American Naturalist* 142, 118–140.

Newberry, R.C. (1993) The space–time continuum and its relevance to farm animals. *Etologia* 3, 219–234.

Nicol, C. and Dawkins, M.S. (1990) Homes fit for hens. *New Scientist* 17 March, 46–51.

O'Brien, P.H. (1988) Feral goat social organization: a review and comparative analysis. *Applied Animal Behaviour Science* 21, 209–221.

Oyama, S. (1985) *The Ontogeny of Information*. Cambridge University Press, Cambridge.

Parker, G.A. and Sutherland, W.J. (1986) Ideal free distributions when individuals differ in competitive ability: phenotype limited ideal free models. *Animal Behaviour* 34, 1222–1242.

Penning, P.D., Parsons, A.J., Newman, J.A., Orr, R.J. and Harvey, A. (1993) The effects of group size on grazing time in sheep. *Applied Animal Behaviour Science* 37, 101–109.

Perez-Barberia, F.J. and Nores, C. (1994) Seasonal variation in group size of Cantabrian chamois in relation to escape terrain and food. *Acta Theriologica* 39, 295–305.

Pitcher, T.J., Magurran, A.E. and Winfield, I.J. (1982) Fish in larger shoals find food faster. *Behavioral Ecology and Sociobiology* 10, 149–151

Price, E.O. (1984) Behavioral aspects of animal domestication. *Quarterly Review of Biology* 59, 1–32.

Pulliam, H.R. (1973) On the advantages of flocking. *Journal of Theoretical Biology* 38, 419–422.

Pulliam, H.R and Caraco, T. (1984) Living in groups: is there an optimal group size? In: Krebs, J.R. and Davies, N.B. (eds) *Behavioural Ecology*, 2nd edn. Blackwell Scientific Publications, Oxford, pp. 122–147.

Pusey, A.E. and Packer, C. (1997) The ecology of relationships. In: Krebs, J.R. and Davies, N.B. (eds) *Behavioural Ecology*, 4th edn. Blackwell Scientific Publications, Oxford, pp. 254–283.

Quenette, P. and Gerard, J. (1992) From individual to collective vigilance in the wild boar (*Sus scrofa*). *Canadian Journal of Zoology* 70, 1632–1635.

Richner, H. and Heeb, P.S. (1995) Is the information-center hypothesis a flop? *Advances in the Study of Behavior* 24, 1–45.

Roberts, G. (1996) Why individual vigilance declines as group size increases. *Animal Behaviour* 51, 1077–1086.

Rowell, T.E. and Rowell, C.H. (1993) The social organization of feral *Ovis aries* ram groups in the pre-rut period. *Ethology* 95, 213–232.

Ruckstuhl, K. E. (1998) Foraging behaviour and sexual segregation in bighorn sheep. *Animal Behaviour* 56, 99–106.

Schaller, G.B. (1972) *The Serengeti Lion*. University of Chicago Press, Chicago, 480 pp.

Scheel, D. and Packer, C. (1991) Group hunting behaviour of lions: a search for cooperation. *Animal Behaviour* 41, 697–709.

Schmidtmann, E.T. and Valla, M.E. (1982) Face-fly pest intensity, fly-avoidance behaviour (bunching) and grazing time in Holstein heifers. *Applied Animal Ethology* 8, 429–438.

Sibly, R.M. (1983) Optimal group size is unstable. *Animal Behaviour* 31, 947–948.

Sinclair, A.G., Edwards, S.A., Hoste, S. and McCartney, A. (1998) Evaluation of the influence of maternal and piglet breed differences on behaviour and production of Meishan synthetic and European White breeds during lactation. *Animal Science* 66, 423–430.

Spinka, M. and Algers, B. (1995) Functional view on udder massage after milk let-down in pigs. *Applied Animal Behaviour Science* 43, 197–212.

Stolba, A. and Wood-Gush, D.G.M. (1989) The behaviour of pigs in a semi-natural environment. *Animal Production* 48, 419–425.

Stricklin, W.R., Zhou, J.Z. and Gonyou, H.W. (1995) Selfish animals and robot ethology: using artificial animals to investigate social and spatial behavior. *Applied Animal Behaviour Science* 44, 187–203.

Trivers, R.L. (1971) The evolution of reciprocal altruism. *Quarterly Review of Biology* 46, 35–57.

Vehrencamp, S. (1983) A model for the evolution of despotic versus egalitarian societies. *Animal Behaviour* 31, 667–682.

Ward, P. and Zahavi, A. (1973) Importance of certain assemblages of birds as information centres for finding food. *Ibis* 115, 517–534.

Waring, G.H., Wierzbowski, S. and Hafez, E.S.E. (1975) The behaviour of horses. In: Hafez, E.S.E. (ed.) *The Behaviour of Domestic Animals*, 3rd edn. Baillière Tindall, London.

Weng, R.C., Edwards, S.A. and English, P.R. (1998) Behaviour, social inter-actions and lesion scores of group-housed sows in relation to space allowance. *Applied Animal Behaviour Science* 59, 307–316.

Wilson, E.O. (1975) *Sociobiology: the New Synthesis.* The Belknap Press of Harvard University Press, Cambridge, Massachusetts.

Wood, G.W. and Brenneman, R.E. (1980) Feral hog movements and habitat use in coastal South Carolina. *Journal of Wildlife Management* 44, 420–427.

Wood-Gush, D.G.M., Duncan, I.J.H. and Savory, C.J. (1978) Observations on the social behaviour of domestic fowl in the wild. *Biology of Behaviour* 3, 193–205.

Wrangham, R.W. (1979) On the evolution of ape social systems. *Social Science Information* 18, 335–368.

Zayan, R. (1985) *Social Space for Domestic Animals.* Martinus Nijhoff Publishers, Dordrecht.

Group Life

2

A. Cecilia Lindberg

Department of Clinical Veterinary Science, University of Bristol, Bristol BS40 5DU, UK

(Editors' comments: If, as has been presented in the previous chapter, the majority of our farm animals live in groups, then it becomes necessary to discuss in more detail life within these groups. This is the aim of Lindberg's chapter and it takes up the basic principles of group life that will appear again several times throughout this book. For example, most people are already familiar with the term dominance hierarchy, but there are many subtle aspects to this seemingly simple concept. These include the effects of group size and available space on hierarchy formation and questions about the importance of individual recognition on the stability of the hierarchy.

Communication is another deceptively simple term, but one which raises many questions as to what information is actually transmitted and what is actually received by individuals. Aspects such as how easy the signal is to detect, how easily it can be discriminated from other signals and how easily it is remembered all influence what is and can be communicated.

The first part of this chapter deals with essential basic information about how groups work. The second part deals with the structure of the group, that is to say how the social dynamics of a group are determined by the types of individuals who make up the group and by the size of the group.)

2.1 Introduction

Chapter 1 discusses *why* animals live in groups and what benefits or disadvantages they might derive from group life, in an evolutionary

©CAB *International* 2001. *Social Behaviour in Farm Animals*
(eds L.J. Keeling and H.W. Gonyou)

sense. In this chapter, I will move on from this to discuss relationships between individuals within the group and consider how the maintenance and successful functioning of the group are achieved through social behaviour.

A definition of a group such as Wilson's (1975), 'any set of organisms, belonging to the same species, that remain together for a period of time interacting with one another to a distinctly greater degree than with other conspecifics', gives us one view of group life. However, in the case of domestic animals, additional factors come into the equation since domestic animal groupings are often controlled entirely by humans. Thus, domestic animal groups may be 'managed', as in an intensive farming system, or they may more closely approach the 'natural' groupings found in their wild ancestors, in the case of free-ranging or feral animals. These situations result in very different types of groups with respect to group size, spacing, dispersal, and age and sex distribution. By confining animals, we also affect the group's habitat and the ability of individuals within the group to choose whether to stay or leave. It is extremely important to consider the species' natural groupings when designing management systems and the most welfare-friendly systems tend to be those where the animals are kept in ways that capture the important features (such as the group size, offspring dispersal patterns and parent–offspring interactions) of natural groups (e.g. the family group system for pigs; Stolba and Wood-Gush, 1984; Wechsler, 1996). Interactions between individuals of groups may be grossly different when kept in inappropriate groupings and many abnormal interactions found under intensive conditions are never seen in free-living groups. For example, the 'buller steer syndrome' is a major problem associated with cattle kept under commercial feedlot conditions, but rarely occurs under pastured conditions where escape is easier (Blackshaw et al., 1997).

In this chapter, I will describe how the behaviour of individuals within the group can result in dominance hierarchies, communication and patterns of spacing. The relevance of these to domestic animal management and husbandry operations is an important consideration. In Europe, there has been considerable behavioural research on laying hens and pigs, partly in response to concerns about their welfare under intensive conditions. Thus, many of the examples in this chapter will come from these two species. Sheep and deer, in comparison, are still farmed relatively extensively and are far more likely to be allowed to determine their own, species-appropriate, ways of living in groups. Cattle are somewhat intermediate, often being kept extensively during the summer and housed intensively during the winter months.

2.2 Maintaining the Group

Once established, a number of mechanisms contribute to maintaining the cohesion of the group. Groups can range from temporary aggregations to highly structured societies. Even within a species, different types of groupings may occur, depending on such factors as seasonal food availability or breeding status. In domestic animals, groups may last longer than they would in a natural situation, since group members are not usually at liberty to decide when to coalesce or to disperse the group. With occasional exceptions, a group is not simply a collection of anonymous animals, but is an actively formed and maintained unit. However, this does not imply that animals act altruistically to maintain the group (see Section 1.2). Rather, a major feature of the group is that each animal acts (or attempts to act) for its own maximum benefit in its interactions with conspecifics, and this may result in 'emergent properties' that affect the group as a whole, such as dominance hierarchies.

2.2.1 Dominance

The concept of social dominance was pioneered by Schjelderup-Ebbe (1935, cited in Syme and Syme, 1979), who was the first to make a scientific study of the 'peck order' in chickens. Dominance relationships are a cornerstone of group life in general and create 'rules' by which social encounters are controlled. The term 'dominance' refers to the predictable relationship between a pair of conspecifics, where one animal has learnt to dominate the other (subordinate), which in its turn tends to avoid confrontations. This is a learned relationship relying on animals recognizing each other and remembering previous social encounters, retaining their relative status during future meetings. Other strategies than individual recognition are also possible, e.g. using a 'rule of thumb' as discussed below in Section 2.3. The sum of all such dominance relationships within a group is known as the dominance hierarchy or peck order. The 'dominance rank' thus represents an individual's relative position with respect to all other animals in the group (Stricklin and Mench, 1987). The dominance order is unique to a particular group and adding or removing individuals will therefore have repercussions through the group, temporarily upsetting the equilibrium until a new dominance order has been established (e.g. Keiper and Sambraus, 1986). The rank of an individual in one group does not tell us what rank it will finally have when it becomes a member of a different group, i.e. dominance ranks are specific to a particular group. Dominance hierarchies do not imply that the difference between each rank is equidistant to any other difference (Fig. 2.1). Beilharz and Zeeb (1982) found that, in three dairy herds studied, there

(a)   (b)

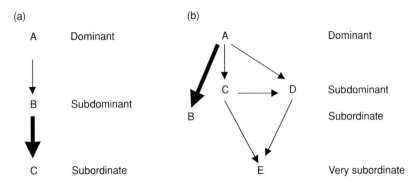

Fig. 2.1. (a) Simple linear dominance hierachy in a small group. Note that dominance behaviour can vary in intensity, e.g. individual B may dominate individual C more strongly than it in its turn is dominated by A, as indicated by the thickness of the arrows. (b) Increasing group size leads to increasing complexity in inter-individual dominance relationships and to variability in the intensity of interactions. Individuals may not fall into the simple dominant/intermediate/subordinate categories found in small groups, while triangular or more complex relationships are common.

was no single cow that was dominant over *all* others, i.e. all cows had to be submissive at some point. Groups can be thought of as 'units of lowered aggressiveness' (McBride, 1971) and other conspecifics are normally attacked if they enter this unit.

Within a group, individuals with different ranks may have different tasks, for example, the alpha individual may not only be dominant to other group members but may also be a leader during migratory movements or be more closely responsible for group defence. Both the dominant and the subordinate status can have advantages and disadvantages. For example, a subordinate might be the recipient of high levels of aggression but might also benefit from group protection and food-locating abilities by more dominant animals. Conversely, dominants may get more matings and priority of access to other resources, but there is also evidence that dominants may be more stressed or fearful (Jones and Faure, 1982). However, this is probably less relevant to the domestic or intensive situation, partly because many tasks (e.g. predator defence, migration) may be unnecessary and partly because domestic animals are often kept in single-sex, same-age flocks. In comparison, in a natural situation there will be a gradual move up the ranks as the animal matures, passing from the juvenile subordinate stage to the more dominant adult stage and possibly taking over from the existing alpha individual. Older animals generally have higher rank than younger ones and adults are almost invariably domi-nant to juveniles. Within the adult group, age effects may be overridden by time of entry to the group. Thus an older adult may be subordinate to a younger but more long-term group member (Beilharz and Zeeb, 1982). Social dominance is established in domestic hens during maturation and may then persist even when the actual ability to win contests with strangers has changed significantly (Lee and Craig, 1981), suggesting that existing hierarchies may not continue to be closely correlated with competitive ability. Small groups of domestic animals of the same sex and approximately the same size often have linear or near linear social organizations, while larger groups usually have more complex organizations (Rushen, 1982; Craig, 1986) (see Section 7.1.6).

The dominance hierarchy functions by the combination of aggression from the dominant and submission from the subordinate, i.e. when the subordinate acts appropriately this inhibits the dominant's aggression. Should the dominant exceed the 'limits' or the subordinate fail to respond in the appropriately submissive way, the interaction may escalate to such an extent that serious injury occurs or the subordinate is forced to leave the group. Alternatively, a 'challenger', e.g. a member of an equine 'bachelor group', could oust the alpha individual. However, in a domestic situation, these controls to the intensity of the interaction may not be effective, which can cause excessive aggressiveness directed at subordinates, since the latter have no way of escaping or leaving the group (Mendl and Newberry, 1997). In flocks of domestic chickens, for example, some individuals are very dominant and some are extremely subordinate with the rest joining in the attack on the subordinates. Thus, a behaviour that is adaptive in the wild, becomes abnormal under intensive management conditions when lack of space means that the dominance hierarchy cannot function correctly.

Although the dominance order does help in allocating resources, it is not an equal distribution dependent on dominance rank (Craig, 1986). More probably, there is plenty for the top-ranking individual and little for the rest (e.g. the most dominant bull or stallion in a herd tends to do most of the mating) or, alternatively, the majority get an equal share but very low-ranking individuals get a poor deal (e.g. Cunningham and van Tienhoven, 1983; the lowest-ranking hen got very little food with resultant poor body condition and the rest of the flock got a larger amount).

The 'avoidance order', whereby more subordinate group members avoid provoking those ranked above them, is equally important. Among domestic animals, this is particularly obvious in, for example, riding horses, which are often extremely fit and fed excess energy; serious injuries are frequently sustained when such horses are not given sufficient space allowance to avoid conflict. Avoidance orders are also observed in sows; thus, when sufficient space was available, subordinates could inhibit aggression in dominant sows by lowering and turning the head away (Jensen, 1982). In contrast, farm animals are often kept at considerably higher density (e.g. poultry) and are frequently less mobile (e.g. turkeys, dairy cows), which may reduce the amount of interaction they can perform.

2.2.2 Aggression/threatening

Aggression is a basic feature during formation of a dominance hierarchy, but although the dominant individuals probably have to be aggressive to achieve their dominant status, they do not need to be aggressive subsequently in all social situations to maintain their position. Indeed, once established, one of the results of a hierarchy is to avoid the need for aggressive and injurious interactions. None the less, when first mixed, it is common in most highly social species to observe heightened levels of aggression, perhaps because there are positive fitness advantages in deterring intruders and retaining priority of access to resources (Fraser and Rushen, 1987). While dominance rank is not exactly the same as frequency of aggressive or submissive acts, these are very important aspects of the rank order. However, if strong antipathies or tolerances exist between animals, this can result in distorted data. For example, an animal of intermediate dominance might be extremely abusive towards the few individuals it does dominate, whereas a very high-ranking animal might only rarely get involved in any obviously agonistic interactions. The level of threatening and aggression is influenced by evolution, which has resulted in ritualization of aggressive behaviour, such that injury and excessive energy wastage can be avoided.

The level of agonistic behaviour in a group is partly dependent on group size; small groups should have a stable and linear or other simple hierarchy and little need for aggression. Slightly larger groups may contain triangular or more complex relationships and there may be more changes or reversals of dominance, and hence slightly higher levels of agonistic interactions. Once the group is very large, e.g. as in intensive poultry systems with thousands of individuals, it seems that aggression gets lower. For example, Hughes et al. (1997) reported that, even when allowed to mix with a previously separate flock, there was no apparent evidence for increased aggression in laying hens. In large groups, Hughes et al. found that birds could be in close proximity without provoking aggression, whereas in small groups moving past or close to a flockmate tended to provoke an agonistic interaction (Grigor et al., 1995). As group size in domestic fowl changes from small to very large, there may be a move from hierarchical organization to an assessment based on body size or other phenotypic factors (Hughes et al., 1997; Pagel and Dawkins, 1997). Alternatively, there may be a threshold beyond which no attempts to form hierarchies are made: if the rate of encountering birds exceeds a certain limit, birds may adopt a non-intervention strategy. Thus, it seems likely that large flocks preclude individual recognition (which may be necessary for a stable hierarchy; McBride, 1964) and that birds do not recognize flockmates as 'familiar' or other birds as 'unfamiliar' (Hughes *et al.*, 1997). Lack of recognition (or social structure) could be a factor in minimizing agonistic interactions between individuals, resulting in a relatively lower level of aggression compared with smaller groups. Evidence from various studies suggests that in domestic hens aggression tends to be low in large flocks and higher but variable in smaller flocks (up to approximately 12 birds). For example, Al-Rawi and Craig (1975) and Hughes and Wood-Gush (1977) report increased aggression with increased group size over a low range of group sizes. Lindberg and Nicol (1996a) also reported an increase in fighting when unfamiliar birds were mixed in a group size of 80. However, when groups become very large, relationships break down and it seems unlikely that hierarchies exist in flocks of several thousand individuals.

Various methods have been used to determine dominance, often involving the creation of competitive situations, for example by food deprivation. Dominance orders based on paired encounters of unfamiliar animals in neutral territories do not tell us much about actual dominance hierarchy that would be established in the home group, but may give some indication of the underlying competitive or aggressive tendencies. These are based on the assumption that a higher social rank always gets priority of access when animals are competing for scarce resources. Syme (1974) was critical of this view and found that different studies using different competitive interactions (e.g. over food, mates, escape from aversive situations) did not correlate well with dominance orders. Therefore we need to validate the competitive order with dominance interactions to use it as an indicator of dominance rank. Another point to note is that different types of competitive interactions might give different results: Banks *et al.* (1979) found that success in competitive encounters for food correlated well with dominance rank in hungry domestic fowl, but no competition occurred in thirsty birds, which simply waited for their turn to drink. Thus, what would seem to be a high priority item for humans may not elicit competitive behaviour in another species (Craig, 1986). Furthermore, the level of priority given to a behaviour will depend on the animal's current motivation. A hen may be unwilling to move past a more dominant individual to reach a nest site, but may be prepared to do so as its nesting motivation increases during the sitting phase of pre-laying behaviour (Freire *et al.*, 1997). Thus, a satiated animal may perform poorly in a competitive test over food, but still be dominant in its home environment. However, if it is sufficiently motivated (e.g. very hungry), an animal might ignore previously established dominance relations.

An alternative method for measuring dominance would be to deprive the animals of a resource, such as food, and then observe agonistic activity that occurs when the group of animals gets access again. This method does not measure the control of the resource *per se* but rather the aggressive and submissive acts elicited by the situation. However, a drawback of this situation is that if deprivation is too long, causing the ensuing competitive situation to become too intense, it can be very difficult to ascertain accurately the hierarchy as numerous peck-order violations and intense aggression may obscure the relationships that normally prevail during less competitive conditions.

2.2.3   Stability of hierarchy

Animals do not usually challenge each other continually; if an animal lost an interaction a few hours ago, then it is likely to lose again now. However, over time, reversals and changes do occur as individuals age or the group composition changes and 'consistently' dominating another individual is not the same as 'permanently' doing so. Managed groups may be unstable because of their constantly changing group membership. For example pigs are mixed and re-mixed during routine husbandry, resulting in high levels of aggression (Jensen, 1994; Erhard and Mendl, 1997; Giersing and Andersson, 1998; Puppe, 1998) (see Section 8.3.3) and domestic horse herds are often in a state of flux, particularly in the case of 'sporting' horses which are removed for varying lengths of time when taken to competitions and similar (see Section 9.1.6). An early experiment on domestic hens by Guhl and Allee (1944) suggested that constant changes of flock members result in increased aggression. They removed a hen from an established group on alternate days and replaced it with a new, unfamiliar, hen, causing disorganization of the flock and resulting in reduced food intake and increased aggression. However, such experiments necessarily also involve repeated intrusion and handling by the human experimenters, which may have played a part in the hens' responses. If the hierarchy does not involve recognition, then stability would depend on all animals consistently following rules such as 'always give way to conspecifics bigger than yourself'. Recognition involves a social memory of the past encounters that initially established social status. In horses, for example, group members can be separated for long periods (6 months or more) and still immediately settle back into the same hierarchy when reunited, retaining previously formed preferences and antipathies for individual herd mates. Conversely, domestic hens appear to have a much more restricted social memory and will treat as a stranger an individual from which they were separated only a few weeks earlier. As discussed by Wiepkema and Schouten (1990), individual recognition may involve 'heavy cognitive efforts', which can vary from simply identifying an individual as a conspecific or a group member to recognizing it as a particular individual, a range of increasing complexity and therefore of increasing cognitive effort. Recognition itself may also have different levels of complexity, including individual recognition of all or only some group members; maintaining superficial or very detailed information on an individual; and at a complex level it may also include information about relationships between other group members (see Section 14.1).

2.2.4 Social requirements for dominance orders to function

Animals need methods of allocating space within (and between) groups. Spacing over a geographical area can be achieved through territoriality, where one animal or group of animals controls an area and its resources by repelling other animals through overt aggression or other signalling (Stricklin and Mench, 1987). Animals very rarely space themselves randomly with respect to other individuals of the same species, which they may be either attracted to or repelled from. Farm animals rarely have the opportunity to defend a fixed area and many species, such as horses, do not usually defend a fixed area even in wild or feral circumstances (see Section 9.1.2). The variation in spacing patterns is related to variation in the distribution and predictability of important resources: territoriality will only be favoured when the benefits of exclusive use of a resource outweigh the costs of defending it, i.e. when the resource is 'economically defendable' (Monaghan, 1990). An alternative (or additional) strategy in such cases is to maintain 'individual distance' or 'social space', which encompasses a 'portable' space that surrounds the animal and moves with it. The size of the required individual space varies, depending on the type of species: some species are known as 'contact species' and frequently spend time in close contact (e.g. horses; Fraser, 1992), whereas for example hens prefer to keep the space around their heads clear of conspecifics (McBride, 1971). It appears that, although pigs do have space requirements and need space for various activities, they do not seem to have a particular requirement for personal space around their bodies and heads that is kept clear of conspecifics. It seems more likely that their tolerance of proximity depends on what behaviour they are currently engaged in and whether they have a particular 'resource' to defend (Baxter, 1985). The space between individuals also depends on the activity currently in progress. Although we can define individual or social space as 'the distance an animal attempts to keep between itself and conspecifics' (Wilson, 1975), this relationship with current behaviour means that it might be better to describe precisely what is

LIVERPOOL JOHN MOORES UNIVERSITY
LEARNING & INFORMATION SERVICES

happening at a particular time. However, proximity can itself be a causal factor for some behaviours (Nicol, 1989). Hens were found to be approximately 3 m apart when walking but only 1.5 m apart during preening (Keeling, 1994) and cattle lie 2–3 m apart but are 4–10 m apart when grazing (Fraser and Broom, 1990). Under intensive conditions, such as battery cages, animals may have little choice but to remain in constant close body contact. Attempts have been made to alleviate this problem by improving the cage environment to include perches and other facilities that allow hens to avoid each other to some extent (Appleby, 1995) and to make better use of the available area.

Careful study of the use of space and what aspects of it are important have aided design of 'escape facilities'. For example, studies of pigs showed that the space around the head was of great importance during aggressive interactions and that simply providing 'popholes' in the wall, in which the pig could hide its head and shoulders from the aggressor, significantly reduced aggression in newly mixed weaned pigs (McGlone and Curtis, 1985). On a more extensive scale, provision of visual cover for farmed red deer was found to have the effect of lowering aggressive behaviour within the herd (Whittington and Chamove, 1995). These studies show that it is possible to use visual barriers to defuse a situation that might otherwise involve a serious violation of personal space and an associated rise in aggression.

It is extremely important to provide sufficient space for animals to establish a hierarchy. Pigs, for example, require sufficient space to be able to fight using the 'parallel/inverse pressing position', whereby two pigs push each other shoulder to shoulder facing in the same or opposite directions, sometimes circling simultaneously (see Section 6.2.1). To perform this, pigs need almost two body lengths of clear space and to allow simultaneous circling may require a clear circle of that diameter. Since this behaviour is a major part of fights, it is most probably important in establishing dominance hierarchies – hence, preventing the behaviour through insufficient space provision might be expected to influence the efficiency with which pigs decide their rank relationships (Jensen, 1982; Baxter, 1985).

Lack of space might also prevent a submissive animal from retreating in defeat. This would mean that submission might not be recognized and fights might be unnecessarily prolonged. The distance that the animal needs to retreat before submission is effective may vary in different species and could also depend on the individual animal's relative rank in the dominance hierarchy. Thus, a very subordinate hen might be chased during its retreat whereas a less subordinate hen could merely be ignored by the dominant. As discussed above, the space required can be influenced by introducing artificial barriers to reduce visibility. For instance, barriers between feeding cows were found to eliminate the differences between dominants and subordinates in time spent feeding since subordinates were no longer forced to withdraw from the communal feeding space on the approach of a more dominant cow (Boissou, 1971, cited in Manson and Appleby, 1990).

2.2.5 Affiliative behaviour

The emphasis on aggression and dominance in this section reflects the amount of work that has been done in this area. Affiliative behaviour has not been investigated as thoroughly in farm animals, although horse owners have long been aware of the importance of preferred associates (e.g. Fraser, 1992). An understanding of the different roles that animals may assume within a group can help us to design husbandry systems that will function adequately and will not excessively strain their capacity to cope. For example, permitting animals to establish 'friendships' with preferred associates or avoiding mixing of unfamiliar individuals whenever possible, is known to reduce aggression in pigs, poultry and horses.

2.3 Communication Within the Group

A signal is a way in which one individual (the 'actor') causes a response in another individual (the 'reactor' or receiver; Krebs and Dawkins, 1984), i.e. it results in some form of action. We can distinguish between cooperative signalling, where the receiver benefits from detecting the signal (e.g. signals to a potential mate), and non-cooperative signalling (Harper, 1991), where it does not (e.g. unintentionally attracting a predator). Signals often develop from existing movements (or structures) that originally may have had no signalling function (the principle of 'derived activities'; Tinbergen, 1952). For example, the juvenile behaviour pattern of food begging is often part of adult courtship behaviour and food passing may appear in response, in the same context. Through the process of 'ritualization', signals have evolved from non-signals and, as a result, a behaviour pattern may have become highly repetitive and exaggerated, making it suitable for its new purpose (Krebs and Dawkins, 1984). Clearly, signals are more likely to evolve from existing behaviour patterns or morphological structures, since they would otherwise have to evolve *de novo* through genetic mutation, necessitating a parallel evolution of an appropriate response in the recipient.

To function in communication, signals need to be *received*: three main factors affect the efficiency with which this occurs (Guilford and Dawkins, 1991). Firstly, the 'detectability', i.e. the physical parameters such as intensity, duration and repetition rate. Such factors may differ depending on the habitat, e.g. birds living in a forest environment tend to have lower frequency songs compared with those living in more open habitats (Morton, 1975). This implies that the signal has become adapted to the environment. In the case of domestic animals kept in man-made environments, we therefore need to consider whether such adaptation is necessary and whether it has had the evolutionary time to occur. Secondly, 'discriminability' or how easily the signal can be discriminated from other stimuli with which it may be confused. In intensive husbandry, the density and social arrangements generally differ greatly from those in natural environments, and it seems reasonable to assume that e.g. excessive crowding might have some influence on discriminability. Thirdly, the 'memorability' of the signal, or how easily it can be remembered or associated with other relevant stimuli. This is particularly important in social contexts, such as during the formation of dominance hierarchies which rely heavily on recognition either of individuals or of some general 'status badge' (Barnard and Burk, 1979).

One of the problems facing signallers is that 'noise' (in the technical sense; Harper, 1991) may impede the detection and recognition of signals by receivers. The receiver may also need to locate the signaller before responding. One of the most common sources of interference is the displays by other signallers attempting to communicate with other receivers nearby (Johnstone, 1997). Breakdown of communication can occur at any stage, whether caused by a complete failure to detect the signal or by the receiver responding to irrelevant stimuli in the environment. Signals can be affected by several different types of noise, such as degradation through space or time, high levels of competing stimulation or even by receivers being highly selected to avoid responding to false alarms.

Different types of signalling modalities have different require-ments, some of which may be violated under normal intensive husbandry conditions. For example, sound signals can be impeded or distorted by high volume levels in the environment or by obstacles. This is particularly likely in the case of small animals (e.g. rodents), which are not physically capable of emitting low-pitched sounds that travel well. Similarly, visual signals may be designed to function in environments lacking physical obstructions or at certain light intensities. To receive a signal, it may also be necessary for the receiver to be located at a particular distance from the signaller. For example, Dawkins (1995, 1996) has shown that hens need to examine conspecifics at very close quarters (less than 30 cm) before they can decide whether a particular individual is familiar and react accord-ingly. The method of signalling is also influenced by factors such as the developmental stage of the signaller or receiver: e.g. initial sow–piglet recognition is by odour (Horrell and Hodgson, 1992), but after weaning recognition becomes increasingly multisensory (Meese and Baldwin, 1975).

Signals are known to be adapted to the environment in which they are used, including the social environment of the group. In some bird species (e.g. Richards, 1981), songs have been shown to include special alerting components, which precede the actual message part of the song. It seems plausible that, as group size and complexity increase (e.g. from jungle fowl in groups of five to eight, to modern poultry kept in groups of several thousand), such alerting components might become increasingly important. Weary and Fraser (1995b) suggested that domestication could lead to increased signalling by enhancing the response to a signal, e.g. in the form of human handlers who can assist an animal that signals its needs. The size of the group is likely to have direct effects on the amount of noise in the environment and hence on the need for repetition and redundancy of signalling components. Conversely, in barren environments, signals might become one of the few stimuli available to animals and could come to be repeated in a stereotypic manner. It has been suggested that vocalizations might be used as indicators of welfare in domestic hens: for example, work on 'gakeln' calls in domestic hens has shown that the number of calls increased during a situation of frustrating non-reward (Zimmerman and Koene, 1997). Vocal behaviour has also been used as an indicator of pain, on the assumption that the calls are reliable signals of the animal's internal state. Thus, piglets were found to vocalize for longer and more loudly during castration than did those handled in an identical way but not subjected to castration (Braithwaite *et al.*, 1995). Similarly, 'non-thriving' and 'non-fed' piglets gave more and higher frequency calls during separation from the sow, suggesting that piglets in greater need of the sow's resources called more and used different calls compared with those in lesser need (Weary and Fraser, 1995a).

2.4 Roles Within the Group

2.4.1 Strategies and models

Different individuals within a group have different roles. I have discussed the dominant/subordinate distinction, but other subdivisions can also be used. For example, Mendl *et al.* (1992) observed pigs in groups to have different propensities to fight. 'High success' pigs were characterized by low levels of inactivity, high aggression and a high involvement in social interactions. A greater number of individuals were classified as 'low success' and were aggressive despite their lack of success, experiencing high levels of aggression from and displacement by other pigs. 'No success' pigs never displaced another pig and were the most inactive, least aggressive, showing low involvement in social interactions. This study thus highlighted that pigs were able to choose whether to fight or not.

In animals reared in family groups, the role assumed by an individual may be influenced to some extent by that of its mother. Several studies described a positive correlation between the female hierarchy in a horse herd and that of their foals, both before and after weaning (Houpt *et al.*, 1978; Houpt and Wolski, 1980; Araba and Crowell-Davis, 1994). Within peer groups, foals obtained ranks comparable with those of their dams among the mares. Interestingly, one study found that the dominance hierarchy among the foals was less stable during the foaling season, when new foals were continuously being introduced to the herd, with the hierarchy stabilizing after weaning (Araba and Crowell-Davis, 1994).

While dominant individuals enjoy various privileges, such as enhanced breeding status, they are always at risk of losing their dominant position. Game theory models can be used to predict what will happen when two animals meet during a pairwise contest. The likelihood of a fight occurring is greatest if the costs of injury are low relative to the value of the resource and a challenge is most likely to materialize when the differences in fighting abilities between the contestants are least (Maynard Smith and Price, 1973). When the risk of defeat is sufficiently great for the dominant animal, it will be advantageous for it to share the resource (e.g. a reproductive opportunity), since sharing will increase the profitability to the subordinate of staying and continuing to cooperate within the group. During repeated encounters in stable groups, with a consistent and detectable difference in resource holding power (RHP) between individuals, the same individual should win each encounter (Pusey and Packer, 1997). However, there may be additional complexities since the outcome of the interaction may be based on traits that are uncorrelated with RHP (e.g. age or seniority). Individuals may even compete for dominance *per se*, rather than a particular resource.

Alliances between individuals may affect hierarchies and can have a destabilizing effect. Thus, if individual X relies on Y to help it defeat Z, X may lose the encounter if Y is absent (Bygott, 1979). Subordinates may also be unreliable in their alliances and support whoever looks most likely to win on a particular day. An alternative to dominance might be the 'ownership' of a resource, such that the first comer 'wins' access. This is more likely if contestants are very evenly matched (e.g. in a group of single-sex, single-age individuals) and the costs of fighting are high relative to the value of the resource (Maynard Smith and Price, 1973).

Within a group, cooperation will be the norm, involving repeated interactions between the same pairs of individuals. Various models, such as the 'Prisoner's Dilemma' (Pusey and Packer, 1997), have explored the evolution of cooperation through some form of reciprocity. There are various ways in which cooperation may arise, such as short- or long-term mutualism. Group living often involves advantages that can be sufficiently strong to overcome the short-term advantages of exploiting each other. An example is cooperative vigilance, and, if individuals do not cooperate, they may risk losing their companions. Another strategy is that of 'producers' and 'scroungers' (Barnard, 1984), where some individuals work to obtain resources and others gain equal access. This strategy can be viable if the scroungers gain equal access to the resource (e.g. food) without paying any extra costs. In this situation, both individuals may achieve similar pay-offs but in different ways; thus, some group members may be good at locating food, while others are good at stealing it once it has been found.

Coerced cooperation is a strategy that may arise in groups with a definite dominance hierarchy. Differences in rank (or in associated RHP) may allow some individuals to 'force' their companions to behave cooperatively, e.g. if the dominant 'punishes' any subordinate that fails to conform. Rather than being expelled from the group, it may be advantageous for the subordinate to cooperate instead of risking further punishment.

2.4.2 Proactive and reactive individuals

Group members may differ in their ways of coping with the environment (see Chapter 12). Work by Koolhaas *et al.* (1997) suggests that the most fundamental difference between group members is the degree to which their behaviour is guided by environmental stimuli. They argue that aggressive individuals develop proactive routines and seem to *anticipate* a situation, whereas non-aggressive individuals *react* to environmental stimuli all the time. This strategy creates a difference in the animals' flexibility and could be the reason aggressive individuals are more successful under stable colony conditions, while non-aggressive ones do better in variable environmental conditions such as migration. Koolhaas *et al.* showed that aggression can be associated with flight behaviour, which one would not predict based on classical motivational theory (Koolhaas *et al.*, 1997). This argument suggests that an animal's control of the environment depends on its capacity to cope; hence an animal with a proactive style of coping might have problems coping with a social environment that is variable and unstable. Stress pathology and development of abnormal behaviour patterns such as stereotypies are likely to be associated with coping style. Koolhaas *et al.* point out that the crucial factor might be the *threat* to control rather than the actual loss of control, such as when a dominant male has difficulties maintaining his dominant position.

It is also helpful to consider why animals may develop various stress-related symptoms such as abnormal behaviour patterns, which are clearly manifestations of a failure to cope with the environment imposed on them. However, by using an arbitrary cut-off point, a unimodal distribution could be made to appear bimodal (Forkman *et al.*, 1995). This means we need to take care in how coping behaviours are recorded to avoid the risk of forcing individuals into bimodal categories when none exist or where a multimodal approach would be more appropriate.

2.5 Group Size

Group size is influenced by a variety of factors, including thermo-regulatory requirements, predation pressure, disease incidence, competition for various resources, dispersal patterns, seasonal effects, and so on. In applied ethology, a fundamental question is: what group size should we keep our animals in for optimal social conditions and welfare? Unfortunately, this apparently simple question has proved difficult to answer, not least because it is difficult to ask animals in an appropriate way. Keeling (1995) discussed what exactly it is we are trying to optimize and pointed out that 'under modern husbandry conditions the word optimum is used to describe the best trade-off between the welfare cost to the animal and the financial benefits to the farmer'. For example, in the case of domestic hens the question has been posed by allowing them to spend time close to particular individuals (Hughes, 1977; Dawkins, 1982), mixing individuals from different groups and observing how they space themselves (Lindberg and Nicol, 1996a) or in the form of a preference test with or without a period of 'experiencing' their choice (Dawkins, 1982; Lindberg and Nicol, 1996a,b; Table 2.1) (see Section 7.2.3).

The most appropriate group size for a particular situation is deter-mined by costs and benefits, which combine to influence survival and reproduction. Optimality models have been used to illustrate factors affecting group size, e.g. Caraco and Pulliam (1980) studying sparrows used time budgets of behaviours such as scanning, fighting and feeding, which are associated with the risk of starvation and predation (Pulliam and Caraco, 1984). Using this model, they predicted that, as scanning for predators decreases and fighting increases with increasing group size, the maximum amount of time could be spent on feeding at intermediate group sizes where these two costs cancel each other out. However, this is a simplistic view since many other factors will also affect group size, such as the dominance status of individuals.

Table 2.1. This table illustrates how aggressive pecking in small (four individuals) and large (44 individuals) groups of domestic hens is affected both by group size and by the level of familiarity of the hens (*P* < 0.001). In very large groups, reduced aggression is often observed.

Group size	Familiarity level	No. of aggressive pecks per hen per 5 min observation ± SE
Small (4)	Familiar	0.2 ± 0.1
Small (4)	Semi-familiar	1.1 ± 0.6
Small (4)	Unfamiliar	1.8 ± 0.5
Large (44)	Semi-familiar	0.1 ± 0.1
Large (44)	Unfamiliar	0.1 ± 0.1

Data from Lindberg (1994).

Dominant and subordinate animals may have different optimal group sizes (Lindberg and Nicol, 1996b; Pagel and Dawkins, 1997), hence the actual group size must necessarily be a compromise. For example, dominant hens had a stronger preference to enter a large group than did subordinates, and once in their chosen groups they behaved differently (Lindberg and Nicol, 1996b). Although we might hypothesize that subordinate hens should prefer a small group where they might more easily assess each other's dominance status and act accordingly, this does not seem to be the case unless the small group is combined with sufficient space. Subordinate hens given a choice between a large group and a small group at the same density, showed a preference for the larger group, perhaps because it is easier to 'blend into' a larger group, either adopting the non-intervention strategy (Section 2.2) or moving away from conflicts. Individuals in different groups could therefore benefit from group living to different extents, but some may be willing to put up with less pay-off than others so long as they could not do better by moving elsewhere (Vehrencamp, 1983). Individuals may also use different strategies for creating social hierarchies depending on the size of the group. This is illustrated by Pagel and Dawkins (1997), whose work on models suggests that as the group gets larger it may no longer pay for the animal to use individual recognition. Instead, there will be a gradual shift to using 'badges of status' (Rowher, 1975; Pagel and Dawkins, 1997) which can be applied to any conspecific without a need to pay a cost of establishing individual recognition. This implies that by following a different type of strategy, group members can adjust to a different group size. Alternatively, individuals may form subgroups within the larger flock, within which hierarchies based on individual recognition may exist (McBride, 1964; Pagel and Dawkins, 1997), although several authors reported weak or no evidence for this in domestic fowl (Hughes *et al.*, 1974; Appleby *et al.*, 1985, 1989; Preston and Murphy, 1989; Widowski and Duncan, 1995).

An important point was made by Sibly (1983): groups of optimal size may be very rare in the wild because, if there was a group of this size, it would pay any solitary individuals to join the group and therefore push it above optimal size. This means that optimal group sizes may be unstable. Hence, in nature we might instead find stable groups which will often be larger than the optimum (Pulliam and Caraco, 1984). The elusiveness of preferred group size (Dawkins, 1982) might be explained by the diversity of factors that play a part in optimality determination. Thus, very rarely will the 'optimum' be the same for all individuals in a flock at all times. This means that, when deciding what group sizes to keep domestic animals in, we need to consider how the group is to function at a practical level: the stability of group membership, density and physical environment may be more important than the size of the group *per se*.

References

Al-Rawi, B. and Craig, J.V. (1975) Agonistic behaviour of caged chickens related to group size and area per bird. *Applied Animal Ethology* 2, 69–80.

Appleby, M.C. (1995) The Edinburgh Modified Cage for laying hens. *British Poultry Science* 36, 707–718.

Appleby, M.C., Maguire, S.N. and McRae, H.E. (1985) Movement by domestic fowl in commercial flocks. *Poultry Science* 64, 48–50.

Appleby, M.C., Hughes, B.O. and Hogarth, G.S. (1989) Behaviour of laying hens in a deep litter house. *British Poultry Science* 30, 545–553.

Araba, B.D. and Crowell-Davis, S.L. (1994) Dominance relationships and aggression of foals (*Equus caballus*). *Applied Animal Behaviour Science* 41, 1–25.

Banks, E.M., Wood-Gush, D.G.M., Hughes, B.O. and Mankovich, N.J. (1979) Social rank and priority of access to resources in domestic fowl. *Behavioural Processes* 4, 197–209.

Barnard, C.J. (1984) *Producers and Scroungers*. Chapman & Hall, New York.

Barnard, C.J. and Burk, T. (1979) Dominance hierarchies and the evolution of 'individual recognition'. *Journal of Theoretical Biology* 81, 65–73.

Baxter, M. (1985) Social space requirements of pigs. In: Zayan, R. (ed.) *Social Space for Domestic Animals*. Martinus Nijhoff Publishers, Dordrecht, pp. 116–127.

Beilharz, R.G. and Zeeb, K. (1982) Social dominance in dairy cattle. *Applied Animal Ethology* 8, 79–97.

Blackshaw, J.K., Blackshaw, A.W. and McGlone, J.J. (1997) Buller steer syndrome review. *Applied Animal Behaviour Science* 54, 97–108.

Boissou, M.F. (1971) Effect de l'absence d'information optiques et de contact physique sur la manifestation des relations hiérarchiques chez bovins domestiques. *Annales de Biologie Animale, de Biochimie et de Biophysique* 11, 191–198.

Braithwaite, L.A., Weary, D.M. and Fraser, D. (1995) Can vocalizations be used to assess piglets perception of pain? In: Rutter, S.M., Rushen, J., Randle,

H.D. and Eddison, J.C. (eds) *Proceedings of the 29th International Congress of the International Society for Applied Ethology.* Universities Federation for Animal Welfare, South Mimms, UK, pp. 21–22.

Bygott, J.D. (1979) Agonistic behaviour, dominance, and social structure in the wild chimpanzees of the Gombe National Park. In: Hamburg, D.A. and McCown, E.R. (eds) *The Great Apes.* Benjamin Cummings, Menlo Park, California, pp. 405–427.

Caraco, T. and Pulliam, H.R. (1980) Time budgets and flocking dynamics. In: *Proceedings of the XVII International Ornithological Congress.* Berlin, Germany.

Craig, J.V. (1986) Measuring social behaviour: social dominance. *Journal of Animal Science* 62, 1120–1129.

Cunningham, D.L. and van Tienhoven, A. (1983) Relationship between production factors and dominance in White Leghorn hens in a study on social rank and cage design. *Applied Animal Ethology* 11, 33–44.

Dawkins, M.S. (1982) Elusive concept of preferred group size in domestic hens. *Applied Animal Ethology* 8, 365–375.

Dawkins, M.S. (1995) How do hens view other hens? The use of lateral and binocular visual fields in social recognition. *Behaviour* 132, 591–606.

Dawkins, M.S. (1996) Distance and social recognition in hens: implications for the use of photographs as social stimuli. *Behaviour* 133, 663–680.

Erhard, H.W. and Mendl, M. (1997) Measuring aggressiveness in growing pigs in a resident–intruder situation. *Applied Animal Behaviour Science* 54, 123–136.

Forkman, B., Furuhaug, I.L. and Jensen, P. (1995) Personality, coping patterns and aggression in piglets. *Applied Animal Behaviour Science* 45, 31–42.

Fraser, A.F. (1992) *The Behaviour of the Horse.* CAB International, Wallingford, UK.

Fraser, A.F. and Broom, D.M. (1990) *Farm Animal Behaviour and Welfare.* Baillière Tindall, London.

Fraser, D. and Rushen, J. (1987) Aggressive behaviour. In: Price, E.O. (ed.) *Farm Animal Behaviour. Veterinary Clinics of North America: Food Animal Practice* 3, pp. 285–305.

Freire, R., Appleby, M.C. and Hughes, B.O. (1997) Assessment of pre-laying motivation in the domestic hen using social interaction. *Animal Behaviour* 54, 313–319.

Giersing, M. and Andersson, A. (1998) How does former acquaintance affect aggressive behaviour in repeatedly mixed male and female pigs? *Applied Animal Behaviour Science* 59, 297–306.

Grigor, P.N., Hughes, B.O. and Appleby, M.C. (1995) Social inhibition of movement in domestic hens. *Animal Behaviour* 49, 1381–1388.

Guhl, A.M. and Allee, W.C. (1944) Some measurable effects of social organization in flocks of hens. *Physiological Zoology* 17, 320–347.

Guilford, T. and Dawkins, M.S. (1991) Receiver psychology and the evolution of animal signals. *Animal Behaviour* 42, 1–14.

Harper, D.G.C. (1991) Communication. In: Krebs, J.R. and Davies, N.B. (eds) *Behavioural Ecology: an Evolutionary Approach*, 3rd edn. Blackwell Scientific Publications, Oxford, pp. 374–397.

Horrell, I. and Hodgson, J. (1992) The bases of sow–piglet identification. 1. The identification of sows by their own piglets and the presence of intruders. *Applied Animal Behaviour Science* 33, 319–327.

Houpt, K.A. and Wolski, T.R. (1980) Stability of equine hierarchies and the prevention of dominance related aggression. *Equine Veterinary Journal* 12, 18–24.

Houpt, K.A., Law, K. and Martinisi, V. (1978) Dominance hierarchies in domestic horses. *Applied Animal Ethology* 4, 273–283.

Hughes, B.O. (1977) Selection of group size by individual laying hens. *British Poultry Science* 18, 9–18.

Hughes, B.O. and Wood-Gush, D.G.M. (1977) Agonistic behaviour in domestic hens: the influence of housing method and group size. *Animal Behaviour* 25, 1056–1062.

Hughes, B.O., Wood-Gush, D.G.M. and Morley Jones, R. (1974) Spatial organization in flocks of domestic fowl. *Animal Behaviour* 22, 438–445.

Hughes, B.O., Carmichael, N.L., Walker, A.W. and Grigor, P.N. (1997) Low incidence of aggression in large flocks of laying hens. *Applied Animal Behaviour Science* 54, 215–234.

Jensen, P. (1982) An analysis of agonistic interaction patterns in group-housed dry sows – aggression regulation through an 'avoidance order'. *Applied Animal Ethology* 9, 47.

Jensen, P. (1994) Fighting between unacquainted pigs – effects of age and of individual reaction pattern. *Applied Animal Behaviour Science* 41, 37–52.

Johnstone, R. (1997) The evolution of animal signals. In: Krebs, J.R. and Davies, N.B. (eds) *Behavioural Ecology: an Evolutionary Approach*, 4th edn. Blackwell Science, Oxford, pp. 155–178.

Jones, R.B. and Faure, J.M. (1982) Tonic immobility in the domestic fowl as a function of social rank. *Biology of Behaviour* 7, 27–32.

Keeling, L.J. (1994) Inter-bird distances and behavioural priorities in laying hens: the effect of spatial restriction. *Applied Animal Behaviour Science* 39, 131–140.

Keeling, L.J. (1995) Spacing behaviour and an ethological approach to assessing optimum space allocations for groups of laying hens. *Applied Animal Behaviour Science* 44, 171–186.

Keiper, R.R. and Sambraus, H.H. (1986) The stability of equine dominance hierarchies and the effects of kinship, proximity and foaling status on hierarchy rank. *Applied Animal Behaviour Science* 16, 120–131.

Koolhaas, J.M., de Boer, S.F. and Bohus, B. (1997) Motivational systems or motivational states: behavioural and physiological evidence. *Applied Animal Behaviour Science* 53, 131–143.

Krebs, J.R. and Dawkins, R. (1984) Animal signals: mind-reading and manipulation. In: Krebs, J.R. and Davies, N.B. (eds) *Behavioural Ecology: an Evolutionary Approach*, 2nd edn. Blackwell Scientific Publications, Oxford, pp. 380–402.

Lee, Y.P. and Craig, J.V. (1981) Agonistic and non-agonistic behaviour of pullets of dissimilar strains of White Leghorns when kept separately and intermingled. *Poultry Science* 60, 1759.

Lindberg, A.C. (1994) The improvement of alternative systems for keeping laying hens: a behavioural approach to the problems of group size and feather pecking. PhD thesis, University of Bristol.

Lindberg, A.C. and Nicol, C.J. (1996a) Effects of social and environmental familiarity on group size preferences and spacing behaviour in laying hens. *Applied Animal Behaviour Science* 49, 109–123.

Lindberg, A.C. and Nicol, C.J. (1996b) Space and density effects on group size preferences in laying hens. *British Poultry Science* 37, 709–721.

Manson, F.J. and Appleby, M.C. (1990) Spacing of dairy cows at a food trough. *Applied Animal Behaviour Science* 26, 69–81.

Maynard Smith, J. and Price, G.R. (1973) The logic of animal conflict. *Nature (London)* 246, 15–18.

McBride, G. (1964) Social discrimination and sub-flock structure in the domestic fowl. *Animal Behaviour* 12, 264–267.

McBride, G. (1971) Theories of animal spacing: the role of flight, fight and social distance. In: Esser, A.H. (ed.) *Behaviour and the Environment: the Use of Space by Animals and Men*. Plenum Press, New York, pp. 53–68.

McGlone, J.J. and Curtis, S.E. (1985) Behaviour and performance of weanling pigs in pens equipped with hide areas. *Journal of Animal Science* 60, 20–24.

Meese, G.B. and Baldwin, B.A. (1975) The effects of ablation of the olfactory bulbs on aggressive behaviour in pigs. *Applied Animal Ethology* 1, 251–262.

Mendl, M. and Newberry, R.C. (1997) Social conditions. In: Appleby, M.C. and Hughes, B.O. (eds) *Animal Welfare*. CAB International, Wallingford, pp. 191–203.

Mendl, M., Zanella, A.I. and Broom, D.M. (1992) Physiological and reproductive correlates of behavioural strategies in female domestic pigs. *Animal Behaviour* 44, 1107–1121.

Monaghan, P. (1990) Social behaviour. In: Monaghan, P. and Wood-Gush, D. (eds) *Managing the Behaviour of Animals*. Chapman & Hall, London, pp. 48–71.

Morton, E.S. (1975) Ecological sources of selection on avian sounds. *American Naturalist* 109, 17–34.

Nicol, C.J. (1989) Social influences on the comfort behaviour of laying hens. *Applied Animal Behaviour Science* 22, 75–81.

Pagel, M. and Dawkins, M.S. (1997) Peck orders and group size in laying hens: 'futures contracts' for non-aggression. *Behavioural Processes* 40, 13–25.

Preston, A. and Murphy, L.B. (1989) Movement of broiler chickens reared in commercial conditions. *British Poultry Science* 30, 519–532.

Pulliam, H.R. and Caraco, T. (1984) Living in groups: is there an optimal group size? In: Krebs, J.R. and Davies, N.B. (eds) *Behavioural Ecology: an Evolutionary Approach*, 2nd edn. Blackwell Scientific Publications, Oxford, pp. 122–147.

Puppe, B. (1998) Effects of familiarity and relatedness on agonistic pair relationships in newly mixed domestic pigs. *Applied Animal Behaviour Science* 58, 233–239.

Pusey, A.E. and Packer, C. (1997) The ecology of relationships. In: Krebs, J.R. and Davies, N.B. (eds) *Behavioural Ecology: an Evolutionary Approach*, 4th edn. Blackwell Science, Oxford, pp. 254–283.

Richards, D.G. (1981) Alerting and message components in songs of rufous-sided towhees. *Behaviour* 76, 223–249.

Rowher, S. (1975) Dyed birds achieve higher social status than controls in Harris' Sparrows. *Animal Behaviour* 33, 1325–1331.

Rushen, J. (1982) The peck orders of chickens: how do they develop and why are they linear? *Animal Behaviour* 30, 1129–1137.

Schjelderup-Ebbe, T. (1935) Social behaviour in birds. In: Murchison, C. (ed.) *Handbook of Social Psychology*. Clark University Press, Worcester, Massachusetts, pp. 947–972.

Sibly, R.M. (1983) Optimal group size is unstable. *Animal Behaviour* 31, 947–948.

Stolba, A. and Wood-Gush, D.G.M. (1984) The identification of behavioural key features and their incorporation into a housing design for pigs. *Annales de Recherches Vétérinaires* 15, 287–298.

Stricklin, W.R. and Mench, J.A. (1987) Social organization. *Veterinary Clinics of North America: Food Animal Practice* 3, 307–322.

Syme, G. J. (1974) Competitive orders as measures of social dominance. *Animal Behaviour* 22, 931–940.

Syme, G.J. and Syme, L.A. (1979) *Social Structure in Farm Animals*. Elsevier Scientific, Amsterdam.

Tinbergen, N. (1952) Derived activities: their causation, biological significance, origin and emancipation during evolution. *Quarterly Review of Biology* 27, 1–32.

Vehrencamp, S.L. (1983) A model for the evolution of despotic versus egalitarian societies. *Animal Behaviour* 31, 667–682.

Weary, D.M. and Fraser, D. (1995a) Calling by domestic piglets: reliable signals of need? *Animal Behaviour* 50, 1047–1055.

Weary. D.M. and Fraser, D. (1995b) Signalling need: costly signals and animal welfare assessment. *Applied Animal Behaviour Science* 44, 159–169.

Wechsler, B. (1996) Rearing pigs in species-specific family groups. *Animal Welfare* 5, 25–35.

Whittington, C.J. and Chamove, A.S. (1995) Effects of visual cover on farmed red deer behaviour. *Applied Animal Behaviour Science* 45, 309–314.

Widowski, T.M. and Duncan, I.J.H. (1995) Do domestic fowl form groups when resources are unlimited? *Applied Animal Behaviour Science* 44, 280.

Wiepkema, P.R. and Schouten, W.G.P. (1990) Mechanisms of coping in social situations. In: Zayan, R. and Dantzer, R. (eds) *Social Stress in Domestic Animals*. Kluwer Academic Publications, Dordrecht, pp. 8–24.

Wilson, E.O. (1975) *Sociobiology, the Modern Synthesis*. Harvard University Press, Cambridge, Massachusetts.

Zimmerman, P.H. and Koene, P. (1997) Differences in reaction to omission of reward in two strains of laying hens *Gallus gallus domesticus*. In: Koene, P. and Blokhuis, H.J. (eds) *Proceedings of the 5th European Symposium on Poultry Welfare*. WPSA, Wageningen, the Netherlands, pp. 71–73.

Parental Behaviour

3

Per Jensen

Department of Animal Environment and Health, Swedish University of Agricultural Sciences, Skara, Sweden

(Editors' comments: A unique social relationship in the animal kingdom is that between parent and offspring. Even though the preceding chapter has discussed relationships between individuals in a group, this special relationship needs to be addressed in a chapter of its own. There are several reasons for this. The first is the great variety of types of parent–offspring relationships that are found in even our most commonly occurring farm animals. Mammals suckle their young, but vary in how they do this depending on the number of young and how well developed the young are when they are born. Birds help their chicks find food and may brood them when they require warmth, but to a large extent chicks must fend for themselves. Fish are at the extreme of this continuum and in the majority of species there is no parental care whatsoever.

Despite this variation, the dominant farm animal species are mammals and this means that at some point in the development of the young, it has to be weaned from its mother's milk. How this weaning is done in commercial practice varies, and is explained more in the species-specific chapters, but the principles underlying it are common to parents and offspring of all species, not just mammals, and are covered in the concept of parent–offspring conflict. This is addressed, very much as in the first chapter, from an evolutionary perspective. The situation where the offspring try to get greater investment from the parent than the parent is willing to give leads to a discussion on the reliability of signals from the offspring.)

©CAB *International* 2001. *Social Behaviour in Farm Animals*
(eds L.J. Keeling and H.W. Gonyou)

3.1   Introduction

This book is on social behaviour, but many animals are thought of as profoundly non-social. A wide range of species live a life more or less on their own, fiercely driving away any intruders from the private territories. Among mammals, we need only think of, for example, mink or hamsters.

However, no matter how non-social an animal appears to be, there are at least two periods in any life history that require extensive social interaction with conspecifics: mating and parental care. For example, the normally solitary female mink spends some 10 weeks every year in daily, intense interactions with its young (Dunstone, 1993), and these interactions contain the same elements as any other social interactions: recognition, cooperation, competition, signalling and so forth. Furthermore, rarely are the fitness costs and benefits of different social behaviour patterns as obvious as in the case of parental behaviour. Young are often helpless and vulnerable to starvation, predation and cold and their survival and fitness rely to a large extent on the ability of the parents to rightly assess their needs and adjust the behaviour accordingly. At the same time, offspring are in fact the fitness units of the parents. If the parent succeeds in its interactions, the offspring will have a higher survival probability and hence the fitness of the parent increases. In case of failure to interact properly, fitness may decrease considerably due to loss of young. So, when observing a given social behaviour between a parent and its offspring we may be more likely to make a correct interpretation of the fitness implications of that behaviour than in many other situations of social interactions between conspecifics.

Given this background, it is surprising that parental behaviour is often overlooked in textbooks on social behaviour. It is true that it does have some unique properties, but, generally, parental behaviour is nothing but one special case of social interaction. It is one of the social behaviours that no species with developed offspring care can fail to express.

This chapter will provide an overview of the social aspects of parental behaviour with a clear emphasis on birds and mammals. It will attempt to highlight the ecological and evolutionary backgrounds to, and the functions of, different types of parental behaviour.

3.2   Parent–Offspring Conflict

Trivers (1974) was the first to suggest that the evolutionary interests of parents and their young were not the same, an idea which has been termed parent–offspring conflict. In short it means that offspring benefit in an evolutionary sense from a higher level of parental care than what is optimal for the parents, since parents have to allocate resources to all offspring, both present and future. The theory has been elaborated and analysed in several mathematical models and been found to be consistent and logical, and the predictions of the theory have been found to hold in a variety of species (Macnair and Parker, 1978, 1979; Lazarus and Inglis, 1986; Godfray, 1995a). Since the parent–offspring conflict theory provides a theoretical framework within which it is possible to understand communication and inter-action, it is necessary to understand some of the details of this theory.

The reason why parents and offspring have different optima for how many resources should be allocated to the young is to be sought in the genetic relatedness between the individuals. As realized by Trivers, in an outbred population of animals, the coefficient of relatedness (r) between a parent and all its offspring is on average 0.5, the same as between any two full siblings. This means that any cost to a parent is only half as big to the offspring (the terms cost and benefit in this chapter refer to fitness costs unless stated otherwise, although this is sometimes indirectly assessed through energy costs). At any given time, the offspring are thus expected to gain in fitness by achieving more care than the parent is willing to give.

A wide array of social interactions between parents and their young can be interpreted within the context of parent–offspring conflict theory, which thus provides a general theoretical framework for this complex of behaviour. In the continuation of this chapter, I will more closely describe some different aspects of parental behaviour and wherever appropriate, I will use parent–offspring conflict theory as a theoretical framework for interpretation.

3.3 Varieties of Parental Behaviour

The forms and means of parental care are as varied as nature itself. Some species, such as salmon (Petersson *et al.*, 1996), lay millions of eggs but provide absolutely no parental care after the young have hatched. Others, such as elephants, give birth only to a single young, which is cared for extensively for many years (Lee, 1987). Many birds, such as great tits and starlings, and all mammals actively feed the young with food that has been caught, prepared or produced, whereas others, for example, pheasants, hens and ducks, do not feed the young at all, but interact with them in order to attract them to food sources and stimulate them to obtain food on their own. Within this diversity of nature, parental behaviour still can be explained by some rather basic evolutionary principles, for example parent–offspring conflict theory, which allow us to interpret the variations in a functional context.

Among birds and mammals, newborn offspring in general belong to one of two different groups: altricial and precocial. Altricial species hatch or are born early in development, and are therefore usually unable to walk or locomote, and they may have undeveloped sensory systems and be blind and deaf. Their capacity for social behaviour is therefore strongly limited in early life and only when they themselves can start to react to signals from the parent can we speak about true interaction. In contrast, precocial young are born or hatched late in development, and are usually capable of locomotion quite early. They may seek and ingest food with little help from the parents and actively

Fig. 3.1. Four examples of species with different reproductive strategies with respect to litter size and developmental stage of the young. (a) Cattle, as most ungulates, are oligotocous with precocial young; (b) pigs are untypical ungulates, being polytocous with precocial young; (c) primates, such as humans, are oligotocous with altricial young; (d) cats and other carnivores are polytocous with altricial young.

initiate social interactions shortly after birth. The early social inter-actions between parents and offspring are therefore strongly dependent on what category the species belong to.

Among mammals, it appears that there is a plausible evolutionary reason for whether offspring are altricial or precocial. Altricial young are typical among carnivores, which might be expected to be hampered in their hunting behaviour if the pregnant mother has to carry its young for a long period. In contrast, typical prey animals such as ungulates are usually precocial, and it appears that there are functional advantages of having newborn young that are fully capable of following the mother soon after birth and having an active and developed anti-predator response. However, this cannot be the only factor of importance, since altricial young are also common among, for example, rodents and primates, which are largely prey animals. A general rule of thumb for mammals seems to be that species with one or few young (oligotocous) generally are precocial and species with larger litters (polytocous) are altricial. There are some noteworthy exceptions to this rule. Pigs and some rodents belonging to Caviomorpha (e.g. guinea-pigs) are polytocous but have extremely precocial young. Primates are usually oligotocous, but the young are best described as altricial. So, whereas the developmental state of the newborn young is likely to reflect an evolutionary adaptation, the selective factors of importance vary, probably in a complex manner.

3.4 Early Parental Care

Parental expenditure starts already before conception, by production of gametes. Some eggs are large and rich in nutrients, and take a considerable expenditure on the part of the mother (Clutton-Brock, 1991). Obviously, parents provide care for the fertilized ovum and for the fetus, and from the parental perspective, pre-birth provisioning may make up a large part of the total investment in offspring. Whereas this is mostly dealt with by means of physiological processes, it may also include some behaviour. For example, most birds and many mammals construct elaborate nests and this activity takes time and energy. A pig nest takes up to about 10 h to construct, and demands a large effort on the part of the sow (Jensen, 1989, 1993). Foxes dig dens for their young, which may be enlarged and improved from year to year, and may represent an impressive cumulative effort (Malm, 1995a). In addition, many mothers spend considerable energy in searching for suitable birth or nest sites and in seeking isolation from other members of the herd. Pigs in a semi-natural enclosure, for example, may walk distances of 5–10 km before selecting a site which is sufficiently protected and isolated (Jensen, 1988, 1989).

Impressive and interesting as such behaviour may be, these aspects of parental expenditure contain little of social interaction. Parental behaviour becomes a social behaviour only when the young are hatched or born and start responding to signals and emit signals (but many birds, for example domestic hens, communicate vocally with the young while they are still in the eggs). Even in the most altricial species, some exchange of information between parents and offspring is evident very early. For example, young of many rodents emit ultrasound vocalizations to communicate during the first days of life (Noirot, 1972). Of course, in precocial species, a rich variety of communication occurs immediately after birth. In mammals, most of it seems to be concerned with the need to quickly form a strong bond between mother and offspring, and a lot goes through the chemical senses, taste and smell, mediated via the licking of the newborn (Baldwin and Shillito, 1974; Townsend and Bailey, 1975; Alexander et al., 1986).

Licking is usually conceived of as a means for the mother to clean and dry the young and to establish a quick, imprinting-like, exclusive bonding to the offspring (Klopfer et al., 1964; Lickliter, 1984). However, it also transfers important nutrients and hormones from fetal fluids and membranes to the mother (Gubernick and Alberts, 1983).

A few mammals do not lick the neonates. Pigs give birth to young which are left completely on their own in order to free themselves of the umbilical cord and the fetal membranes, and are never licked by the mother (Signoret et al., 1975; Fraser, 1984). The same appears to be the case with camelids (Vila, 1992). It is not known whether the behaviour has never developed in these animals, or whether it has been dropped during evolutionary history (wild boars, the ancestors of domestic pigs, do not lick the young either; Gundlach, 1968).

3.5 Mother–Offspring Bonds

The quick forming of stable bonds between mother and offspring is essential. With respect to the early social response of the precocial ungulate neonates, the species form two different groups, which have been named hiders and followers (Lent, 1974). The hider species are those where the young remain at or near the birth site for some time after parturition, usually many days, while the mother may be away for long time periods. Followers are those with offspring which immediately start moving around with the mother and are seldom or never away from her. The young of hiders can afford to have a low degree of social responsiveness, since the mothers actively seek them out at feeding times, whereas followers must be fully sensitive and interactive from birth.

At first glance it may seem that followers would have a greater need for establishing a strong social bond, but the same actually goes for hiders. Unless the mother forms a strong bond to the newborn, the motivation to leave the grazing areas and return to the hide may be weak. Also, as most mammals give birth in a relatively synchronized manner, there will usually be a number of different young hiding around the feeding areas of the adults, and it is necessary for the mother to be able to distinguish her own young with certainty (Green and Rothstein, 1993).

Hiders and followers are often thought of as distinct categories of ungulates. However, there may be a considerable flexibility in the response pattern of the young within a species. For example, cattle are usually conceived of as a hider species, but studies of cattle under semi-natural conditions have shown that calf behaviour may range from a typical hider strategy to a follower strategy, depending on ecological factors (Lidfors and Jensen, 1988; Lidfors *et al.*, 1994a). It is not clear to what extent the behaviour of the mothers affect the behaviour of the calves in this context, but it seems likely that signals from the dam may modify offspring behaviour (see Section 5.1.4). It has been observed in pigs, a profound hider species, where piglets usually remain in the nest for about a week post-partum, that by emitting vocal signals the sow can attract the piglets and make them move to another nest site if the conditions (for example climatic) are unfavourable in the first one (P. Jensen, unpublished observations). This indicates that the mother may affect whether the young remain in hiding or not (see Section 6.1.6).

Mechanisms of the formation of maternal bonds have been the focus of much research. In mammals, the process involves mainly the chemical senses, and only later hearing and vision are used in mutual recognition (Klopfer *et al.*, 1964; Horrell and Eaton, 1984; Lickliter, 1984; Hepper, 1987; Romeyer, 1993). In birds, the imprinting phenomenon is well known, and it involves mainly vision and hearing (Bateson, 1965; Immelman, 1972; Gallagher, 1977). Sometimes imprinting is treated as a special phenomenon in ethology texts. However, it has few unique features, and can very well serve as a model for learning in general, and social recognition in particular (Johnson *et al.*, 1992). For example, imprinting seems to depend on predispositions. Animals will not imprint equally well on any arbitrary visual stimulus, but rather have some very specific set-ups of preconceptions about what constitutes meaningful stimuli for social recognition. Chickens, for example, seem to have an innate crude picture of an archetypical hen head, which strongly enhances rapid imprinting of the mother's appearance (Johnson *et al.*, 1992). Other important aspects of imprinting, which have implications for the social relationships between mothers and offspring, are the stable character of the learned recognition ('irreversibility') and the fact that it can usually only take place during a limited ontogenetic phase ('sensitive period'). This means that maternal bonds (which involve social recognition) can be created only at a certain developmental time and that the recognition and associated bonds remain for a long time, sometimes into adulthood (Hepper, 1994).

3.6   Types of Food Provisioning

Although some animals use considerable energy and take great risks in defending nestlings and young (Carl and Robbins, 1988; Hogstad, 1993), food provisioning clearly stands out as the largest part of parental expenditure. Consequently, it is probably the resource causing most communication and conflicts between parents and offspring.

Many birds actively feed their young. Often, as in most insectivorous and carnivorous birds, the food is caught by the parent and brought to the young, where it is either fed actively by placing it directly into the beak or the throat of the nestling or (in the case of larger prey) is torn in pieces by the parent and offered to the young (Bustamante, 1994; Hamer and Hill, 1994).

A seemingly more energy-demanding provisioning is performed by many birds that transport food a long distance, often from sea or lakes, and where the food is provided to the young in a pre-digested form. Here the parents eat the food, e.g. fish, and store it in the stomach, where it is sometimes partly processed by digestive hormones. As the parent comes to the nest, the nestlings signal their urge for food and the parent regurgitates it directly into the beaks of the young. The well-known pecking of young gulls towards the red spot on the beak of the parent is part of the signalling from the offspring in this connection (Tinbergen and Perdeck, 1950).

An even more demanding feeding method among birds is found among doves, where both parents use the food they have ingested to produce a substance rich in nutrients, called crop milk, which in turn can be regurgitated to the young (Lehrman, 1964).

Among mammals, feeding of young has taken one further step. As in doves, the food ingested by the mother is processed to provide energy and nutrients for herself, but also for producing a special, nutritious and homogeneous food for the young, the milk. However, in contrast with the doves, all placental mammals produce and provide the milk in specially adapted organs, consisting of mammary tissue, alveoli, milk ducts, milk cisternae and teats. The actual transfer of the milk to the young is done during discrete sucklings. In order to achieve a suckling, the mother and the offspring need to communicate extensively. They need to maintain proximity before the suckling and they need to coordinate their behaviour to make the mother stand or lie in the right position for the young to get full access to the teats. At the same time, the mother needs to ascertain that she is providing for her own young, so identity signalling is common. The occurrence of intense communication is also reflected in the fact that, in many species, actual milk flow constitutes a very small part of the total suckling time: for example, in pigs only about 20 s out of about 3–10 min of interactions at every suckling (Fraser, 1980), and in cattle about 1–2 min out of 10–15 min of interaction (Lidfors *et al.*, 1994a). The rest of the time is used mainly for communication.

3.7 Communication During Suckling

The communication in connection with suckling can take various forms. In species with one or a few offspring, it is sometimes sufficient with a brief nose contact before actual suckling starts in order to establish the individual identity, followed by the mother and the young assuming the suckling position. This is what happens in sheep and, while in the reverse parallel suckling position, the ewe sniffs the anogenital region and the lamb wags its tail intensely, presumably to spread pheromones more efficiently to the mother (Ewbank, 1967; Festa-Bianchet, 1988). In this species, the milk ejection happens fast and the suckling is rapid, whereas, for example in cattle, the actual suckling bout lasts for some time and includes additional communication. Before milk ejection, there is about 1 min of sucking and butting and, after the milk ejection, there is a period of several minutes of continued sucking from different teats and some butting (Lidfors *et al.*, 1994a). According to Lidfors *et al.* (1994a), the sucking after actual milk ejection may partly be a way for the calf to stimulate further milk production, and thereby communicate its nutritive needs to the dam.

The most complex interaction pattern described in connection with suckling is the one found in pigs (Fraser, 1980). During a suckling, the sow emits regular deep grunts at a rate of about one grunt every 2 s in the period preceding milk ejection. About 20 s before milk ejection, the grunt rate suddenly increases to a peak of about two grunts per second, and then wanes. It has puzzled some researchers that the grunt peak occurs some time before the milk ejection, since one could expect that the sow would communicate precisely this ejection to the piglets. However, it has been found that the grunt peak coincides with the release of oxytocin (the hormone which triggers the milk ejection) from the pituitary into the bloodstream (Ellendorff *et al.*, 1982; Castrén *et al.*, 1989). It therefore appears that the sow actually communicates the hormone release, not the milk ejection. The time from hormone release until milk is ejected is quite constant and depends only on the speed the substance moves with the blood to the mammary tissue. Once it reaches its target, the reaction is a stereotyped all-or-nothing reaction, where the smooth muscles of the milk alveoli contract for a fixed period of time (about 20 s). This means that once the hormone is released, there is no way for the piglets to alter the outcome until the actual milk ejection is over.

Information concerning the oxytocin release is essential for the piglets, because they spend a substantial amount of energy in massaging the teats by rubbing the noses against the udder while they await the milk ejection. They have to do this for about 1 min in order to trigger the oxytocin release, and it is necessary that this is a concerted action on the part of the litter, otherwise no milk will be ejected. It has been suggested that the grunt pattern is a way for the sow to make sure that the milk is allocated to all piglets in the litter and a method to prevent some individuals in the litter from obtaining a disproportionately large part of the milk (Algers, 1993). Whereas this is probably one part of the truth, piglets get very uneven rations anyhow. The milk output from different teats along the udder of any individual sow may differ by 200% or more (Algers, 1993), and, since piglets have an almost exclusive teat order (Fraser, 1980), it means that different piglets in the litter get quite different amounts. The consequence of this system is that no piglet can obtain milk on its own, whereas the sow still has the possibility of varying investment in different young within the same litter, perhaps in accordance with their specific needs.

It may seem obvious that the sow gruntings serve to communicate with the piglets (and not with, for example, other sows in the vicinity), but what is the actual evidence? This was examined in a study where the grunts were masked by playing back loud ventilator fan noise at a level which overshadowed most of the grunts of the sow (Algers and Jensen, 1985). Piglets exposed to this situation lost the within-litter coordination and massaged less during the period before milk ejection, and they also tended to obtain less milk (Algers and Jensen, 1991), which strongly indicates that the grunting has a communication function.

However, the interactions do not stop once the milk is consumed after the approximately 20 s milk ejection. Directly following this, the piglets resume massaging and may continue for several minutes. This so-called post-massage has been the target of some discussion and research, since it is unclear which function it fills. One early suggestion was that the piglets used this behaviour to scent mark their own teats by rubbing facial glands against the udder (McBride, 1963), and indeed piglets have greater problems in locating their own teats if the udder is washed with a detergent between sucklings (Jeppesen, 1982). However, such scent marking may occur at any time during a suckling (presumably already during the first massage movements before milk ejection), and the question still remains why the piglets are so persistent after the milk has been consumed. The most likely explanation seems to be that piglets, by the intensity and duration of their massage activities, actually communicate to the sow their degree of satiation in relation to their needs and affect the future milk production of their own teat. In one experiment, piglets deprived of food massaged more intensely after milk ejection for several sucklings after the deprivation ended, and the sow reacted by increasing the milk output in teats which received more massage (Jensen *et al.*, 1998). This communication on the part of the piglets seems to affect milk production in both the shorter (over the next few sucklings) and longer term (over several days) (Spinka and Algers, 1995).

3.8 Regurgitation

Even though lactation is the main way for mammals to supply food to their young, other ways do exist. Among carnivores, many species show regurgitation similar to the behaviour seen in some birds (described earlier). For example in wolves and dogs, this behaviour is typically seen when the young have passed the neonatal period (about 3–4 weeks of age of the pups), but well before the weaning (Malm, 1995b).

Again, this behaviour requires intense social interaction (Malm and Jensen, 1992). When the adult returns from a foraging trip, the pups will gather around it, whine and lick and butt the adult in the corner of the mouth. This behaviour causes the adult to regurgitate the stomach contents, consisting of food which has been partly digested by stomach hormones and juices, and the pups can ingest it. This behaviour can be performed not only by the mother, but also by the father, and other adult members of the pack as well. In any case, it depends on the performance of mouth-licking by the pups, and the fact that many dog breeds today do not perform regurgitation may possibly be attributed to an indirect effect, caused by a decrease of mouth-licking during selection (Malm, 1995b).

3.9 Begging

In the previous section, we have seen a number of examples where offspring seemingly have to use quite demanding signals in order to obtain resources from their parents. Why has evolution shaped behaviour that apparently leads to the young spending a lot of energy on obtaining food from the parents?

This question relates closely to the parent–offspring conflict theory, and has been approached in a series of theoretical and empirical research papers on the evolution of begging in young. Begging is of course only a special case of signalling, and a lot that has been revealed about the evolution of begging signals applies to any signal among animals. Begging can simply be defined as a signal emitted by young animals in order to increase the provisioning behaviour of adults (usually parents, but not always). Since parents have an evolutionary interest to restrict the feeding of any young (to save resources for other, equally valuable offspring), while offspring have the evolutionary interest of obtaining more from the parent than what is optimal from the parent's perspective, a parent would gain fitness from being able to detect whether a signal reflects the actual need of the young. The evolutionary interest of the young is to emit a signal which causes the parent to provide resources at a level which corresponds with its particular optimum. This situation has evoked the question of whether begging signals are honest. In this context, an honest signal can be defined as a signal which reliably reflects an underlying state of the sender.

So, is begging honest? The problem for the parent is that it cannot directly assess the need of an offspring, and so has to rely on the signals it emits. Those young in worse condition will derive a relatively higher fitness gain from parental resources. The crux of the game is therefore to make sure that only those young that are really in need beg strongly, which requires that the begging carries a fitness cost. If a begging signal carried with it no costs to the emitter, the field would be open for cheating to evolve.

There has been considerable theoretical effort to examine under what conditions honest begging might have evolved. In general, evolutionary models suggest three predictions which would all be found in a system of honest begging (Harper, 1986; Godfray, 1995):

1. The intensity and duration of a signal should be closely correlated to a measurable need in the offspring. For example, the intensity of food begging should be correlated to time since last food intake.
2. Accordingly, the parent should reply to increased intensity by increased care, for example by increasing food supply to young that beg more intensely.
3. The signal should have a fitness cost to the emitter. In particular, the cost should increase with increasing intensity of signal emission.

In general, there is plenty of evidence that begging signals increase in intensity when offspring become more hungry. In birds, several experiments have manipulated time since last feeding and subsequently found that nestlings adjust their begging effort in direct relation to their needs (Stamps et al., 1989; Redondo and Castro, 1992; Christe et al., 1996; Leonard and Horn, 1996). In mammals few examples have been investigated. In one experiment, piglets responded as expected with an increase in post milk ejection massage (a likely example of begging, as described earlier) when milk provision was reduced, but did not reduce their massage as expected when milk provision was increased (Jensen *et al.*, 1998).

Concerning the second prediction, in general parents seem to respond as expected by theory by increasing provisioning when begging increases, in birds (Stamps *et al.*, 1989; Price and Ydenberg, 1995; Christe *et al.*, 1996; Leonard and Horn, 1996; Ottosson *et al.*, 1997) and in pigs (Jensen *et al.*, 1998). It has been suggested that post milk ejection butting and sucking in cattle serves the same function, but this has so far not been experimentally investigated (Lidfors *et al.*, 1994b).

When it comes to the last (and maybe most critical) prediction of the theory, that begging has to be costly, the evidence is less clear. Even if begging of bird nestlings seems to be quite demanding to the young, energetic measurements have failed to demonstrate any substantial cost in terms of energy expenditure (McCarty, 1996, 1997). In pigs, meta-bolic expenditure increases several times during massage compared with basal metabolism, but on a diurnal time scale, massaging is responsible for a quite small proportion of the total energy expenditure of piglets, and this probably reflects the situation in most mammals (Spinka and Algers, 1995).

However, even if energy is not important, it appears that begging may carry other costs. In birds, an increased predation risk has been demonstrated as a consequence of begging calls being emitted by nestlings (Ryan, 1988). For newborn piglets, the most dangerous part of the world is the mother, and sometimes up to about one in three piglets is killed by crushing when the sow rolls over or when she lies down and the piglets get caught under her (Fraser *et al.*, 1995). It appears that increased begging in the form of udder massaging increases the risk of being killed in this way, and such increased risk is of course a substantial cost to the piglet (Weary *et al.*, 1996). So, weighing together all the evidence, it appears that much offspring begging constitutes good examples of honest signalling systems.

Young animals may solicit other forms of parental care than feeding. For example, young American white pelicans perform specific begging behaviour (shivering, vocalizing) when they are in need of warmth, which is provided by the mother by covering the young with her wings. This behaviour follows all the predictions of honest begging theory (Evans, 1994). Correspondingly, piglets call more intensely and loudly when they are hungry, but also when they are cold and need the warmth of the mother for maintaining body temperature (Weary *et al.*, 1996). Furthermore, piglets in pain call more intensely than others, presumably to solicit maternal protection (Weary and Fraser, 1996).

The fact that offspring signalling is a reliable and honest reflection of the state and needs of the signaller has led some researchers to suggest that such signalling may be a good indicator of animal welfare (Weary and Fraser, 1995, 1996). It has also been used to determine, for example, the responses of piglets to weaning at different ages (Weary and Fraser, 1997).

3.10   Factors Affecting Parental Care

The amount of care that a parent provides for its offspring depends on a number of different factors. As a general rule, the parent seeks to optimize its investment, in line with the parent–offspring conflict theory. This leads to a huge amount of possible factors that may affect levels of parental care, and an exhaustive treatment of all those is beyond the scope of this chapter. Rather, I will consider a selection of different factors as examples, and describe how evolution may have shaped parents to respond to these.

The first factor to be considered is the size of the litter or the brood. In general we would expect that, when comparing parents of the same species, parental effort should increase with increased number of offspring, and this also seems to be the case (Clutton-Brock, 1991). However, the relationship is usually not linear, so, in larger litters, each young receives less care. This is well known in animal production, where piglets in large litters have a lower average growth than in small litters (Donald, 1937; Fraser et al., 1995). In house mice, mothers produce more milk and higher quality milk when litters are larger, but each young receives a smaller amount and therefore grows more slowly (König et al., 1988). The reason for this non-linear relationship is the fact that parental care is a limited resource, and a linear increase of care would seriously decrease the possibility for the parents to care for future offspring.

Another factor which might affect the way parents behave towards their young is offspring age. From an evolutionary perspective, it is not self-evident what relationship we would expect. In one sense, young animals need greater care since they are more vulnerable. This may lead us to assume that parental care would decrease with offspring age; however, older young represent a higher value to the parents (more has been invested in them and their reproductive prognosis is better), which would give the prediction that care would increase with age (Clutton-Brock, 1991). It appears that parental care is often the result of a complicated weighing of both these factors: offspring need and value. In most mammals, suckling frequency decreases rather linearly with the age of the offspring (Reiter et al., 1978; Berger, 1979; Gauthier and Barrette, 1985; König and Markl, 1987; Lawson and Renouf, 1987;

Jensen and Stangel, 1992; Packard *et al.*, 1992; König, 1993; Malm and Jensen, 1996). However, milk production often shows a curvilinear relationship to offspring age, increasing to a peak less than halfway through lactation and then decreasing again (Arman *et al.*, 1974; Days *et al.*, 1987; Doreau, 1994; Malm and Jensen, 1996).

Also the age of the parent appears to be important. In general, younger parents seem to have a higher residual fitness cost for reproduction and a poorer probability of raising young successfully, and tend to show less parental care (Clutton-Brock, 1991). Hence, parental care is likely to increase with parent age, and a particular prediction following this is the theory of terminal investment: where an animal has a reduced probability of surviving one more reproductive cycle, selection should favour animals increasing their parental efforts correspondingly (Clutton-Brock, 1991). In the extreme case, where the parent's probability of survival beyond the present offspring is nil, it should devote all available resources to the young it is caring for at the moment.

The amount of parental care is also affected by whether other adults help in caring for the young or not. In birds where both parents care for the young, the parental effort of each parent is reduced in relation to when one parent is alone (Clutton-Brock, 1991). In mammals, care by others than the mother is also not uncommon; in fact alloparental care (including non-offspring nursing, adoption and other forms of care for non-offspring) has been reported in well over 100 species (Riedman, 1982; Packer *et al.*, 1992).

The last factor affecting parental care to be covered here is the sex of the offspring. The potential future reproductive success of a given young may often depend on its sex. For example, due to competition, some males may often be fathers of a disproportionately large number of young, whereas the variation in reproductive success may be smaller among females. In species where reproductive variation is larger in one sex, and where this variation is due to size and strength of the members of that sex, parents would be expected to invest more in offspring of the sex with higher reproductive variation (Willson and Pianka, 1963). This is a way to increase the chance that any investment will subsequently pay off in the form of more grandchildren. There is some experimental evidence that this happens in many mammalian species (Clutton-Brock, 1991).

3.11 The Transition to Offspring Independence

Social bonds between parents and offspring may last for long times, sometimes throughout life. For example, in mammals groups are often made up of closely related females. Such groups are usually formed by female offspring remaining with the mother, a phenomenon often referred to as natal philopatry (Greenwood, 1980). In some species, mainly primates, young have also been reported to 'inherit' their social position in the group from their mothers (Bernstein, 1981).

Even when social relations last for a long time, with respect to parental care, the weaning represents a distinct breaking point. In certain species, this is literally true: for example, in some seals, mothers nurse the young while these are still confined to a mostly terrestrial life at the seashore, and then suddenly one day simply abandon the pups and leave them for good (Reiter et al., 1978; Lawson and Renouf, 1987). More commonly, the weaning is a prolonged process, where it is often difficult to determine exactly when milk transfer definitely ceases.

Weaning is often referred to as the process leading up to the termination of lactation, and hence does not refer to a specific point in time (Counsilman and Lim, 1985). Since not only milk transfer but other parental care also decreases gradually during this process, it has been suggested that weaning should be defined as the period where the drop in parental investment per time unit is largest (Martin, 1984). With this definition, weaning is not a process unique to mammals, but also birds reducing feeding rates and general care would be said to wean the young. Therefore, even if the rest of this section will be concerned with mammals, it is likely that the general principles which are described are applicable also to other groups of animals showing parental care.

Since lactation represents such a heavy part of parental care in mammals, and since we know that there is an evolutionary conflict between parents and offspring over the allocation of this care, one might expect the young to make considerable efforts to obtain more than the parent is prepared to provide. Weaning might therefore be expected to be a process signified by overt conflicts, such as intense begging efforts from the young and aggressive rejection from the mother. Indeed, when the theory of parent–offspring conflict was first presented, Trivers suggested looking at weaning in order to find the overt signs of this conflict (Trivers, 1974). However, usually very little conflict behaviour is obvious in this period. Weaning has been studied under more or less natural conditions in a range of species, and mostly the findings are similar: there is a gradual and slow decline in maternal care and suckling, but rarely do mothers forcefully reject their young or show aggressive behaviour towards them; for example this has been reported for bighorn sheep (Berger, 1979), wolves (Packard et al., 1992), horses (Duncan, 1979), deer (Gauthier and Barrette, 1985), bison (Green et al., 1989), house mice (König and Markl, 1987), dogs (Malm and Jensen, 1996) and pigs (Jensen and Recén, 1989).

So, if the mother does not forcefully reject her young, how then is weaning brought about? A common observation in all the species mentioned before is that the mother usually initiates fewer sucklings and terminates more, as lactation goes on. In addition, mothers are often observed to make suckling more tedious to the young, in the sense that they have to work harder for the milk. This is accompanied by a general decrease in milk production and in milk quality, so that the pay-off to the young in the form of the milk they manage to obtain becomes smaller. For example, in both pigs and dogs, it has been demonstrated that the mothers always nurse in a lying position early in the lactation, but use a standing position to a higher and higher extent as weaning proceeds; presumably, milk ingestion is more difficult when the mother is standing (Jensen and Recén, 1989; Malm and Jensen, 1996). Mothers also tend to keep a longer distance between themselves and the young during weaning, which forces the young to use time and energy to maintain contact with the mother (Jensen, 1995).

Findings such as these led to the suggestion that weaning is controlled by the mother only to the extent that she affects the energetic costs and benefits which the young obtain from continued sucking (Jensen and Recén, 1989). Young animals can be assumed, just like any animal, to obtain food in an energy-optimal manner by choosing a strategy that maximizes net intake, i.e. to follow the general rules of optimal foraging (Charnov, 1976). In nursing animals, at least three possible strategies could be distinguished: (i) to obtain all nutrients by suckling only; (ii) to combine suckling with solid food intake; and (iii) to refrain from suckling and only feed on solid food. As nutritional requirements of the young increase, it will change from the first to the second strategy. From that point, it will be expected to continuously weigh the energy benefits from continued sucklings against the energy costs. When the marginal benefit of continued suckling compared with feeding only solids is below zero, the young will wean itself.

As suggested by Jensen and Recén (1989), the behaviour of the mother may act so as to affect this decision of the young. By continuously increasing the costs of suckling and decreasing the benefit (less milk production, less nutrient-rich milk), the young will at some point cease suckling simply because it does not pay off any more. This would make weaning a flexible and dynamic process and open the possibility for a mother to wean different offspring at different ages, depending on their needs and the ecological conditions.

There are some examples which lend support to this hypothesis. In bighorn sheep, weaning time and weaning abruptness are affected by the availability of alternative food for the young (Berger, 1979). In free-ranging pigs, weaning occurs later during winter when it is harder for the young to find solid food (Jensen and Recén, 1989). Furthermore, in pigs, the pre-massage time, i.e. the time needed to perform massage in order to release a milk ejection (representing an energy cost of obtaining milk), increases during lactation. As already mentioned, a higher proportion of nursings are performed when the sow is standing, which also may be considered a more energy-costly way for the young to obtain the milk (Jensen and Recén, 1989).

In dogs and other canids, the earlier described regurgitation behaviour may also be interpreted in this framework: not only is it a less energy-demanding feeding method for the mother compared with using the food for milk production, it may also be a more favourable alternative for the pups, thereby facilitating a smooth weaning (Malm and Jensen, 1992). In favour of this suggestion is the observation that weaning may be more associated with aggressive behaviour from the mother in dog breeds where regurgitation for some reason is absent (Malm, 1995b).

The weaning examples above demonstrate that communication and social interaction between parents and offspring may be subtle and unobvious. However, signalling is often defined as the behaviour of one animal affecting the behaviour and/or the strategies of another, and, according to this definition, weaning is a typical case of social signalling, even if the signals may be well hidden in the form of physiological processes.

A lesson from parental behaviour to all students of social behaviour could therefore be to broaden the general idea of what constitutes a signal and an interaction, and to look for the hidden methods which animals may use to affect and even manipulate each other's behaviour.

References

Alexander, G., Poindron, P., Le Neindre, P., Stevens, D., Lévy, F. and Bradley, L. (1986) Importance of the first hour post-partum for exclusive maternal bonding in sheep. *Applied Animal Behaviour Science* 16, 295–300.

Algers, B. (1993) Nursing in pigs – communicating needs and distributing resources. *Journal of Animal Science* 71, 2826–2831.

Algers, B. and Jensen, P. (1985) Communication during suckling in the domestic pig. Effects of continuous noise. *Applied Animal Behaviour Science* 14, 49–61.

Algers, B. and Jensen, P. (1991) Teat stimulation and milk production during early lactation in sows: effects of continuous noise. *Canadian Journal of Animal Science* 71, 51–60.

Arman, P., Kay, R.N.B. and Goodall, E.D. (1974) The composition and yield of milk from captive red deer (*Cervus elaphus*). *Journal of Reproduction and Fertility* 37, 67–84.

Baldwin, B.A. and Shillito, E.E. (1974) The effects of ablation of the olfactory bulbs on parturition and maternal behaviour in Soay sheep. *Animal Behaviour* 22, 220–223.

Bateson, P.P.G. (1965) Changes in chicks' responses to novel objects over the sensitive period for imprinting. *Animal Behaviour* 12, 479–489.

Berger, J. (1979) Weaning conflict in desert and mountain bighorn sheep (*Ovis canadensis*): an ecological interpretation. *Zeitschrift für Tierpsychologie* 50, 188–200.

Bernstein, I.S. (1981) Dominance: the baby and the bathwater. *The Behavioral and Brain Sciences* 4, 149–157.

Bustamante, J. (1994) Family break-up in Black and Red Kites *Milvus migrans* and *M. milvus* – is time of independence an offspring decision? *Ibis* 136, 176–184.

Carl, G.R. and Robbins, C.T. (1988) The energetic cost of predator avoidance in neonatal ungulates: hiding versus following. *Canadian Journal of Zoology* 66, 239–246.

Castrén, H., Algers, B., Jensen, P. and Saloniemi, H. (1989) Suckling behaviour and milk consumtion in newborn piglets as a response to sow grunting. *Applied Animal Behaviour Science* 24, 227–238.

Charnov, E.L. (1976) Optimal foraging, the marginal value theorem. *Theoretical Population Biology* 9, 129–136.

Christe, P., Richner, H. and Oppliger, A. (1996) Begging, food provisioning, and nestling competition in great tit broods infested with ectoparasites. *Behavioural Ecology* 7, 127–131.

Clutton-Brock, T.H. (1991) *The Evolution of Parental Care.* Princeton University Press, Princeton, New Jersey.

Counsilman, J.J. and Lim, L.M. (1985) The definition of weaning. *Animal Behaviour* 33, 1023–1024.

Dawkins, R. and Krebs, J.R. (1978) Animal signals: information or manipulation. In: Krebs, J.R. and Davies, N.B. (eds) *Behavioural Ecology.* Blackwell Scientific Publishers, Oxford, pp. 282–309.

Days, M.L., Imakawa, K., Clutter, A.C., Wolfe, P.L., Zalesky, D.D., Nielsen, M.K. and Kinder, J.E. (1987) Suckling behavior of calves with dams varying in milk production. *Journal of Animal Science* 65, 1207–1212.

Donald, H.P. (1937) The milk consumption and growth of suckling pigs. *Empire Journal of Experimental Agriculture* 5, 349–360.

Doreau, M. (1994) Mare milk production and composition: particularities and factors of variation. *Lait* 74, 401–418.

Duncan, P. (1979) Time-budgets of Camargue horses. II. Time-budgets of adult horses and weaned sub-adults. *Behaviour* 72, 26–49.

Dunstone, N. (1993) *The Mink.* T. & A.D. Poyser, London.

Ellendorff, F., Forsling, M.L. and Poulain, D.A. (1982) The milk ejection reflex in the pig. *Journal of Physiology* 33, 577–594.

Evans, R.M. (1994) Cold-induced calling and shivering in young American white pelicans: honest signalling of offspring need for warmth in a functionally integrated thermoregulatory system. *Behaviour* 129, 13–34.

Ewbank, R. (1967) Nursing and suckling behaviour amongst Clun Forest ewes and lambs. *Animal Behaviour* 15, 251–258.

Festa-Bianchet, M. (1988) Nursing behaviour of bighorn sheep: correlates of ewe age, parasitism, lamb age, birthdate and sex. *Animal Behaviour* 36, 1445–1454.

Fraser, D. (1980) A review of the behavioural mechanism of milk ejection of the domestic pig. *Applied Animal Ethology* 6, 247–255.

Fraser, D. (1984) The role of behaviour in swine production: a review of research. *Applied Animal Ethology* 11, 317–339.

Fraser, D., Phillips, P.A., Thompson, B.K., Pajor, E.A., Weary, D.M. and Braithwaite, L.A. (1995) Behavioural aspects of piglet survival and growth. In: Varley, M.A. (ed.) *The Neonatal Pig – Development and Survival.* CAB International, Wallingford, pp. 287–312.

Gallagher, J.E. (1977) Sexual imprinting: a sensitive period in Japanese quail (*Coturnix coturnix japonica*). *Journal of Comparative Physiology* 91, 72–78.

Gauthier, D. and Barrette, C. (1985) Suckling and weaning in captive white-tailed and fallow deer. *Behaviour* 94, 128–148.

Godfray, H.C.J. (1995a) Evolutionary theory of parent–offspring conflict. *Nature* 376, 133–138.

Godfray, H.C.J. (1995b) Signaling of need between parents and young: parent–offspring conflict and sibling rivalry. *American Naturalist* 146, 1–24.

Green, W.C.H. and Rothstein, A. (1993) Asynchronous parturition in bison – implications for the hider–follower dichotomy. *Journal of Mammalogy* 74, 920–925.

Green, W.C.H., Griswold, J.G. and Rothstein, A. (1989) Post-weaning associations among bison mothers and daughters. *Animal Behaviour* 38, 847–858.

Greenwood, P.J. (1980) Mating systems, philopatry and dispersal in birds and mammals. *Animal Behaviour* 28, 1140–1162.

Gubernick, D.J. and Alberts, J.R. (1983) Maternal licking of young: resource exchange and proximate controls. *Physiology and Behavior* 31, 593–601.

Gundlach, H. (1968) Brutfürsorge, Brutpflege, Verhaltensontogenese und Tagesperiodik beim Europäischen Wildschwein (*Sus scrofa*, L.). *Zeitschrift für Tierpsychologie* 25, 955–995.

Hamer, K.C. and Hill, J.K. (1994) The regulation of food delivery to nestling cory's shearwaters *Calonectris diomedea* – the roles of parents and offspring. *Journal of Avian Biology* 25, 198–204.

Harper, A.B. (1986) The evolution of begging: sibling competition and parent–offspring conflict. *American Naturalist* 128, 99–114.

Hepper, P.G. (1987) The amniotic fluid: an important priming role in kin recognition. *Animal Behaviour* 35, 1343–1346.

Hepper, P.G. (1994) Long-term retention of kinship recognition established during infancy in the domestic dog. *Behavioural Processes* 33, 3–14.

Hogstad, O. (1993) Nest defence and physical condition in fieldfare *Turdus pilaris. Journal für Ornithologie* 134, 25–33.

Horrell, R.I. and Eaton, M. (1984) Recognition of maternal environment in piglets: effects of age and some discrete complex stimuli. *Quarterly Journal of Experimental Psychology*, Sect. B, 36B, 119–130.

Immelman, K. (1972) Sexual and other long-term aspects of imprinting in birds and other species. *Advances in the Study of Behaviour* 4, 147–174.

Jensen, P. (1988) Maternal behaviour and mother–young interactions during lactation in free-ranging domestic pigs. *Applied Animal Behaviour Science* 20, 297–308.

Jensen, P. (1989) Nest site choice and nest building of free-ranging domestic pigs due to farrow. *Applied Animal Behaviour Science* 22, 13–21.

Jensen, P. (1993) Nest building in domestic sows: the role of external stimuli. *Animal Behaviour* 45, 351–358.

Jensen, P. (1995) The weaning process of free-ranging domestic pigs: within- and between-litter variations. *Ethology* 100, 14–25.

Jensen, P. and Recén, B. (1989) When to wean – observations from free-ranging domestic pigs. *Applied Animal Behaviour Science* 23, 49–60.

Jensen, P. and Stangel, G. (1992) Behaviour of piglets during weaning in a seminatural enclosure. *Applied Animal Behaviour Science* 33, 227–238.

Jensen, P., Gustafsson, G. and Augustsson, H. (1998) Massaging after milk ejection in domestic pigs – an example of honest begging? *Animal Behaviour* 55, 779–786.

Jeppesen, L.E. (1982) Teat-order in groups of piglets reared on an artificial sow. II. Maintenance of teat-order with some evidence for the use of odour cues. *Applied Animal Ethology* 8, 347–355.

Johnson, M.H., Bolhuis, J.J. and Horn, G. (1992) Predispositions and learning – behavioural dissociations in the chick. *Animal Behaviour* 44, 943–948.

Klopfer, P.H., Adams, D.K. and Klopfer, M.S. (1964) Maternal 'imprinting' in goats. *Proceedings of the National Academy of Sciences USA* 52, 911–914.

König, B. (1993) Maternal investment of communally nursing female house mice (*Mus musculus domesticus*). *Behavioural Processes* 30, 61–74.

König, B. and Markl, H. (1987) Maternal care in house mice. I. The weaning strategy as a means for parental manipulation of offspring quality. *Behavioral Ecology and Sociobiology* 20, 1–9.

König, B., Riester, J. and Markl, H. (1988) Maternal care in house mice (*Mus musculus*): II. The energy cost of lactation as a function of litter size. *Journal of Zoology* 216, 195–210.

Lawson, J.W. and Renouf, D. (1987) Bonding and weaning in harbor seals, *Phoca vitulina*. *Journal of Mammalogy* 68, 445–449.

Lazarus, J. and Inglis, I.R. (1986) Shared and unshared parental investment, parent–offspring conflict and brood size. *Animal Behaviour* 34, 1791–1804.

Lee, P.C. (1987) Allomothering among African elephants. *Animal Behaviour* 35, 278–291.

Lehrman, D.S. (1964) The reproductive behaviour of ring doves. *Scientific American* 211, 4854.

Lent, P.C. (1974) Mother–infant relationships in ungulates. In: Geist, U. and Walther, F. (eds) *The Behaviour of Ungulates and its Relationship to Management*. New series, no. 24, IUCN Publisher, pp. 14–55.

Leonard, M. and Horn, A. (1996) Provisioning rules in tree swallows. *Behavioural Ecology and Sociobiology* 38, 341–347.

Lickliter, R.E. (1984) Behaviour associated with parturition in the domestic goat. *Applied Animal Behaviour Science* 13, 335–345.

Lickliter, R.E. and Heron, J.R. (1984) Recognition of mother by newborn goats. *Applied Animal Behaviour Science* 12, 187–192.

Lidfors, L. and Jensen, P. (1988) Behaviour of free-ranging beef cows and calves. *Applied Animal Behaviour Science* 20, 237–247.

Lidfors, L., Jensen, P. and Algers, B. (1994a) Suckling in free-ranging beef cattle – temporal patterning of suckling bouts and effects of age and sex. *Ethology* 98, 321- 332.

Lidfors, L., Moran, D., Jung, J., Jensen, P. and Castren, H. (1994b) Behaviour at calving and choice of calving place in cattle kept in different environments. *Applied Animal Behaviour Science* 42, 11–28.

Macnair, M.R. and Parker, G.A. (1978) Models of parent–offspring conflict. II. Promiscuity. *Animal Behaviour* 26, 111–122.

Macnair, M.R. and Parker, G.A. (1979) Models of parent–offspring conflict. III. Intrabrood conflict. *Animal Behaviour* 27, 1202–1209.

Malm, K. (1995a) Behaviour of parents and offspring in two canids. Report 37. Thesis (Doctoral), Department of Animal Hygiene, Swedish University of Agricultural Sciences, Skara.

Malm, K. (1995b) Regurgitation in relation to weaning in the domestic dog: a questionnaire study. *Applied Animal Behaviour Science* 43, 111–122.

Malm, K. and Jensen, P. (1992) Regurgitation as a weaning strategy – a selective review on an old subject in a new light. *Applied Animal Behaviour Science* 36, 47–64.

Malm, K. and Jensen, P. (1996) Weaning in dogs: within- and between-litter variation in milk and solid food intake. *Applied Animal Behaviour Science* 49, 223–235.

Martin, P. (1984) The meaning of weaning. *Animal Behaviour* 32(4), 1257–1259.

McBride, G. (1963) The 'teat-order' and communication in young pigs. *Animal Behaviour* 11, 53–56.

McCarty, J.P. (1996) The energetic cost of begging in nestling passerines. *Auk* 113, 178–188.

McCarty, J.P. (1997) The role of energetic costs in the evolution of begging behavior of nestling passerines. *Auk* 114, 135–137.

Noirot, E. (1972) Ultrasounds and maternal behaviour in small rodents. *Developmental Psychobiology* 5, 371–387.

Ottosson, U., Backman, J. and Smith, H.G. (1997) Begging affects parental effort in the pied flycatcher, *Ficedula hypoleuca*. *Behavioural Ecology and Sociobiology* 41, 381–384.

Packard, J.M., Mech, L.D. and Ream, R.R. (1992) Weaning in an arctic wolf pack – behavioral mechanisms. *Canadian Journal of Zoology – Revue Canadienne de Zoologie* 70, 1269–1275.

Packer, C., Lewis, S. and Pusey, A. (1992) A comparative analysis of non-offspring nursing. *Animal Behaviour* 43, 265–281.

Petersson, E., Järvi, T., Steffner, N.G. and Ragnarsson, B. (1996) The effect of domestication on some life history traits of sea trout and Atlantic salmon. *Journal of Fish Biology* 48, 776–791.

Price, K. and Ydenberg, R. (1995) Begging and provisioning in broods of asynchronously hatched yellow-headed blackbird nestlings. *Behavioural Ecology and Sociobiology* 37, 201–208.

Redondo, T. and Castro, F. (1992) Signalling of nutritional need by magpie nestlings. *Ethology* 92, 193–204.

Reiter, J., Stinson, N.L. and Le Boeuf, B.J. (1978) Northern elephant seal development: the transition from weaning to nutritional independence. *Behavioural Ecology and Sociobiology* 3, 337–367.

Riedman, M.L. (1982) The evolution of alloparental care and adoption in mammals and birds. *Quarterly Review of Biology* 57, 405–435.

Romeyer, A. (1993) Ontogeny and selectivity of the mother–young bond in sheep and goats. *Revue d'Ecologie – La Terre et La Vie* 48, 143–153.

Ryan, M.J. (1988) Energy, calling and selection. *American Zoologist* 28, 885–898.

Signoret, J.P., Baldwin, B.A., Fraser, D. and Hafez, E.S.E. (1975) The behaviour of swine. In: Hafez, E.S.E. (eds) *The Behaviour of Domestic Animals.* Baillière Tindall, London, pp. 295–329.

Spinka, M. and Algers, B. (1995) Functional view on udder massage after milk let-down in pigs. *Applied Animal Behaviour Science* 43, 197–212.

Stamps, J., Clark, A., Arrowood, P. and Kus, B. (1989) Begging behaviour in budgerigars. *Ethology* 81, 177–192.

Tinbergen, N. and Perdeck, A.C. (1950) On the stimulus situation releasing the begging response in the newly hatched herring gull chick (*Larus argentatus argentatus* Pont). *Behaviour* 3, 1–39.

Townsend, T.W. and Bailey, E.D. (1975) Parturitional, early maternal and neonatal behaviour in penned white-tailed deer. *Journal of Mammalogy* 56, 347–362.

Trivers, R.L. (1974) Parent–offspring conflict. *American Zoologist* 14, 249–264.

Vila, B.L. (1992) Mother–offspring relationship in the vicuna, *Vicugna vicugna* (Mammalia: Camelidae). *Ethology* 92, 293–300.

Weary, D.M. and Fraser, O. (1995) Signalling need: costly signals and animal welfare assessment. *Applied Animal Behaviour Science* 44, 159–169.

Weary, D.M. and Fraser, D. (1996) Calling by domestic piglets: reliable signals of need? *Animal Behaviour* 50, 1047–1055.

Weary, D.M. and Fraser, D. (1997) Vocal response of piglets to weaning: effect of piglet age. *Applied Animal Behaviour Science* 54, 153–160.

Weary, D.M., Pajor, E.A., Thompson, B.K. and Fraser, D. (1996) Risky behaviour by piglets: a trade off between feeding and risk of mortality by maternal crushing? *Animal Behaviour* 51, 619–624.

Willson, M.F. and Pianka, E.F. (1963) Sexual selection, sex ratio and mating systems. *American Naturalist* 97, 405–407.

The Evolution and Domestication of Social Behaviour

4

W. Ray Stricklin

Department of Animal and Avian Sciences, University of Maryland, MD 20742, USA

(Editors' comments: Major climatic and ecological changes at the end of the last ice age, around 30,000 years ago, led animals and humans into closer contact and to the eventual domestication of certain species. Initially, this social contact would have been symbiotic, with both parties benefiting from the association. Later, humans selected animals for ease of handling and this inadvertently influenced their social behaviour, favouring individuals that stayed together in a herd and tolerated the close proximity of humans. The development and widespread use of semen collection and artificial insemination made it possible to exert even greater selection pressure and dramatically influence production traits, and this only came into effect in the last 50 years.

In this chapter, Stricklin presents evidence that social behaviour is an evolutionary adaptation and that it has played a major role in domestication. He does this by first discussing domestication in general, emphasizing that domestication led to increased genetic fitness of the animals. He then speculates on how this combination of evolutionary and molecular genetic approaches could lead to improved productivity and welfare of farm animals. He concludes with a statement that selection for production traits, independent of concern for behavioural traits, has resulted in animals that are no longer adapted to domestication and even raises the question whether we have genetically altered animals to the degree that they are 'beyond domestication').

4.1 Introduction and Overview

All domestic animals are social animals! In a discussion about the domestication of farm animals, acknowledging social behaviour as a trait common to *all* domestic animals is possibly the most universal – most descriptive – and probably the most important statement that can be made.[1] Recognition of the importance of social behaviour to domestication is not a new idea. Indeed, in 1875 Darwin wrote that '. . . complete subjugation generally depends on an animal being social in its habits, and on receiving man as the chief of the herd or family'.

After such a strong opening statement, one finds few if any other statements that apply to all behavioural, morphological or physiological traits among domestic animals. There are, however, general trends or tendencies. Many of the other common traits, however, are also behavioural characteristics linked or tied to social behaviour, possibly genetically.

The traditional classification system of sorting animals into species, genus, family, etc. relies on grouping individuals according to their having common morphological and behavioural traits. This system is of relevance to this discussion because a strong argument can be made for viewing domestication as similar to speciation (Darwin, 1859, 1875; Spurway, 1955). However, classifying domestic animals presents a problem because the one trait, social behaviour, thus far identified as being common to all domestic animals is a trait also common to many wild animal species. Therefore, formulating a scientific class for all domestic animals (or groups of domestic animals) on a trait or genetic basis is a problem, as is finding a scientific definition of domestication.

So how does one define domestication? Or, more precisely, how does one provide a scientific definition of a 'domestic animal'? One could say that it is common or public knowledge that the dog, cow, pig, horse, sheep, goat and chicken are domestic animals. These animals have in common a significant feature – their lives are more closely or directly linked to human activity than are those of animals that are said to be living naturally in a wild or feral state. Hale (1962) stated that domestication was a condition wherein the breeding, care and feeding of animals are, to some degree, subject to continuous control by humans. Price (1998) discussed in detail the problem of defining domestication. His contention was that domestication is best viewed as a process wherein animals adapt to living with humans and to the environments provided by humans.

A limitation of these definitions is of course that they emphasize either a condition or a process relative to human actions and not the specific genetic consequences or changes experienced by the population of animals. The definitions are sound, pragmatic and appropriate when writing about the broad topic of domestication. The problem is that the current level of scientific understanding of domestication is limited. As such, it results in domestic animals being placed into a classification category that is dependent on what humans do, not what is different about the animals. Possibly there is no solution to this problem because there may be no truly distinguishing features or traits that are unique to domestic animals. However, defining and recognizing the problem in itself should be useful and could contribute to the identification of uniqueness associated with domestication.

This discussion is not intended to be another thorough review of the topic of domestication, as was recently presented by Price (1998). Nor is the following information intended to be a review and discussion of behavioural genetics as it relates to farm animals, for Hohenboken (1987) has previously made this contribution.

The purpose of this chapter is to demonstrate that social behaviour as an evolutionary adaptation played a critical role in the domestication process and today continues to be a trait of great importance in commercial production systems. This will be undertaken by first discussing the process of domestication (in general) and then how the process has influenced social behaviour in our farm animals. This discussion will also attempt to go beyond viewing domestication as simply a process and will attempt to raise questions about possible DNA-level changes resultant from the domestication process. Conjecture will be presented as to how a combined evolutionary and molecular-level understanding of domestication as it pertains to social behaviour could ultimately improve both productivity and welfare of farm animals.

4.2   A Brief History of Time – Relative to Domestication and Modern Farm Animals

Anthropologists and evolutionary biologists have determined that initiation of domestication of the most commonly known farm animals occurred some 8000–12,000 years ago (Clutton-Brock, 1981, 1999; Zeder and Hesse, 2000). This means that domestic animals have existed only during the current, most recent period of geological time known as the Holocene, the period since the last major glaciation activity. This equates to chickens, sheep, pigs and cattle, as we know them today, not existing before the end of the last ice age. Dogs, which are generally recognized as the first species to have been domesticated, have existed for only about 14,000–15,000 years.[2] Considering the age of life on this planet or even the length of time that the major mammalian and avian species have existed, domestic animals arrived comparatively recently and are products of only a relatively few generations of selection

Table 4.1. Chart representation presenting the relative time of some factors related to the development of domestic animals.

Years ago (approx.)	Events	Relative time
4.5 to 5 billion (10^9)	Origin of Earth	1 Jan.
2 to 3.5 billion (10^9)	Life begins (genetic replicators), algae and bacteria	19 Apr.
600,000,000	Invertebrates: jellyfish, worms, etc.	17 Nov.
500,000,000	Vertebrates	28 Nov.
400,000,000	Land plants, fishes, amphibians and insects	1 Dec., evening
310,000,000	Reptiles appear, amphibians dominant	
220,000,000	Dinosaurs	17 Dec., 10:00 am
180,000,000	Appearance of birds and mammals	19 Dec., evening
135,000,000	Beginning of extinction of dinosaurs	
35,000,000	Appearance of most modern genera of mammals	
5,500,000	*Homo erectus*	
1,000,000	First human societies	
250,000	*Homo sapiens*	31 Dec., 11:30 pm
14,000	Dog domesticated	
7000 to 11,000	Sheep, goats, pigs, cattle and chickens domesticated	
5000	Written records begin	
2000	Birth of Christ	31 Dec., 11:59 pm
200	Beginning of the confinement of livestock into fenced areas, not commons around manor houses, leading to more control over mating and breeds as we know them	
	Robert Bakewell – first publications on animal breeding	
150	Era of Darwinian evolutionary theory begins with publication of *Origin of the Species* (1859)	
	Mendel publication	
100	Start of many herd books and breed associations	
	Beginning of the study of genetics as a science	
	Polled Hereford breed developed, one of the first applications of Mendelian laws	
75	Development of formula for calculating inbreeding, heritability and genetic correlations, leading to the use of selection techniques that could greatly increase the rate of change in production traits	
	Beginning of the study of ethology	

Continued

Table 4.1. *Continued.*

| 50 | Start of modern confinement housing systems employing greater use of technology to reduce labour costs, leading to more animals per caretaker. Development and application of artificial insemination Watson and Crick (1953) | |
| Present, year 2000 | Era of information revolution and genetic engineering | 31 Dec., midnight |

(Table 4.1). The relative times of events and ages of species strongly imply that the primary moulding of the gene structures of modern farm animals occurred well before domestication. The majority of the genetic structure of current farm animals occurred as a consequence of natural selection, in the long evolutionary period that preceded the relatively few generations that have been influenced by humans through artificial selection.

Even though humans have been present for a relatively long period, their living in association with other animals is more recent. It was probably around 30,000 years ago, during the Pleistocene, when important changes started that eventually led to domestication (Geist, 1971; Coppinger and Smith, 1983). There were relatively rapid and dramatic changes in the environment, the most significant of which was the subsiding of glaciers exposing vast new areas that were habitable by humans and other animals. The simultaneous expansion of humans and animals into these new habitats is thought to be related to the eventual domestication of species (Coppinger and Smith, 1983).

The initial contact between humans and the pre-domesticants probably began with their simply living in close spatial proximity (Coppinger and Smith, 1983). This first social relationship was probably a more symbiotic condition, and not a situation whereby humans controlled the animals with fences or buildings (Budiansky, 1992). Initially, humans probably used procedures such as castration and hobbling techniques to control behaviour, combined with the common practice of shepherding or herding livestock. The widespread use of fencing is relatively recent, with its common use in Britain dating back only 200–300 years. Prior to this time livestock were kept outside on grounds referred to as 'commons'. Many of the major breeds of livestock as we know them owe their origin to the British practice of fencing individually owned livestock (Briggs and Briggs, 1980).

There has been some artificial selection pressure on animals as a consequence of human intervention, dating back at least to the first use of castration to control behaviour of the 'wilder' more aggressive males (Briggs and Briggs, 1980). Some of the selection pressure on the early

domestic, or semi-domestic, groups of animals was probably a consequence of the genetically less tractable animals simply escaping back to
the wild. Additionally, in the early stages and in some species continuing until today, humans routinely chose to kill and eat the animals
whose behaviour was difficult to control before they had a chance to
breed. Thus, for the vast majority of time that humans have controlled
domestic animals, the selection pressure exercised has probably been
related to the animals' behaviour (Belyaev, 1979). Initially, the selection for more tractable behaviour probably occurred more as a consequence of human attempts to control behaviour, not planned conscious
acts by humans to produce tamer, more tractable animals (Darwin,
1859; Budiansky, 1992). This unintentional selection no doubt also
influenced social behaviour in that individuals with a tendency to stay
together in herds and flocks would be favoured, as would individuals
that tolerated the close proximity of humans. Today, in reindeer culling
there is discrimination between animals that stay in the herd centre
and those that stay on the perimeter of the group, and similar selection
pressure was probable in the early stages of domestication of other
species. Castration of more aggressive males made them more tractable,
but it also contributed to lowered behavioural libido in domestic males.
Other techniques employed since early stages of domestication, such as
cross-fostering of young and early weaning, also affected maternal
behaviour of livestock and correspondingly influenced social behaviour. Domestic hens are known to be more likely to accept chicks
hatched from eggs of other species than are wild birds (Lorenz, 1965).

An understanding of the genetic basis for selection to directly 'fix'
traits has developed only in the last 100 years (Lush, 1947; Zirkle,
1952). Previously, and in some cases even today, humans believed the
behaviour, coat colour, reproduction, etc., of animals was controlled
by gods or spirits (Briggs and Briggs, 1980; Campbell, 1991; Olesen
et al., 2000). As a consequence of widespread myth and superstition
about animals and inheritance, when all farm animals are considered,
very little effective or intentionally directed selection on production
traits occurred in livestock before the 20th century. Prior to widespread
use of automobiles in much of North America and Europe, livestock
roamed fields, woodlands, and even in cities with fences constructed
to keep animals outside areas that were used for crop and vegetable
production (Briggs and Briggs, 1980). Today there are still major
regions in the world, such as parts of Australia, where livestock are
kept under range or extensive conditions with only minimal contact
with humans. Under these conditions, natural selection remains a
major factor and matings are only partly controlled by humans.

However, for a significant portion of the world's livestock, a major
shift in the level of confinement began just after World War II. Over the
last 50 years there has been a trend towards larger scale farms utilizing

more intensive confinement housing systems. New artificial selection techniques were developed by animal breeders, and computers made it possible to store and access records and plan matings, leading to implementation of these selection systems (Stricklin and Swanson, 1993). The development and widespread use of semen collection, storage and artificial insemination made it possible to exert even greater selection pressure on desired traits (Olesen *et al.*, 2000).

One of the most dramatic consequences of the move to modern farming techniques is related to social behaviour. The social group size (see Section 2.5) for farm animals today is sometimes several thousand times larger than the group sizes found prior to 1950. The ability of domestic animals to live in groups is a behavioural trait that made intensive livestock systems possible. However, the consequences of crowding and large group sizes on the welfare of farm animals are being questioned (Dawkins, 1998). The natural agonistic behaviours, including pecking, biting, butting, chasing and fighting, common among social animals, account for many animal deaths and lower rates of gain among closely confined animals. Some behavioural actions that cause damage to other animals (such as buller syndrome in steers, feather pecking in hens and tail-biting in pigs), while not necessarily true social behaviours are nevertheless a consequence of group living. Sometimes the term social stress is used to cover the negative aspects of group-related activities of animals (Stricklin and Mench, 1987).

In concluding this section, it must be acknowledged that there have been dramatic changes in traits such as growth rate, milk and egg production, even litter size, in some domestic animals, with much of this change occurring during the past 50 years (Stricklin and Swanson, 1993). Additionally, the domestication process has influenced the behaviour of farm animals. However, the most significant factor in the formation of the genome of modern farm animals was natural selection, which preceded and has continued to act through the domestication process. Significantly, the modern farm animal of today compared with the pre-domestication ancestor has a full behavioural repertoire. No behaviours have been added or deleted through the domestication process. It is only the level of needed stimuli associated with the initiation of a given behaviour and the corresponding final level of expression for the behaviour that have been shown to be modified (Hale, 1962; Wood-Gush, 1983).

4.3. A Cursory Overview of the Evolutionary Basis of Social Behaviour

Development of social behaviour arose in part because animals that share genes in common mutually increase their genetic fitness values as a result of cooperation. The genetic basis for this relationship was not defined until Hamilton (1964) wrote a two-part article entitled 'The genetical evolution of social behaviour'. These papers formed the theoretical basis for the explanation of an individual's behaviour that appears to benefit the group at the expense of the individual's fitness. Hamilton mathematically formulated a predictive model for the probability of an altruistic act by a given animal based on the number of genes it shared in common with the animal(s) that would benefit from the act. Having more genes in common was predicted to increase the probability of an altruistic act because the action increased the 'inclusive fitness' of the individual, even though the act could be negative to the immediate survival or well-being of the individual. The differential reproduction of one's genes as a consequence of cooperation with close relatives was named kin selection. These aspects are discussed in more detail in Chapter 1.

An important aspect of kin selection, as developed by Hamilton and investigated by others, is that individuals benefit most when they can distinguish their close kin from other individuals. The term for such behaviour is kin recognition and is known to exist in a few species, but is not known to be common in the domestic species. However, kin selection (independent of kin recognition) is also demonstrated to be effective when individuals behave differentially with respect to those animals that share a common close association, especially if this association begins at birth. The basis for kin selection functioning under these conditions, of course, is that typically the animals in close proximity under natural conditions are close relatives. The family structure of the wild ancestors of current domestic animals was predominantly a matrilineal-based group (Reinhardt and Reinhardt, 1981; Stricklin, 1983). The extended family of related females thus shared genes in common, as did their offspring. It would therefore be expected that behaviours would evolve that promoted social behaviour.

Some production systems may act to cause non-genetically related adult farm animals to behave as if they are relatives. For example in dairy heifer replacement systems and chick rearing systems for laying hens, individuals share a common and close association in their early development. This early association may lead the animals as adults to behave socially toward each other as if they are kin-related (Stricklin, 1983). This early association in development has considerable implications for the later social behaviour of domestic farm animals. The traditional family structure wherein the offspring remains with the mother from birth until natural weaning occurs somewhat infrequently among current farm animals. However, farm animals developed from species that evolved to behave differentially towards those individuals with whom they were in close association following birth and during their early development (Reinhardt and Reinhardt, 1981; Graves, 1984).

Today, it is possible – one might even suggest probable – that dairy heifers behave towards their age peers, with whom they shared a common early development, as if they were genetic relatives, even as if they were sisters or cousins, when they are not.

The model presented by Hamilton was built upon and/or used concepts that originated from the work of animal and plant breeders. Jay L. Lush, who is considered by Americans to be the founder of animal breeding, stated as early as 1947 that the degree of being helped or hindered among animals was probably proportional to their genetic relatedness. This expression by Lush (1947) is basically a summation of the kin selection theory, which was later formulated by Hamilton.

Interestingly, in his discussion of the evolution of social behaviour as it relates to group selection, E.O. Wilson in *Sociobiology* (1975) acknowledged that:

> One of the principal contributions to theory was provided by Jay L. Lush (1947), a geneticist who wished to devise a prescription for the choice of boars and gilts for use in breeding. It was necessary to give each pig 'sib credits' determined by the average merit of its littermates. A quite reliable set of formulas was developed which incorporated the size of the family and the phenotypic correlations between and within families. This research provided a useful background but was not addressed directly to the evolution of social behavior in the manner envisioned by Darwin.

Lush is known to have been a strong advocate for the study of behaviour of farm animals building from an evolutionary foundation. However, traditional animal breeders have continued these investigations to only a limited extent. But research of the type advocated by Lush on social behaviour and genetics became the domain of workers in a discipline recognized today as behavioural ecology, a discipline that traces a major part of its foundation to the work of Hamilton.

There is some support for the theory that animals under domestication conditions adopt a more energy conservation strategy of behavioural activity, at least relative to optimal foraging patterns, as was proposed by Beilharz *et al.* (1993). Behaviours with energy costs that are high, such as extensive foraging and social interactions, were found to decrease in frequency in domestic birds. When domestic pigs were compared with domestic swine–wild boar hybrids, the domestic pigs used a less costly foraging strategy (Gustafsson *et al.*, 1999). One implication of this work is that domestic animals are not necessarily less adapted compared with their wild counterparts, as has been implied or stated (Ratner and Boice, 1969; Coppinger and Smith, 1983). Instead, it can be argued that domestication is an evolutionary strategy that tends to favour animals that adopt energy-conserving behavioural foraging strategies (Gustafsson *et al.*, 1999). If this theory applies more universally to the behaviour of domestic animals, then lower levels of agonistic behaviour might be expected among animals under domestic conditions relative to their wild counterparts (Fig. 4.1 illustrates different body types in pigs).

4.4   The Role of Behaviour in Domestication

Social behaviour between humans and animals (see Chapter 13) was the foundation on which the domestication process was built (Zeuner, 1963). Docility is recognized as the primary and essential trait for domestication (Kretchmer and Fox, 1975). Docility by definition has to do with the behaviour of an animal while in the presence of humans – and thus at a minimum is inclusive of the degree of sociality exhibited between the animal and humans.

Fig. 4.1.   Illustration of the differences in body type for (a) a European wild boar sow and piglets; (b) a feral boar from Issabaw Island, Georgia, USA; (c) a common domestic breed – Yorkshire – sow and pigs in a farrowing crate, and (d) a Meishan sow. Note that the domestic Yorkshire pigs are white, which tends to be a common trait among domestic animals, especially those kept in close confinement such as laboratory animals. Also, the Meishan sow shows considerable morphological neoteny in facial and body features, and pigs of this breed reach puberty at less than 3 months of age, while common breeds reach puberty by about 6 months of age. Meishan, a Chinese breed, are noted for having very large litter sizes (15 pigs or more), and were developed independently from European domestic breeds. Note also that the feral boar (b) shows a body type that is much more similar to the European wild ancestor (a) compared with the body characteristics of domestic pigs, from which it is more recently descended.

In *The Behaviour of Domestic Animals*, edited by Hafez, E.B. Hale (1962) wrote a chapter entitled 'Domestication and the evolution of behaviour'. In this chapter, Hale outlined traits he considered favourable and unfavourable to domestication. The original chart has been reproduced in other discussions about domestication (e.g. Kretchmer and Fox, 1975; Price, 1984), but the information is considered worthy of presentation again in the current discussion (Table 4.2).

Hale's general argument was that there were behavioural traits common to the progenitors of domesticated species that favoured their transition from the wild to the domesticated state. Tennessen and

Table 4.2. Behavioural characteristics which favour domestication and those which do not favour domestication of a species. (After Hale, 1962.)

Favourable characteristics	Unfavourable characteristics
1. Group structure	
(a) Large social groups (flock, herd, pack), true leadership	(a) Family groupings
(b) Hierarchical group structure	(b) Territorial structure
(c) Males affiliate with female groups	(c) Males in separate groups
2. Sexual behaviour	
(a) Promiscuous matings	(a) Pair-bond matings
(b) Males dominant over females	(b) Males must establish dominance over or appease female
(c) Sexual signals provided by movements or posture	(c) Sexual signals provided by colour markings or morphological structures
3. Parent–young interactions	
(a) Critical period in development of species bond (imprinting, etc.)	(a) Species bond established based on species characteristics
(b) Female accepts other young soon after parturition or hatching	(b) Young accepted on basis of species characteristics
(c) Precocial young	(c) Altricial young
4. Response to humans	
(a) Short flight distance to humans	(a) Extreme wariness and long flight distance
(b) Not easily disturbed by humans or sudden changes in environment	(b) Easily disturbed by man or sudden changes in environment
5. Other behavioural characteristics	
(a) Omnivorous	(a) Specialized dietary habits
(b) Adapt to a wide range of environmental conditions	(b) Require a specific habitat
(c) Limited agility	(c) Extreme agility

Hudson (1981) used a form of cluster analysis to evaluate Hale's classification of favourable and unfavourable traits and concluded that his system is not highly accurate in its predictive ability. However, Hale (1962) acknowledged that the progenitor of a domesticated species need not have possessed all the indicated adaptations and stated that there are notable exceptions, including many species that possessed traits favourable to domestication but remained in the wild state. For example he speculated that almost any of the family Bovidae would be easily domesticated, yet the majority remain wild. He also argued that it would be extremely difficult to domesticate a species that possessed only the unfavourable traits. More importantly, Hale argued that species originally possessing unfavourable characteristics come under selection pressures that move them towards developing the corresponding favourable characteristics.

Hale stated that perhaps the most remarkable aspect of early domestication was the disproportionate contribution of a single order of mammals, the Artiodactyla, to the successful domesticants. This order includes swine, sheep, goats, European cattle, Zebu cattle, Indian buffalo, yak, camels, llamas, alpacas and reindeer. He noted a similar but less extreme situation in birds with the order Galliformes contributing a large number of species including chickens, pheasants, peafowl, guinea-fowl and turkeys. Hale contended that the traits favourable to domestication (Table 4.2) are highly descriptive of many ungulates and gallinaceous birds, helping to explain their major contribution to domesticated species.

It is worth noting in this discussion that Hale started his list of traits favourable to domestication with social behaviour. Hale listed first large social groups, and, in the light of the current group sizes used in some production systems, this remains a highly important trait. He argued that a hierarchical group structure could benefit domestication through tending to reduce fighting to a minimum, thus subjecting the animals to less social stress. It should be noted (Table 4.2) that, in making these statements, Hale was comparing animals that are hierarchical in social group structure (and possess a system of portable personal space) with animals that have a territorial-based social structure (and defend a fixed and definable physical space against intrusion by other animals).

In summary, the use of Hale's traits favourable to domestication may be useful in answering the question – What is a domestic animal? According to Hale, a domestic animal is one that is a member of a population whose behaviour either fits the traits listed in Table 4.2 or which is in a condition wherein there is selection pressure that is moving the group's behaviour towards fitting the favourable traits. Others, including Coppinger and Smith (1983) and Belyaev (1979), have stated or implied that behaviour may be the best method of defining a domestic animal.

4.5   Domestication as an Adaptive Trait Leading to Increased Genetic Fitness

Geneticists define fitness as the contribution an individual makes to the gene pool of subsequent generations relative to the contributions of other individuals in the population (Wilson, 1975; Dawkins, 1976). Inherently, the process of natural selection leads eventually to the prevalence of the genotypes with the highest fitness.

The most numerous bird on earth is the domestic chicken. The increase in the number of domestic animals relative to their wild counterparts has been argued to be evidence of the adaptive benefit of having traits favourable to domestication (Budiansky, 1992). This argument is one based on Darwinian fitness. Traits from genes that lead to offspring in greater number are said to be adaptive (Dawkins, 1976). Behavioural traits leading to domestication ultimately produced greater numbers of animals and therefore are more adaptive. Thus, by using a strict Darwinian argument, the behaviour of domestic animals can be said to be more adaptive than that of their wild counterparts who are now relatively fewer in number or in some cases extinct. Domestic animals have differentially reproduced their genes through exploiting a strategy that is dependent on humans providing food, shelter, etc. (Budiansky, 1992). This argument contends that the groups of animals that 'chose' domestication as an evolutionary strategy were more successful than were the species that adopted other evolutionary strategies that culminated with their extinction or that now place them in jeopardy.[3]

Natural selection in wild populations involves selection for *greater individual viability* (competitive ability, longevity, etc.) and for *greater reproductive success* (Tchernov and Horwitz, 1991). The two strategies were termed *K-selection* and *r-selection*, respectively, by MacArthur and Wilson (1967). In population biology, the symbol 'r' is used to represent the intrinsic rate of increase or growth in the number of individuals of a population, and the symbol 'K' is used to represent the carrying capacity of the environment (upper reproductive limit). Using these distinctions, r-strategists and K-strategists have been reported to move towards acquiring different behavioural and morphological traits (Wilson, 1975), as outlined in Table 4.3. The r-strategists are species that emphasize colonization of short-lived environments, rapid population increase and full utilization of resources. In contrast, K-strategists adapt to stable, predictable environments in which rate

Table 4.3. Some characteristics of r-strategists and K-strategists.[a] (After Pianka, 1970; Wilson, 1975; and Dewsbury, 1978.)

Correlate	r-strategists	K-strategists
Climate	Variable or unpredictable	Relatively constant and predictable
Population size	Variable over time with wide fluctuations, usually below carrying capacity of environment	Relatively constant over time and often near carrying capacity of environment
Intraspecific and interspecific competition	Variable, often lax	Usually keen
Development	Rapid	Slower
Age of reproduction	Young	Delayed
Body size	Small	Relatively large
Frequency of reproduction	Once	Repeated
Emphasis in energy utilization	Productivity	Efficiency
Lifespan	Short	Long
Colonizing ability	Substantial	Minimal
Social behaviour	Weak, mostly schools or aggregations	Frequently well developed

[a]It should be noted that few, if any, species fit entirely all criteria of either of these two strategies. However, this conceptual model is considered useful in presenting the expected correlated responses for a species that is moving towards adaptation of either strategy.

of population growth is unimportant but formation of stable social patterns is stressed.

Species that are r-strategists are sometimes said to be 'opportunistic species' and the K-strategists are 'specialists'. K-strategists exist at or near the maximal carrying capacity of an environment and, therefore, are subjects of a strategy wherein they evolve towards maximizing their genetic ability to utilize the resources in their environment in the most efficient manner. Such behavioural traits as increased social behaviour, leading to greater cooperation, and behaviour such as greater investment in, or at least extended, parental care would be expected in these specialized species.

These two forms of natural selection, of course, are not mutually exclusive. Species are subject to varying levels of selection pressure towards one or the other of these two strategies, and there is considerable debate concerning the trade-off between selection for increased viability and increased reproduction (Tchernov and Horwitz, 1991).

Tchernov and Horwitz (1991) argued that domestication is characterized by a shift to a more r-selected strategy, even though domestic ungulates (with one or few offspring, parental investment, etc.) tend not to fit the criteria outlined for r-selection[4] (Pianka, 1970). However, Tchernov and Horwitz (1991) argued that the environmental conditions associated with domestication eventually lead to a shift that is predominantly an r-selection strategy. These authors build an argument that the decrease in body size of domesticants relative to their wild progenitors was a consequence of their being shifted to a greater r-selection strategy and not a consequence of intentional, directed selection by humans. They further speculate that many traits, including limited agility, decreased aggressiveness, a more omnivorous diet, more precociality and earlier maturity, may all be the consequence of a shift to an r-selection strategy, and not the consequence of human-directed genetic selection.

Tchernov and Horwitz (1991) were attempting to use the r- and K-selection theory to obtain a better understanding of the early stages of the domestication process. Their contention was that a change in environment drove the animals to adopt an r-selection strategy. Interestingly, they did not discuss the impact of a shift to an r-selection strategy on social behaviour. However, from the trend outlined in Table 4.3 one could predict that the animals would become less social and move towards becoming an aggregation as opposed to being true social animals (see Section 2.5).

Wilson (1975) defined an aggregation as a group of individuals of the same species gathered in the same place but not internally organized or engaged in cooperative behaviour. He made a distinction between an aggregation and a true social organization (his term was society), which he defined as a group of individuals organized in a cooperative manner with reciprocal communication and interactions extending beyond mere sexual activity. The term aggregation as defined here seems to describe many groupings of modern farm animals that show lower levels of agonistic behaviour, especially groups of an extremely large size such as broiler flocks. (See Chapter 7 for a discussion of chickens and group size and Chapter 10 for information on fish and aggregations.)

If domestication drives animals toward an r-selection strategy and if r-selection tends to result in behavioural traits as listed in Table 4.3, then modern farm animals may be moving away from being truly social and towards becoming aggregations. Today's artificial selection strategies that emphasize growth and reproduction may additionally exacerbate the shift towards traits associated with r-selection. In some cases these modern selection criteria are being imposed on animals that no longer live in a context where there is social interaction between themselves and humans,[5] further lessening or eliminating selection pressure for social behaviour. Such a combination of selection pressures could produce dramatic genetic changes in social behaviour and other traits in a relatively short time period.

Hohenboken (1987) suggested that in some cases domestic animals in modern, highly mechanized, intensive farming systems face changes to their physical and social environment that are possibly more dramatic than those of their ancestors. He further suggests that these changes may be occurring much more rapidly than is the ability of populations to genetically adapt. Dairy cows that are kept in large groups and must pass through milking parlours two or more times daily may be an example of animals confronting such rapid and dramatic social and physical environmental changes (see Section 5.2.3). The system may force cows to move more and thus come into social contact with a large number of other cows on a daily basis. An essential component of a true social organization is that members are able to recognize other group members. Theoretically, there is an upper limit to the ability of a cow to individually recognize (see Section 14.1) and remember other cows. Once this limit is exceeded, then group stability would be expected to be compromised (Wilson, 1975). Within groups that are heterogeneous for age, weight and sex, when group stability breaks down, excessive aggression and even aberrant behaviours may result (Calhoun, 1963). However, among groups that are homogeneous in composition, which is true for most farm animals, then a different strategy may result – one with group members showing lowered levels of agonistic encounters. For example, a group of 1000 3-year-old heifers that were mixed frequently into smaller groups and placed in smaller areas exhibited low rates of agonistic behaviour (Stricklin and Kautz-Scanavy, 1984). Lower rates of agonistic encounters may be an indication that under some circumstances farm animals move from being social groups towards formations that are simply aggregations.

4.6   Is Domestication a Single-gene Trait?

No doubt, in the broad sense, domestication is the consequence of numerous genes. However, while still speculation, there may be indirect support for an argument that the primary or major event associated with 'true' domestication is the consequence of a gene complex, wherein one gene controls or regulates a number of other genes, all of which influence traits associated with domestication. In this discussion, it is proposed that there is merit in examining the possibility that a major contribution to domestication came from a rather simple DNA change – a single gene.

As an overview, it is suggested that initially, animals that genetically possessed traits favourable to domestication lived in close proximity to humans. In some animals, a relatively simple DNA change (mutation) that regulated the action of a large number of other genes occurred, leading to a rapid change in behaviour, morphology and physiology. This rapid and dramatic change led toward traits more favourable to domestication. Finally, there is a period of 8000–12,000 years commonly associated with domestication of livestock. During this period artificial selection has led to gradual rates of change in behaviour, growth, reproduction and production traits, with the most dramatic changes occurring within the past 50 years (and which may be significant enough to be considered a separate stage itself). However, it is the suggested early phase possibly involving a simple DNA change based on a single mutation that raises the question – Is domestication a single-gene trait?

If a rapid and dramatic genetic change occurred, the intriguing possibility is that this was a consequence of a single 'master gene' that switched on and off a large number of other genes or else regulated the rate of their expression. But, before discussing this hypothesis further, the traditional view supporting genetic gradualism as the basis for domestication should be addressed.

4.6.1 The argument for gradualism

The predominant view of domestication implies that traits, including behaviour, in animals have changed gradually over large numbers of generations as a consequence of selection that is directed by humans, or at times inadvertently caused by humans (Hale, 1962; Price, 1984; Ratner and Boice, 1969; Ricker *et al.*, 1987). This view implies that, through differential rates of reproduction (i.e. selection) of those animals that were tame in their behaviour, humans gradually created the domestic farm animals of today. Further, if one were to chart the proposed rate of change, it would show a pattern of gradual transition from the wild type to the modern domestic farm animal (Briggs and Briggs, 1980).

This model suggests that all traits found in domestic animals, for example tameness or tractability, changed only gradually and as a consequence of manipulation of the additive genetic variation that originally existed in the gene frequency of wild progenitors to modern livestock. In this model it is generally assumed that there are a large number of genes related to tractability (Hohenboken, 1987). Selection by humans caused an increase in frequency for the genes producing tractable behaviour and a decrease in the genes related to fearfulness and flightiness.

Population geneticists would predict that the expected response to selection pressure is relative to the amount of additive genetic variation in the population – which is to say that highly heritable traits (ones with more additive genetic variation) show more rapid rates of change when under selection (Wilson, 1975). However, selection pressure (either natural or artificial) on a given trait leads to less additive genetic variation in the population, causing the heritability estimate to decrease. Some traits such as those related to reproduction typically are said to be poorly heritable because there is little associated additive genetic variation, presumably because natural selection has acted and continues to act to keep the variation at low levels.[6]

However, in studies of populations of domestic animals, traits that reflect tractability have typically been found to be moderately to highly heritable (Hohenboken, 1987). Intuitively, it would seem that if domestication is the result of selection for more tractable animals, then artificial selection would have acted to decrease the amount of additive genetic variation as the animals became domesticated. However, this does not seem to be the case. Further, feralization investigations wherein animals are returned to a natural state, though limited in number, indicate that tractability is quickly lost and that the animals revert to 'wild type' behaviour in only a few generations (Wood-Gush, 1983). The fact that tractability retains relatively high levels of additive variation raises the possibility that domestication is not simply a process of gradualism – other factors may also be influencing domestication.

4.6.2   The argument against gradualism

Darwin proposed that species originated as a consequence of gradual genetic change over long periods of time.[7] But the fossil record does not fit this pattern. Instead the fossil record shows abrupt appearance and extinction of species (Eldredge and Gould, 1972). According to Belyaev (1979), domestic animals differ from their wild counterparts, and from each other, much more than do some species and even genera. He further argued that there is no evidence of any period in the history of evolution where a similar magnitude of variability occurred compared with domestication.

Eldredge and Gould (1972) proposed that the pattern of rapid species development could be described by a model they called 'punctuated equilibrium'. They used this term to explain situations whereby morphological features of a species remained unchanged for long periods of time but originated from rapid change concentrated in a geologically short time period.[8] The implication of this theory is that, if species can arise quickly, then animals with domestic-like behaviour could also develop rapidly.

Belyaev (1979) and Belyaev and co-workers (1981) presented research evidence to support the notion that domestication was a rapid process. He selected silver foxes on the basis of a single trait, short flight distance to humans. He found that tame foxes were obtained in a relatively few generations (less than 20), but he also found significant changes in other traits as well. The foxes behaved very much like dogs, responding to humans by licking their hands, whining and wagging their tails, and snarling fiercely at one another as they seek the favour of human handlers (Trut, 1999). He also found dramatic changes in coat colour and pattern, with some closely resembling that of dogs. Belyaev (1979) concluded that 'The process of domestication in all animal species seems to have resulted in the same kinds of homologous variations as a result of selection for the single important characteristic of tame behavior.' There is some support, though more anecdotal, for the suggestion that the domestication of fallow deer was also a consequence of rapid change and occurred in the manner Belyaev suggested (Hemmer, 1990).

Neoteny has been suggested as the genetic process that produced the rapid change (Coppinger and Smith, 1983; Shea, 1989). Neoteny is generally defined as the retention of juvenile traits into adulthood (Gould, 1977). This retention of juvenile traits is thought to function via a mechanism that controls the timing of trait development, possibly through a single gene or single-gene complex (Shea, 1989). Thus, a minor change in a gene mechanism that controls (switches on or off) other genes could produce dramatic physical and behavioural changes. The resultant juvenile-like traits due to neoteny would perhaps have produced an animal that had 'looser' ties to adult drives (including traits such as species-specific recognition), increased tameness and more dependent, care-soliciting behaviour. In short, neotenization could very much enhance social behaviour.

Neoteny has been suggested to affect the domestic animal not only behaviourally, but also morphologically. Changes that have been suggested to occur via neoteny are based on the juvenile form. Specifically, animals that are neotenized resemble juvenile forms as adults. A flatter face, rounder head and shorter extremities are the major characteristics of a juvenile and also a neotenous adult (Lorenz, 1965; Gould, 1980; Campbell, 1982). Geist (1971) provided very strong support for sheep becoming neotenized. Some of the Chinese breeds of pigs appear to be highly neotenized, compared with the European wild boar and even some of the modern domestic breeds (Fig. 4.1).

Support for the idea that neoteny influences the process of domestication has been reported for the dog (*Canis familiaris*). Evidence of behavioural neoteny has been documented by Coppinger *et al.* (1987) where evidence of 'selected differential retardation (neoteny)' in motor pattern development was found when compared with ancestral species. Frank and Frank (1982) found that there was a breakdown in ritualized aggression in domestic dogs when compared with timber wolves (*Canis lupus lycaon*), which they stated was a consequence of neotenization.

Structural changes supporting the morphological consequences of neoteny were reported by Morey (1994). Morey measured snout length and cranial width in the prehistoric dog, modern wild canid and modern dog. Snout length did not vary greatly between wild canids and modern dogs, but the width of the palate and cranial vault did vary. While the ratio of snout length to total skull length was similar between wolves and dogs, the cranial width to total skull length proportion was not. Instead, these proportions in dogs closely resembled those of juvenile wolves. It was concluded that this supported the idea that dogs are neotenous forms of their wolf ancestors. Interestingly, similar changes in facial morphology were found in the comparison of captive and wild alligators (Meers, 1996).

Neoteny has been proposed as the basis for increased brain size in humans compared with other primates, and this is possibly both a cause and effect of social behaviour (Shea, 1989). In fact humans are highly neotenized, possibly the most neotenized of all animals (Montegu, 1989). This observation raises a question that is of interest, but only incidentally related to the current topic, as to whether or not humans are domestic animals. Many persons, including Darwin and Lorenz, have discussed this question, and the general agreement is that a strong case can be made.[9] Human behaviour fits rather well with Hale's list of traits favourable to domestication (Table 4.2). Budiansky (1992) argued that the domestication of humans and animals occurred simultaneously. In fact one could argue that, through the process of domesticating plants and animals, humans have domesticated themselves.

Paradoxically, fairly dramatic reduction in the brain sizes of domestic animals (up to 30%) has been reported, compared with their wild counterparts (Hemmer, 1990). This seems contradictory from the viewpoint that it has been argued that neoteny was influential in the development of both domestic animals and humans. A contended major aspect of neoteny is that head size relative to body size, as found in neonates, remains relatively high in neotenized adults (Gould, 1980), producing larger brains. It is generally accepted that neoteny produced an increase in brain size in humans (Montagu, 1989; Shea, 1989). It therefore seems contradictory that neoteny could produce the behavioural and morphological changes listed by Hemmer (1990) and also result in a decrease in brain size of the magnitude purported by Hemmer. It would seem that, if there was a punctuated effect that occurred in humans that is related to neoteny and if animals were similarly neotenized, then this would be expected to lead to increased brain size in both.

4.6.3   Possible molecular basis for a gene related to domestication

Although the evidence is far from being direct or conclusive, there is some theoretical support for the contention that domestication occurs rapidly, possibly as a result of a regulator gene, or master gene, that controls a large number of other genes. This epistatic genetic relationship (non-allelic gene interactions) viewed at a molecular level could be a form of a homeobox gene. Homeobox genes have been found to play crucial roles in a wide range of organisms (Duboule, 1994). A homeobox gene functions by auto-regulatory and cross-regulatory gene interactions, with homeobox gene complexes linked into genetic networks. Through this action, it is theoretically possible that a simple mutation could have a profound pleiotropic effect. Homeobox genes have been found to be influential from the earliest steps in embryogenesis (such as setting up an anterior–posterior gradient in a fruit fly egg) to the latest stages of cell differentiation (such as the differentiation of neurons). These gene complexes are phylogenetically widely distributed, having been found in all vertebrate species investigated (Duboule, 1994). The implication of a homeobox gene for this discussion is that a very simple mutation can possibly mediate the expression of a gene complex that then has a major impact on the morphological, physiological and behavioural development of an organism. Belyaev's (1979) relatively rapid development of domestic-like foxes may be the product of gene change that involved a homeobox, which would be the equivalent to the stage of rapid and dramatic change associated with domestication.[10] While his work preceded the discovery of homeoboxes, Belyaev (1979) seems to have had an awareness of the underlying probable cause of the rapid changes he observed. He wrote, 'In a genetic and biochemical sense, what may be selected for are changes in the regulation of genes – that is, the timing and the amount of gene expression rather than changes in individual structural genes.'

As indirect support for the suggestion that domestic animals experienced morphological changes rapidly rather than gradually, one can also look to the archaeological evidence (Clutton-Brock, 1981, 1999; Zeder and Hesse, 2000). The discussions by Belyaev (1979), Budiansky (1992), Coppinger and Smith (1983), Geist (1971) and Hemmer (1990) about rapidly occurring morphological and behavioural changes in animals, in association with neoteny, are also indirectly supportive of the hypothesis that a homeobox complex plays the dominant role in the genetic basis of domestication.

If a master gene for domestication could be identified through molecular biology techniques, then at least a better understanding of domestication should result. However, such a finding could additionally result in the development of animals that are optimally suited to the conditions experienced by modern domestic farm animals.

4.7   Impact of Production Trait Selection on Social Behaviour

Typically, when animal scientists list the traits that are considered to be of economic importance, the list is inclusive of traits such as rate of body weight gain, efficiency of conversion of feed to body weight, number of eggs produced, volume or weight of milk produced, fleece weight, etc. Seldom are behavioural characteristics included as being important production traits in animal breeding or animal production textbooks. Yet having animals with the appropriate behavioural traits is highly important to the production of food and fibre from animals. Specifically, the behaviour(s) that allow animals to live together and to live in association with, or under the control of, humans are of primary significance to efficient, indeed profitable, animal production systems (Stricklin and Mench, 1987). From this viewpoint, social behaviour is a highly significant production trait, but one that is not typically included in genetic selection schemes.

Social behaviour is also linked, at least to some extent, with a number of other behavioural traits that are important in animal production systems. These behavioural traits include especially sexual and maternal characteristics of farm animals (see Chapter 3). Additionally, growth is primarily a trait associated with young animals. However, growth in farm animals has been extended into older ages. Budiansky (1992) suggested that this is a consequence of neoteny and is linked to a complex of traits, a major one being social behaviour. Thus, these traits are interrelated with, or are a part of, social behaviour to such an extent that, when one is genetically altered, the others are also influenced.

Social behaviour also includes bonding across species between animals and humans (see Section 13.2). Because domestic animals are more social and/or tractable, humans can keep them in groups, restrain, handle and transport them, etc., whereas the lack of tractability makes game ranching difficult, even impractical, for some species of wild animals. The importance of sociability and tractability of domestic animals is sometimes overlooked, but they are characteristics that have allowed the development of many practices common in animal agriculture. In the words of Darwin, it is these behaviours that lead animals to 'receive humans as the chief'.

4.8   Beyond Domestication?

The genetic changes that define a domestic animal relative to its wild progenitor or contemporary counterpart have not been defined, especially at the molecular level. Not understanding these relationships could result in our selecting farm animals in a manner contrary to that of 'true' domestication, especially if a single production trait becomes the primary focus of the selection process. As a consequence of removing selection pressure on traits that led to domestication, including social behaviour, we may have moved or will move animals to undomesticated genetic states – or a state beyond domestication. In other words it is conceivable that animals kept and selected under close confinement conditions could genetically revert to a pre-domestication state or transcend to a state that is considerably different from that originally described by Hale (1962) in his categories of behaviour favourable to domestication. This could especially be true for social behaviour. Under current production conditions it is conceivable that some farm animals through r-selection type pressure are being shifted towards having group structures that are aggregations rather than true social organizations. Additionally, by not having a clear understanding of domestication at the molecular level we may not be selecting animals that are genetically the most suited to the conditions experienced by farm animals – including adaptations to confinement that are welfare positive and also compatible with optimal levels of production.

Notes

1   The cat is possibly a noted exception or maybe even an anomaly. Unlike true social animals that typically live in large groups, cats tend to be solitary. Only in the special circumstances of sexual activity or maternal litter care is prolonged social contact typically found in free-ranging cats (Rosenblatt and Schneirla, 1962). Cats also tend to allocate space using temporally separated but spatially overlapping home ranges, and some of their wild relatives are truly territorial. This is in contrast to true social animals, which employ portable space, with each individual in the gregarious formation having an area around their head or body whose intrusion by conspecifics they tend to defend against (Wilson, 1975). The behaviour and morphology of the domestic cat is little changed compared with its wild ancestors, leading some persons to suggest that the cat lives more in cohabitation with humans rather than having been truly domesticated (Davenport, 1910; Daniels *et al.*, 1998). Some recent semi-domesticated animals such as foxes and mink are also from species that are not truly social.

2   This estimate is based on archaeological evidence. Vila *et al.* (1997) reported that mitochondrial DNA analyses indicate the genetic divergence of dogs from wolves began 100,000 or more years ago. These researchers suggested that association between humans and the precursors to the domestic dog may also have occurred quite

early. Thus, associations with other species may also have occurred earlier than reported dates of domestication.

3   It should be noted that the word 'chose' as used by Budiansky (1992) and others does not imply any cognitive process on the part of individual animals. It is used in reference to the adoption of an evolutionary strategy by a group of animals. Some animals chose domestication in the same sense that the ancestors of *Homo sapiens* chose to become bipedal as an adaptive strategy in their evolutionary development.

4   Some traits of domestic ungulates do, of course, fit into r-selection. Pigs tend to fit some criteria. However, horses and members of the family Bovidae (cattle, sheep, goats, etc.) tend not to fit r-selection strategy criteria.

5   Farm animals continue to have contact with humans, but the contact is much less social than was true for past production systems. The modern intensive confinement, mass production systems employ greater technology in substitution for human labour for animal care (Stricklin and Swanson, 1993). This has led to considerably less frequent and less direct contact between humans and farm animals. Hens in battery cages, gestating sows, pigs in finishing pens, broilers in open floor settings, steers in large feedlot settings and other types of modern production systems result in animals having only limited contact with their human caretakers.

6   While this statement is the classical population genetics view, it is only partly true – primarily when the environment is constant. Houle (1992) in an analysis of published data on this topic demonstrated that often fitness traits in fact have high additive genetic variance components but that the residual variance is also high. Thus, the heritability estimate in the narrow sense, which is a ratio of additive to total variance, is low.

7   Darwin (1859) acknowledged that with domestic animals 'some variations useful to [humans] have probably arisen suddenly, or by one step . . .'. However, he viewed gradual and accumulated change, over many generations, as the primary and most important factor contributing to domestication and breed development.

8   Because the topic is evolution maybe it is not surprising that there are some persons who disagree with this theory (see Hoffman, 1982, for example). However, even if Eldredge and Gould (1972) were not correct as to exactly what produced the abrupt changes in the fossil record, the 'punctuated' changes remain. In the current discussion, it is proposed that molecular genetics will eventually determine the likely causes of these changes, and it is further suggested that Eldredge and Gould will be demonstrated to be more correct than wrong on this topic. Additionally, it is proposed that

the same processes will be found to have caused the early genetic changes associated with domestication.

9   There is not, however, total agreement on whether or not humans should be considered as domesticated. The term 'domestic' has a negative connotation in modern language and in some cases the contention that humans are domestic animals is intended to carry this meaning. In fact Lorenz in his early writings argued that crowding of humans into cities resulted in behavioural and moral degradation, which he also viewed as being a consequence of domestication (Lerner, 1992), and Lorenz (1974) returned to this same theme in his final book, *Civilized Man's Eight Deadly Sins*. In a contrary view but still with a negative attitude to farm animals, Spurway (1955) wrote, 'Evidence from social behaviour strongly contradicts the suggestion that civilization and domestication are processes which have anything in common. Civilization has not resulted in human beings having barnyard morals.'

10 The onset of domestic dog-like behaviour did not occur immediately in all foxes selected. The foxes considered to exhibit the highest level of domestic behaviour progressed approximately as follows: 18%, 35% and 75% for generations 10, 20 and 40, respectively (Trut, 1999). However, this is a remarkably rapid change compared with the thousands of years generally considered to have been necessary to produce domestic-type behaviour. These data also suggest that the underlying genetic change occurred in only some animals but was capable of remaining intact across generations when under selection pressure. Trut (1999) reports that 35% of the variation in the tameness response of foxes was genetic, which is the same as saying that, in their population, the trait had a heritability estimate of 0.35.

References

Beilharz, R.G., Luxford, B.G. and Wilkinson, J.L. (1993) Quantitative genetics and evolution: is our understanding of genetics sufficient to explain evolution? *Journal of Animal Breeding and Genetics* 115, 439–453.

Belyaev, D.K. (1979) Destabilizing selection as a factor in domestication. *The Journal of Heredity* 70, 301–308.

Belyaev, D.K, Ruvinsky, A.O. and Trut, L.N. (1981) Inherited activation–inactivation of the star gene in foxes: its bearing on the problem of domestication. *The Journal of Heredity* 72, 267–274.

Briggs, H.M. and Briggs, D.M. (1980) *Modern Breeds of Livestock*, 4th edn. Macmillan, New York, 802 pp.

Budiansky, S. (1992) *The Covenant of the Wild: Why Animals Chose Domestication*. William Morrow and Company, New York, 190 pp.

Calhoun, J.B. (1963) The social use of space. In: Mayre, W.V. and van Gelder, R.G. (eds) *Physiological Mammalogy.* Academic Press, New York, pp. 2–187.

Campbell, J. (1982) *Grammatical Man: Information, Entropy, Language, and Life.* Simon and Schuster, New York.

Campbell, J. (1991) *The Masks of God: Primitive Mythology.* Arkana, New York, 504 pp.

Clutton-Brock, J. (1981) *Domesticated Animals from Early Times.* British Museum/Heinemann, London.

Clutton-Brock, J. (1999) *A Natural History of Domesticated Mammals,* 2nd edn. Cambridge University Press, 232 pp.

Coppinger, R.P. and Smith, C.K. (1983) The domestication of evolution. *Environmental Conservation* 10, 283–292.

Coppinger, R.P., Glendinning, J., Torop, E., Matthay, C., Sutherland, M. and Smith, C.K. (1987) Degree of behavioral neoteny differentiates canid polymorphs. *Ethology* 75, 89–108.

Daniels, M.J., Balharry, D., Hirst, D., Kitchener, A.C. and Aspinall, R.J. (1998) Morphological and pelage characteristics of wild living cats in Scotland: implications for defining the 'wildcat'. *Journal of Zoology* 244, 231–247.

Darwin, C. (1859) *On the Origin of Species* (reprinted 1950). Watts and Co., London.

Darwin, C. (1875) *The Variation of Plants and Animals under Domestication,* 2nd edn. John Murray, London.

Davenport, E. (1910) *Domesticated Animals and Plants.* Ginn and Company, Boston, Massachusetts.

Dawkins, M.S. (1998) *Through Our Eyes Only? The Search for Animal Consciousness.* Oxford University Press, 208 pp.

Dawkins, R. (1976) *The Selfish Gene.* Oxford University Press, New York, 352 pp.

Dewsbury, D.A. (1978) *Comparative Animal Behavior.* McGraw-Hill Book Company, New York, 452 pp.

Duboule, D. (1994) *Guidebook to the Homeobox Genes.* Oxford University Press, Oxford.

Eldridge, N. and Gould, S.J. (1972) Punctuated equilibria: an alternative to phyletic gradualism. In: Schopf, T.J.M. (ed.) *Models in Paleobiology.* Freeman, Cooper, San Francisco, pp. 82–115.

Frank, H. and Frank, M.G. (1982) On the effects of domestication on canine social development and behavior. *Applied Animal Ethology* 8, 507–525.

Geist, V. (1971) *Mountain Sheep: a Study in Behavior and Evolution.* The University of Chicago Press, Chicago, 383 pp.

Gould, S.J. (1977) *Ontogeny and Phylogeney.* Belknap Press of Harvard University Press, Cambridge, Massachusetts.

Gould, S.J. (1980) *The Panda's Thumb.* Norton, New York.

Graves, H.B. (1984) Behavior and ecology of wild and feral swine (*Sus scrofa*). *Journal of Animal Science* 58, 482–492.

Gustafsson, M., Jensen, P., deJonge, F.H. and Schuurman, T. (1999) Domestication effects on foraging strategies in pigs (*Sus scrofa*). *Applied Animal Behaviour Science* 62, 305–317.

Hale, E.B. (1962) Domestication and the evolution of behaviour. In: Hafez, E.S.E. (ed.) *The Behaviour of Domestic Animals*, 2nd edn. Baillière, Tindall & Cassell, London, pp. 22–42.

Hamilton, W.D. (1964) The genetical evolution of social behaviour (I & II). *Journal of Theoretical Biology* 7, 1–16 and 17–52.

Hemmer, H. (1990) *Domestication: the Decline of Environmental Appreciation.* Cambridge University Press, Cambridge, 208 pp.

Hoffman, A. (1982) Punctuated versus gradual mode of evolution: a reconsideration. *Evolutionary Biology* 15, 411–436.

Hohenboken, W.D. (1987) Behavioral genetics. In: Price, E.O. (ed.) *Farm Animal Behavior*, Vol. 3, *Veterinary Clinics of North America: Food Animal Practice.* W.B. Saunders, Philadelphia, pp. 217–229.

Houle, D. (1992) Comparing evolvability and variability of quantitative traits. *Genetics* 130, 195–204.

Kretchmer, K.R. and Fox, M. (1975) Effects of domestication on animal behaviour. *The Veterinary Record* 96, 102–108.

Lerner, R.M. (1992) *Final Solutions: Biology, Prejudice, and Genocide.* Pennsylvania State Press, University Park, Pennsylvania, 238 pp.

Lorenz, K. (1965) *Evolution and Modification of Behavior.* University of Chicago Press, Chicago.

Lorenz, K. (1974) *Civilized Man's Eight Deadly Sins.* Methuen, London.

Lush, J.L. (1947) Family merit and individual merit as bases for selection, I, II. *American Naturalist* 81, 241–379.

MacArthur, R.H. and Wilson, E.D. (1967) *The Theory of Island Biogeography.* Princeton University Press, Princeton, New Jersey.

Meers, M. (1996) Three-dimensional analysis of differences in cranial morphology between captive and wild American alligators. (http://itech.fgcu.edu/faculty/mmeers/res/gatorheads.html)

Montegu, A. (1989) *Growing Young*, 2nd edn. Bergin & Garvey Publishers, New York.

Morey, D. (1994) The early evolution of the domestic dog. *American Scientist* 82, 336–347.

Olesen, I., Groen, A.F. and Gjerde, B. (2000) Definition of animal breeding goals for sustainable production systems. *Journal of Animal Science* 78, 570–582.

Pianka, E.R. (1970) On 'r' and 'K' selection. *American Naturalist* 104, 592–597.

Price, E.O. (1984) Behavioral aspects of animal domestication. *The Quarterly Review of Biology* 59, 1–32.

Price, E.O. (1998) Behavioral genetics and the process of animal domestication. In: Grandin, T. (ed.) *Genetics and the Behavior of Domestic Animals.* Academic Press, San Diego, pp. 34–35.

Ratner, S.C. and Boice R. (1969) Effects of domestication on behaviour. In: Hafez, E.S.E. (ed.) *The Behaviour of Domestic Animals.* Williams and Wilkins, Baltimore, Maryland, pp. 3–18.

Reinhardt, V. and Reinhardt, A. (1981) Cohesive relationships in a Zebu cattle herd (*Bos indicus*). *Behaviour* 77, 121–151.

Ricker, J.P., Skoog, L.A. and Hirsch, J. (1987) Domestication and the behavior–genetic analysis of captive populations. *Applied Animal Behaviour Sciences* 18, 91–103.

Rosenblatt, J.S. and Schneirla, T.C. (1962) The behaviour of cats. In: Hafez, E.S.E. (ed.) *The Behaviour of Domestic Animals*. Williams and Wilkins, Baltimore, Maryland, pp. 453–488.

Shea, B.T. (1989) Heterochrony in human evolution: the case for neoteny reconsidered. *Yearbook of Physical Anthropology* 32, 69–101.

Spurway, H. (1955) The causes of domestication: an attempt to integrate some ideas of Konrad Lorenz with evolution theory. *Journal of Genetics* 53, 325–362.

Stricklin, W.R. (1983) Matrilineal social dominance and spatial relationships among Angus and Hereford cows. *Journal of Animal Science* 57, 1397–1405.

Stricklin, W.R. and Kautz-Scanavy, C.C. (1984) The role of behavior in cattle production: a review of research. *Applied Animal Ethology* 11, 359–390.

Stricklin, W.R. and Mench, J.A. (1987) Social organization. In: Price, E.O. (ed.) *Farm Animal Behavior*, Vol. 3, *Veterinary Clinics of North America: Food Animal Practice*. W.B. Saunders, Philadelphia, 307–322.

Stricklin, W.R. and Swanson, J.C. (1993) Technology and animal agriculture. *Journal of Agricultural and Environmental Ethics* 6, 67–80.

Tchernov, E. and Horwitz, L.K. (1991) Body size diminution under domestication: unconscious selection in primeval domesticates. *Journal of Anthropological Archaeology* 10, 54–75.

Tennessen, T. and Hudson, R.J. (1981) Traits relevant to the domestication of herbivores. *Applied Animal Ethology* 7, 87–102.

Trut, L.N. (1999) Early canid domestication: the farm-fox experiment. *American Scientist* 87, 160–169.

Vila, C., Savolainen, P., Maldonado, J.E., Amorim, I.R., Rice, J.E., Honeycutt, R.L., Crandall, K.A., Lundeberg, J. and Robert, K.W. (1997) Multiple and ancient origins of the domestic dog. *Science* 276, 1687–1689.

Wilson, E.O. (1975) *Sociobiology: the New Synthesis*. Harvard University Press, Cambridge, Massachusetts, 697 pp.

Wood-Gush, D.G.M. (1983) *Elements of Ethology: a Textbook for Agricultural and Veterinary Students*. Chapman & Hall, London, 240 pp.

Zeder, M.A. and Hesse, B. (2000) The initial domestication of goats (*Capra hircus*) in the Zagros Mountains 10,000 years ago. *Science* 287, 2254–2257.

Zeuner, F.E. (1963) *A History of Domesticated Animals*. Harper & Row, New York, 560 pp.

Zirkle, C. (1952). Early ideas on inbreeding and crossbreeding. In: Gowen, J.W. (ed.) *Heterosis: a Record of Researches Directed Toward Explaining and Utilizing the Vigor of Hybrids*. Iowa State College Press, Ames, Iowa, pp. 1–13.

The Social Behaviour of Domestic Species

It has been suggested that two approaches to studying animal behaviour are: (i) to choose a topic and study it across species or (ii) to choose a species and learn about all facets of its behaviour. In a sense this book has attempted to combine the two approaches. In Part II of the book we examine the details of social behaviour within several farmed species. It is a recognition that each species has its own unique social behaviour, and that management methods must reflect these differences.

It has been our philosophy that, to understand social behaviour in the commercial context, one must first understand the social characteristics that have enabled that species to survive for thousands of generations in its evolutionary environment. Thus, we have asked the authors to begin their chapters with a description of the basic social characteristics of their species. They have then moved from the general to the applied, as they describe the social behaviour of their species under commercial conditions. Finally, they have examined issues in the management and welfare of the animals arising from social behaviour. It is our intent that, once a reader is familiar with one species covered in this section, they could easily find comparable information on another species in another chapter. To facilitate this we have organized the sections within each chapter to be as similar as possible. However, different issues arise for each species, and so the emphasis within each chapter may vary. In some cases, topics have not been addressed because they are considered of lesser importance for that species.

We have included chapters on the most significant food-producing species: cattle, pigs, poultry and sheep. Even though management systems for cattle are distinctly different for beef and dairy production, animals in both systems stem from the same wild stock. This chapter then provides a look at how the same species exists under two different management systems. In a similar way, the chapter on poultry examines several species and how they adapt to systems for meat production, and one species (chickens) and how it adapts to both meat and egg management.

Horses were included in this book because of their major role as a recreational and sporting animal. Management in this case is often more focused on individuals, perhaps because of their greater value and interaction with humans. It is interesting that many of the problems associated with managing horses arise from this individualization of what is normally a social animal.

The final chapter in Part II examines the social behaviour of several species of fish, which have only recently been subject to domestication and farming. This chapter must address social behaviour in a relatively general manner as the species included evolved in different habitats, and are now farmed in quite different ways. This chapter also allows us to examine the interaction of social behaviour and animal management from a different perspective from the one those of us who are most familiar with terrestrial animals are accustomed to. We would encourage you, once you have read a chapter on a more conventionally farmed species, to examine this chapter on fish. We think it will surprise you how many of the topics and issues are common to both chapters.

In a sense, by spending time on examining the interaction between social behaviour and management in several important species, we return to the concept that many aspects of social behaviour are common to all. After moving from the general to the specific, we then return to the general again in the final part of the book.

The Social Behaviour of Cattle

Marie-France Bouissou,[1] Alain Boissy,[2] Pierre Le Neindre[2] and Isabelle Veissier[2]

[1]*Laboratoire d'Etude du Comportement Animal, INRA, 37380 Nouzilly, France;* [2]*Unité de Recherches sur les Herbivores, INRA, 63122 Saint-Genès-Champanelle, France*

(Editors' comments: Female feral and free-ranging cattle live in herds, which may include mature males, or the males may run in small male groups or as solitary individuals. Under extensive conditions the herds have home ranges. Visual, auditory and olfactory cues are used in communication within these herds. Within herds the relationships within a family, and affiliative relationships with other animals are strong and long-lasting.

The development of separate meat and milk industries has resulted in divergent social paths for managed cattle. Extensively raised beef cattle exist in near natural social groupings, at least until weaning takes place at several months of age. In contrast, intensively raised dairy cattle are often isolated at an early age, or kept in all-calf groups. Many of the socially mediated problems in cattle are due to this early experience. An interesting point that the authors make is that the social relationships within a herd are often beneficial in reducing the effects of stressful conditions.

5.1   Basic Social Characteristics

Domestic cattle belong to the Bovidae family. This family comprises 14 subfamilies, among them the Bovinae, Ovinae and Caprinae, certain species of which have been domesticated. Bovinae are the most recent and advanced of the bovid tribe and are not territorial. Major features of their social organization include the integration of males and females into mixed herds, precocial young, group defence, social licking and minimal social distance (Estes, 1974). Groups are of variable size but generally consist of around 20 individuals. In certain circumstances the gathering of several groups leads to very large herds of up to hundreds or even thousands of animals (e.g. bison, Lott and Minta, 1983; African buffalo, Sinclair, 1977). Outside the rutting period, males are either solitary or in male groups of from two to ten 3- or 4-year-old individuals. These groups are less cohesive than female groups.

The only representatives of the genus *Bos* are the domesticated species *Bos taurus* and *Bos indicus*. However the observation of feral cattle can provide a basic understanding of the social structure and behaviour of the ancestral wild species, namely the auroch (*Bos primigenius*, Bojanus), which became extinct in 1627. Few populations of cattle in the world are really feral. These include the feral cattle of Amsterdam, an island southwest of Madagascar (Daycard, 1990), the Maremma cattle in Italy (Lucifero *et al.*, 1977), a population of 140 animals running in the south of Spain (Lazo, 1994) and a herd in the Orkney islands (Hall and Moore, 1986). A special case is that of the Chillingham cattle kept in a closed park, in northern England, for more than 700 years with minimal human interference (Hall, 1986). Among observations of domestic cattle that are free ranging for at least part of the year are those of the Camargue cattle (Schloeth, 1961), a herd in Utah (Howery *et al.*, 1996) and a small mixed herd on the Isle of Rhum (Clutton-Brock *et al.*, 1976).

5.1.1 Composition and structure of social groups

On Amsterdam Island, females were often associated with two calves: one recently born and a yearling. Apart from this basic structure, Daycard (1990) observed three main types of groups: (i) groups of females of all ages with some sub-adult males (mean group size 10.5 animals); (ii) adult and sub-adult males, of which a high proportion were solitary adult males (mean group size 3.5 animals); and (iii) mixed groups of adult males and females mainly during the mating season (mean group size 18 animals). The young males were often associated with other sub-adult males or with adult males, whereas the females were more often associated with the adult females (Lésel, 1969; Daycard, 1990; Berteaux and Micol, 1992).

In the Chillingham herd, adult cows, heifers and young bulls form mixed groups, whereas males more than 4 years old live in male groups of up to three animals. However, during the winter period, when hay is provided at one point of the park only, the various bull groups are forced to come into contact (Whitehead, 1953; Hall, 1986). Large mixed groups of males and females are the rule in the herd in the south of Spain (Lazo, 1994) but the number of males in this case was very low.

5.1.2   Use of space

On Amsterdam Island, males and females occupy different home ranges (Lésel, 1969). Specific home ranges have been described for cattle in the south of Spain (Lazo, 1994) and Utah (Howery *et al.*, 1996). This strategy could increase the global feeding efficiency, but Swona cattle foraged as a single group in winter and there was no sign of home ranges (Hall and Moore, 1986). Inter-herd dominance relationships have been described (Lazo, 1994) and affects the use of specific parts of the environment. When a home range exists, it is learned by calves at an early age (Howery *et al.*, 1998).

5.1.3   Communication (see Section 2.3)

Visual communication

Visual signals are one of the most important means of communication in cattle. Grazing mammals have wide-set eyes and panoramic vision, an adaptation for survival as prey animals. Their angle of vision is approximately 320°. Cattle have only 1/22 to 1/12 the visual acuity of humans (Entsu *et al.*, 1992). Colour vision has been demonstrated by operant experiments (Soffié *et al.*, 1980; Gilbert and Arave, 1986; Riol *et al.*, 1989).

Body language or visual signals may involve movements of the entire body or only parts of it. Facial expressions are poor in cattle compared with horses. In contrast, the mobility of the head allows displays in which its position with respect to the body plays an important role, e.g. in aggressive or submissive displays (Schloeth, 1958). Tail position does not seem to play an important role in communication in cattle, with the possible exception of cows in oestrus. The position of the tail is mostly an indicator of a cow's mood and activity (Kiley-Worthington, 1976; Albright and Arave, 1997). Bulls are attracted by the sight of oestrous females mounting each other (Kilgour *et al.*, 1977; Baker and Seidel, 1984/85).

Vocal communication

Cattle like other gregarious grazing mammals use vocalizations to communicate although to a lesser extent than do forest-dwelling species. Eleven different acoustic signals have been reported for Camargue cattle by Schloeth (1961). Using sonograms, Kiley (1972) described six different types of calls; however, they are more like a continuum. In cattle, vocalizations do not seem to be specific to the situation but to the degree of excitement and interest in the stimulus (Kiley, 1972). Most are related to frustration and stress (for example, an isolated animal seeking conspecifics or animals anticipating a pleasurable event such as water, food or milking).

Olfactory communication

The considerable number of odoriferous glands (interdigital, infra-orbital, inguinal, sebaceous glands, etc.) present in cattle suggests the importance of olfaction in their social life. Olfactory cues are important in social, sexual and maternal behaviour. Both the main olfactory system (olfactory bulbs) and secondary olfactory system (vomeronasal organ) are used. The flehmen response, in which the animal presents a special facial expression, allows the animal to put odours into direct contact with the vomeronasal organ (for a review see Albright and Arave, 1997).

Olfaction is of importance in social relationships as it contributes to individual recognition. Cattle can be trained to individually discriminate between conspecifics through olfactory cues alone (Baldwin, 1977). The role of olfaction in the determination and maintenance of social rank has been studied in groups of unfamiliar cows deprived of the sense of smell by surgical removal of the olfactory bulbs. Neither the establishment of rank order nor its maintenance differed from

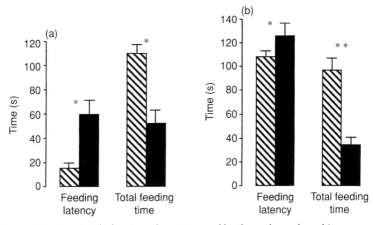

Fig. 5.1. (a) Mean (SE) behavioural reactions of heifers when placed in an unfamiliar arena in the presence of a non-stressed (hatched bars) or stressed (black bars) conspecific (*n* = 8 per treatment). (b) Influence of the odour of urine from non-stressed (hatched bars) or stressed (black bars) conspecifics on the feeding behaviour of heifers. Urine samples were placed under the food in an unfamiliar arena (*n* = 10 per treatment) (means and SE; **P < 0.01; *P < 0.05). (From Boissy *et al.*, 1998.)

that in intact animals (Mansard and Bouissou, 1980; Bouissou, 1985). However, steers whose vomeronasal organs have been plugged and cauterized were more aggressive and tended to attain higher social rank than controls (Klemm *et al.*, 1984). On the other hand, spraying lactating cows with aniseed oil reduces the frequency of aggressive acts and alleviates the reduction in milk yield following regrouping (Cummins and Myers, 1991).

Finally, animals can communicate their psychological state, especially when frightened or stressed, by means of pheromones. Heifers are slower to learn a task if they are in the presence of stressed conspecifics, and their latency in approaching a bucket containing food is increased in the presence of urine of stressed peers (Boissy *et al.*, 1998; see Fig. 5.1).

Tactile communication

This type of communication has been far less documented, although it is important in sexual and maternal behaviour, in establishing rank order, in affiliative relationships (allogrooming) and in human–animal relationships.

5.1.4   Intra-group interactions

Male–male

The social system of the wild Bovinae is characterized by a dominance hierarchy between adult males (Estes, 1974). In feral cattle this hierarchy has been observed between adult males and also in the bachelor groups where sub-adult males live together. Dominance relationships between males are less stable from year to year than those between females, and middle-aged males (3–5 years) tend to be dominant. In the Chillingham herd an older male is dominant and is responsible for all mating. In a herd of Camargue cattle Schloeth (1961) stated that among ten bulls only the two highest-ranking ones were observed copulating.

Aggressive behaviour is the most prevalent of the social activities recorded for both bulls and steers maintained together on pasture (Kilgour and Campin, 1973; Hinch *et al.*, 1982a). There is also a high incidence of sexual behaviour. Until 18 months of age bulls and steers have similar spatial requirements (Hinch *et al.*, 1982b), although, when older, bulls become more 'territorial' (Kilgour and Campin, 1973).

Female–female

The interactions between cows have rarely been described for feral animals. In the Chillingham herd, no particular associations between individuals have been found, but strong affinities exist among classes, e.g. high-ranking females tend to associate with high-ranking males (Hall, 1986). Clutton-Brock *et al.* (1976) concluded that the hierarchical organization of free-ranging cows on the Isle of Rhum is similar to that of animals raised under human control.

Male–female interactions

Lésel (1969) observed that each day the adult males shift from their inland higher altitude areas to the lowland ones, where the females are concentrated, but that they do not maintain close contact with any specific group of females and there is no 'harem' behaviour. On the other hand, in southern Spain, Lazo (1994) observed several adult bulls living all year round in each of the four herds.

Parent–offspring (see Chapter 3)

Only a few descriptions of the mother–young relationships of feral cattle are available. In the Chillingham and Maremma herds, cows isolate themselves as far as possible from the herd to give birth. The calf remains hidden for some days before joining the herd with its mother (Hall, 1986; Vitale *et al.*, 1986). Cows suckle their young until the next calf is born, but sometimes calves can be suckled after the birth of a second one, thus leading to starvation of the younger calf (Bilton, 1957).

For some days before calving, domestic cows running free on pasture are restless (Hafez, 1974) and may isolate themselves from the group (Wagnon, 1963; Edwards, 1983). Apparent isolation could be the result of reduced mobility arising from the cow remaining near the amniotic fluids, which are very attractive for her (George and Barger, 1974). A large percentage of cows stay in sheltered areas during calving, especially when climatic conditions are harsh (Scheurmann, 1974; Le Neindre, 1984).

During the postnatal period (see Section 3.4), cows display protective behaviour and may attack dogs, foxes (Rankine and Donaldson, 1968) or humans coming close to their calves (P. Le Neindre, unpublished results). During the first few days after birth cows stay close to their calves although, after this postnatal period, cows begin to spend more time away from their calves and integrate progressively with the herd. The postnatal period is essential for the establishment of the bond between the calf and its dam (Poindron and Le Neindre, 1975). When cows without previous maternal experience have no contact with any calves during the first 24 h after calving, they will not later accept being suckled by a calf (Le Neindre and Garel, 1976; Hudson and Mullord, 1977). However, in contrast to ewes, cows with previous maternal experience do accept calves even when their first contact with them is not until the day after birth (Le Neindre and D'Hour, 1988). After calving beef cows are very selective and suckle only their own calves.

Generally the calf suckles within the first hour although some calves were not even suckled for 4 h (Selman *et al.*, 1970a, b). When cows give birth to twins there are usually no problems in the establishment of the mother–young relationship (Price *et al.*, 1981; Owens *et al.*, 1985).

Licking is the other important activity of the cows towards their calves. After birth the cow licks her young for a long time, basically until it is dry, and the number of lickings per day remains high for more than 10 months after birth. About 56% of the licking bouts are associated with suckling. The specific relationship between the dam and her calf is long-lasting (Le Neindre, 1984; Veissier *et al.*, 1990a). When cows do not have a new calf the following year, calves still suckle three times a day at 10 months of age and 1.5 times per day at about 400 days. However, this relationship is not dependent on the possibility of the calf obtaining milk (Veissier and Le Neindre, 1989). After the birth of a new calf, the young of the previous year still have preferential contact with their dams. Reinhardt and Reinhardt (1982b) observed that in African zebu, despite the cows actively weaning their calves between the age of 7 and 14 months, the dam remains the preferred partner.

Among juveniles

Based upon the relationship between the dam and offspring in the days following the birth, calves can be considered as 'hiders' (Lent, 1974). When not with their mother, they stay alone, concealed in the vegetation, usually lying down for long periods. Le Neindre (1984) observed that calves had no neighbour in 12% of the scans made when they were between the age of 2 and 5 days. However, this period of hiding was rather short and 3 weeks after birth calves spent most of their time with other calves even if the number of interactions with them was limited. This 'creche' behaviour has been observed by several authors and reaches a peak between the 11th and 40th day of life (Vitale *et al.*, 1986; Sato *et al.*, 1987), and some have observed dams remaining within the proximity, which they have called 'cow guards'. Several functions of this tendency for the young to cluster have been hypothesized, e.g. it could have an anti-predatory function, it may decrease the negative influence of flies or allow the socialization of the calves.

The number of interactions of the young with other members of the herd increases slowly with age and is similar to those observed in adults. However, they are largely non-agonistic before the age of 2 months. Activities resembling fights occur as soon as 2 weeks of age, described as 'mock fighting' (Reinhardt and Reinhardt, 1982a) although they are displayed in a different social context from real fights. They usually start with a solicitation from one of the partners and usually end abruptly with no evident consequences for either of the two animals. During the first few months of life, female calves are more often engaged in play fighting than males. But after 6 months of age it is the males who initiate more play fights and are more attractive as partners (Reinhardt *et al.*, 1978; Vitale *et al.*, 1986). Mock fights are associated with running games and playful mountings. Other types of cattle play include gambolling, bucking, kicking, prancing, butting, vocalizing, head shaking, sporting, goring and pawing. Schloeth (1961) even described a play-call, a play-specific tail position of calves and specific play areas of cattle on free range.

During the first months after birth, male and female calves have similar relationships with their mothers and with the other members of the herd. However, 10-month-old males have many more interactions with the other calves (see Section 3.11), especially other male calves, and with other cows than do the female calves (Reinhardt and Reinhardt, 1982b; Le Neindre, 1984; Vitale *et al.*, 1986). Young males progressively form bachelor groups (Schloeth, 1956). At the same age, females remain very close to their mothers (Kimura and Ihobe, 1985). This probably facilitates the building of a matriarchal relationship with specific affinities between females (Lazo, 1994), and may be the first step to the social structure observed in feral cattle.

5.2   Social Behaviour Under Commercial Conditions

5.2.1   Social groupings

Cattle are raised all over the world under very different management systems. In developing countries, cattle are often associated with pastoralism. The cattlemen drive their animals so as to find food and guard them. In developed countries two main types of management are found. In suckling herds, the cows raise their calves for 4 to 9 months. During the summer season the cows graze with their calves and usually there is one bull per herd. When not used for reproduction the males are gathered together after weaning into fattening units or feedlots. In contrast, in the modern dairy industry, calves are isolated from their mothers soon after birth, and in most cases before 3 days of age (see Section 11.2.3). The female calves are artificially reared and raised in groups of females until they join the cow herd. The males can be used as veal calves or raised in fattening units. Groups of dairy cows are often assembled based on their production and physiological status (lactating, dry). Artificial insemination is commonly practised.

Different types of cattle are used under these different management types. They are not only different in their ability to produce meat or milk, but also in their behaviour. For example, Salers cows, which are used for beef production, are more selective than Holsteins and actively reject alien calves. The calves are also different as Salers calves try to maintain closer contact with their mothers than do Friesians and have more problems in adapting to artificial rearing management (Le Neindre, 1989). These two types of animals appear to be adapted to specific environments.

5.2.2　Social structure

Social interactions

Social interactions (see Section 2.4) can be roughly divided into agonistic (including aggressive acts and the responses to aggression, mainly avoidance reactions) and non-agonistic (including in particular allogrooming and sexual behaviour).

Several types of threat can be described in cattle according to the extent of ritualization, from a simple swing movement of the head to more sophisticated patterns (Schein and Fohrman, 1955; Schloeth, 1958, 1961; Bouissou, 1985) (Fig. 5.2). In the lateral display, which is also common in other bovids, the animal presents itself laterally, with the head down, the back arched and the hind legs drawn forward, showing its largest profile. It also often turns slowly so that it is always presenting the flank (Schloeth, 1956; Fraser, 1957) (Fig. 5.2f). This threatening stance is ambivalent and may express a tendency both to attack and to withdraw. If the threatened animal is slow to submit or fails to notice the threat, the dominant animal butts. A butt (also called a bunt) is a blow with the forehead directed at the opponent's side or rump, which can seriously injure the victim, especially if the dominant animal has horns.

However, when dominance relationships are well established, threats generally induce the retreat of the threatened animal at the slightest movement of the dominant (Fig. 5.2c). A retreat or withdrawal is often accompanied by a submissive (or appeasement) posture, in which the head is low and directed away from the opponent (Fig. 5.2d). Generally speaking, avoidance reactions or withdrawals follow an aggression, but, in well-established groups, most of these behaviours happen without any overt aggression from the dominant animal (Fig. 5.2e). Such retreats can be provoked from significant distances. These 'spontaneous withdrawals' constitute, in some groups, more than 90% of all agonistic acts recorded.

Before the establishment of dominance relationships a threat can provoke another threat in return. In this latter case, fighting generally occurs. Fighting among cattle is head to head followed by head to neck (Schein and Fohrman, 1955; Bouissou, 1985) (Fig. 5.2a and b). There is much manoeuvring for position as each combatant strives for a flank rather than a frontal attack. The duration of fights is highly variable, from a few seconds to nearly 1 h. However, most of them are of short duration: 80% last less than a minute (Bouissou, 1974a). In prolonged

Fig. 5.2. Agonistic interactions among cattle. (a) Fights between Hérens cows at pasture. (b) Exhibition of fighting between Hérens cows. (c) Threat (no. 88) and avoidance reaction (no. 38) in Hérens cows. (d) Submissive posture (right) following a threat (left). (e) Spontaneous withdrawals (not provoked by an aggression). (f) Lateral threat display (no. 88). Photographs by M.-F. Bouissou.

fights, a behaviour called the 'clinch' has been described by Schein and Fohrman (1955). In this, one participant allows the opponent to gain a flank advantage as it pushes its muzzle between the opponent's hind leg and udder. No cow can attack from such a position, and the clinch is an ideal way for combatants to rest safely. According to Schein and Fohrman (1955) and Schloeth (1961), under herd conditions agonistic interactions are much more common between neighbours in social rank (up to three rank positions apart) than between animals widely separated on the social scale.

Allogrooming, which has a communicative and social function, consists mainly in licking of the head, neck and shoulder areas (Sambraus, 1969; Bouissou, 1985). Anogenital and rump lickings, as well as licking of the penis in bulls, are more associated with sexual behaviour. Lickings are often preceded by a solicitation to be licked, including the adoption of a special posture with the head and neck lowered and often with slight bunts under the neck or chest (Bouissou, 1985; Sato *et al.*, 1991). According to Sato (1984) all animals are groomed, but only 75% of individuals perform grooming. Depending on the authors, grooming is mainly performed by the subordinate member of a pair (Fraser and Broom, 1997), more often initiated by the dominant (Sambraus, 1969), or is independent of the social position of the groomer (Schloeth, 1961; Bouissou, 1985). More obvious are the preferences for particular partners. Preferred partners are often animals of similar age, of neighbouring rank or are related animals (Wood, 1977; Reinhardt, 1981). Familiarity has a significant effect on licking, whose frequency increases with the length of cohabitation (Bouissou, 1985; Sato *et al.*, 1991). Allogrooming may function as a tension-reducing effect, reinforcing social bonds and the stabilization of social relationships (Sato *et al.*, 1993).

Leadership

Leadership (see Section 2.4) is the ability of an animal to influence the movements and activities of its group mates. According to Meese and Ewbank (1973) it is defined as 'a form of unequal stimulation, acting possibly through animals of a low threshold of response to a given environmental change, i.e. certain animals may react faster than others to environmental change and these may stimulate their fellows'. All herding animals show leadership–followership behaviour under various social circumstances. Leadership can be qualified as 'social', concerned with control of aggression and altruistic aspects such as protection of other members when the group is faced with a danger, or 'spatial', concerning group movements (Syme and Syme, 1979).

In spatial leadership, the leader is generally said to be the animal at the head of the movement. However, it has been suggested by Leyhausen and Heinemann (1975) that, in fact, the animals responsible for movements could well be at the rear of the group, thus pushing the other animals. The capacity to push would be correlated with dominance, whereas the capacity to incite other animals to follow, or to pull them along, would be more related to social attractiveness.

Voluntary or forced leaderships (that is, movements forced by a human) have been studied in cattle in a wide variety of situations: grazing, during movements to or from the milking parlour, entering the milking parlour, squeeze chute or crushes, and in various other management situations, and highly repeatable orders are commonly observed. However, correlations are low between these various leadership orders (Beilharz and Mylrea, 1963; Dickson *et al.*, 1967; Reinhardt, 1973; Arave and Albright, 1981). Reinhardt (1973) reported a correlation of 0.41 between dominance and milking order. However, according to Dietrich *et al.* (1965), entrance into a milking parlour is unrelated to body weight, age or dominance, but higher yielding cows do enter first (Rathore, 1982).

There is generally little correlation between leadership and dominance order (McPhee *et al.*, 1964; Dietrich *et al.*, 1965). However, during voluntary movements, middle-ranking cows are generally at the front of group, high-ranking cows are in the middle and low-ranking cows at the rear (Kilgour and Scott, 1959; Arave and Albright, 1981). During forced movements subordinate cows are in front (Beilharz and Mylrea, 1963).

Dominance relationships

Dominance relationships in cattle have been studied extensively since the pilot works of Schein and Fohrman (1955) and Guhl and Atkeson (1959). Various methods have been proposed to assess dominance (e.g. dominance value) (Beilharz and Mylrea, 1963; Bouissou, 1970, 1985; Reinhardt, 1973; Arave *et al.*, 1977; Soffié and Zayan, 1978; Le Neindre and Sourd, 1984; Wierenga *et al.*, 1991). Although the mere observation of a group can be sufficient to determine relationships among animals, this may constitute an arduous and tedious task for large groups or groups with low social activity. For these reasons several methods based on competition for a desired resource have been proposed (Bouissou, 1970; Friend and Polan, 1978; Stricklin *et al.*, 1985; Rutter *et al.*, 1987) (Fig. 5.3). (For a recent review of dominance establishment and assessment see Albright and Arave, 1997.)

In wild ungulates, the young animal is integrated into the pre-existing social structure based partly on its age and mother's social position. Schloeth (1961) observed in the Camargue herd, under semi-natural conditions, that dominance relationships develop gradually and are still somewhat unstable at 12 months of age. In zebu cattle, dominance relationships are apparent in the fifth month (Reinhardt and Reinhardt, 1982b).

At a very young age, especially among artificially reared dairy calves, agonistic interactions tend to be bidirectional and bunts often fail to elicit withdrawal (Bouissou, 1977, 1985; Canali *et al.*, 1986). However, social hierarchy can be established at a young age, depending on the animals' experience and the social context. Dam-reared calves establish dominance relationships at an earlier age than artificially reared ones (on average 4–5 months vs. 9 months), and they learn the significance of social interactions such as threat displays at an earlier age (Bouissou, 1985). The appearance of dominance among familiar animals often corresponds to the time of first oestrus, when there are more social interactions (Bouissou, 1977). The establishment of dominance between ovariectomized heifers is delayed (Bouissou,

1985). In the case of unfamiliar animals, 6-month-old heifers are able to establish clear and stable dominance relationships (Bouissou and Andrieu, 1977).

Under natural conditions, the introduction of new members into groups is rare among females, but it is a common practice in animal husbandry. The most striking feature is that dominance relationships are established extremely rapidly, in most cases without a fight, and even without physical contact (Bouissou, 1974b, 1985). The comparison of groups of previously unacquainted 'naive' (no prior contact with strangers from birth) or 'experienced' heifers (one to five previous encounters with strangers) demonstrated that social experience profoundly affects the speed and the means by which dominance relationships are established. Experienced animals fought less, established dominance relationships much more quickly and these relationships were more stable (Bouissou, 1975) (Fig. 5.4a and b).

Dominance relationships among adult females are extremely stable and may persist for several years. Neither oestrus, ovariectomy nor pregnancy modifies social rank (Bouissou, 1985). Rank reversals are less than 10% per year (Beilharz and Mylrea, 1963; Reinhardt and Reinhardt, 1975; Sambraus *et al.*, 1978). When reversals do occur, they generally happen suddenly and are difficult to explain. Dominance relationships are less stable among young animals (Reinhardt and Reinhardt, 1975) and among males (Bouissou, 1985).

It is very difficult to modify dominance relationships experimentally either by manipulation of social experience (being dominant

Fig. 5.3. Food competition tests between heifers: the dominant animal (left) controls the food during the entire period. The subordinates often do not try to get access to food. (From Bouissou, 1970.)

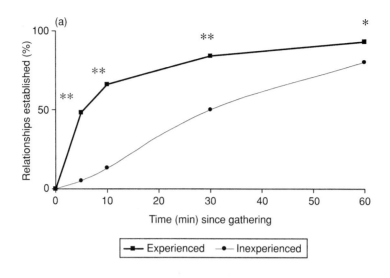

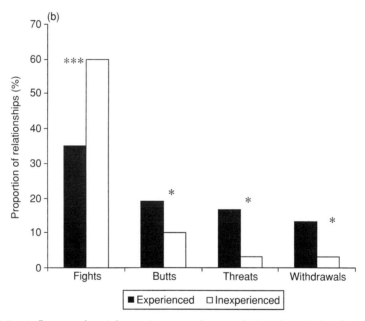

Fig. 5.4. Influence of social experience on the speed (a) and mode (b) of establishment of dominance relationships in ten groups of four previously unacquainted heifers (five in which the animals had never been exposed to strangers, five in which they had had such experience). (a) Cumulative percentage of established relationships during the first hour following the regrouping of the animals. (b) The bars indicate the proportion of relationships (30 per treatment) decided after each type of interaction (fights, butt, threats or withdrawals); in the remaining cases it was impossible to determine which interaction was responsible for the establishment of the relationship. *$P < 0.05$; **$P < 0.01$; ***$P < 0.001$.

or submissive in other groups), or by modifications of physical appearance (colour or odour modifications, adorning with larger false horns, etc.) (Bouissou, 1985). However, Bouissou (1978, 1990) was able to completely change the social order of established groups of cows by treating half of the animals with testosterone propionate or oestradiol benzoate. Treated animals invariably became dominant over non-treated ones without modifying the relationships between themselves, and such rank reversals are long lasting even after the cessation of treatment.

Fights are limited to the first days, even the first hours, after meeting. Thereafter relationships are generally maintained by threats from the dominant and avoidance reactions from the subordinate, without physical contact. Therefore, non-contact senses such as vision and olfaction seem to be of importance in the maintenance of relationships. However, dominance relationships are not modified by deprivation of sight, and can be revealed by food competition tests under controlled conditions (Bouissou, 1971) (Fig. 5.5). Moreover, 80% of dominance relationships were maintained in a group of anosmic cows after they had been blindfolded (Mansard and Bouissou, 1980). On the contrary, even if not necessary, physical contact should remain possible. Different types of separations have been used to improve feeding time of subordinate cows and it has been shown that the protection of the head is of major importance (Bouissou, 1970).

According to most authors, age is important in determining social position (Cummins and Myers, 1991; Kabuga, 1992a, b). However, it is

Fig. 5.5. Food competition test between cows deprived of sight. (a) Control situation: dominant cow (right) prevents subordinate from feeding. (b) The same interaction when the animals are deprived of sight by a blindfold. (c) The subordinate cow has no access to food and stands aside. (From Bouissou, 1971.)

almost impossible to control for other factors, as age is often associated with seniority in the group, weight and experience. Many authors have reported a correlation between social rank and weight or height at the withers (Schein and Forhman, 1955; Guhl and Atkeson, 1959; McPhee *et al.*, 1964; Dickson *et al.*, 1967; Arave and Albright, 1976, 1981; Sambraus *et al.*, 1978; Kabuga, 1992a, b). However, the existence of a correlation between weight and rank does not necessarily prove the influence of weight, for weight could instead be a consequence of a high social rank. Horns have been thought to be of importance in determining rank, and farmers often believe that removing horns will modify the social order. Although the presence of horns confers a significant advantage at the time when the social order is being established, the removal of horns in a well-established group modifies only a few relationships, if any (Bouissou, 1985).

Early experience, including rearing conditions, influences social position in adulthood. Warnick *et al.* (1977) found that group-reared calves were dominant over individually reared and isolated calves. Salers heifers that had been bucket-fed were less dominant than those reared by a foster mother, whereas the rearing conditions had no effect on Friesian heifers (Le Neindre and Sourd, 1984). Bouissou (1985) compared dominance ability of calves reared in complete isolation from birth to 6 months of age, then first regrouped with calves reared in the same conditions, and finally with group-reared calves. There was no difference in dominance ability between isolated and group-reared calves. The differences between these results and those of other studies could be explained by the fact that the animals had first been grouped among themselves, and thus had gained social experience, whereas, in other studies, they were grouped just at the end of individual rearing or isolation. It seems that the consequences of social deprivation have no long-lasting effect in cattle (Bouissou, 1985; Arave *et al.*, 1992).

The social status of the young does not appear to be determined by its dam (Le Neindre, 1984; Bouissou, 1985). In a study of groups of cows including granddam–dam–daughter families, Stricklin (1983) concluded that granddam and dam were always dominant over their adult offspring.

Genetic influences on dominance have been demonstrated in several studies (Stricklin, 1983; Le Neindre, 1984; Mench *et al.*, 1990; Kabuga *et al.*, 1991). Dominance among heifers raised separately from their mothers from birth has been reported to be poorly heritable (Dickson *et al.*, 1970). However, identical twins or clones are difficult to rank with respect to each other, having the same rank in a larger group. It is possible to exchange individual twins between the groups without modifying the social structures (Bouissou, 1985; Purcell and Arave, 1991). It may be that the other members of the group are unable to distinguish one twin from the other (Bouissou, 1985).

Temperament, including emotional reactivity or fearfulness, is probably one of the most important factors in determining social position. High-ranking animals withdraw less often from their own superiors, and future dominant calves also withdraw less often before relationships are established. When anosmic, blindfolded, unfamiliar cows first meet, some animals consistently withdraw as soon as they perceive another, and thus become subordinate. Thus, fearfulness seems to play a critical role in the establishment of the relationships iissou, 1985). In a series of experiments aimed at modifying iinance relationships in stable groups of cows, or influencing future al rank of individuals (either calves or adults) using androgen tment, Bouissou (1978), Bouissou and Gaudioso (1982) and Boissy Bouissou (1994) clearly demonstrated that the higher dominance ity consistently attained by treated animals was the result of a lower : of conspecifics, as well as a lower general reactivity.

'liative relationships (see Section 2.4)

ə basis of social organization in most ungulates is a matriarchal ıup, in which aggressive behaviour is rare and dominance relation- ps difficult to reveal. This suggests that preferential relationships ıst between members of these groups and are responsible for their hesion. In cattle, affinities include spatial proximity, reduced aggres- ʾeness, enhanced positive interactions and tolerance in competitive uations. Such relationships can remain stable for several years tween dams and their offspring or even unrelated animals einhardt, 1981). Twins often present strong affinities, but the same is ıe for unrelated calves reared together (Ewbank, 1967). Heifers reared the same group from birth were less aggressive among themselves, ʿchanged more non-agonistic interactions, remained spatially associ- ed during feeding and resting and were more tolerant in a food- ɔmpetitive situation than heifers of the same herd but with which they ad not been reared (Bouissou and Hövels, 1976a, b). The period from irth to 6 months of age is the most suitable for the complete develop- ıent of such preferential relationships, and could be limited by the evelopment of dominance relationships around puberty in Friesian ɪeifers (Bouissou and Andrieu, 1978). The existence of such relation- hips is of economic importance as it reduces the unfavourable conse- quences of the existence of dominance relationships for subordinates.

5.2.3   Effects of group size and space allowance on social behaviour

High social density (minimal space allowance per individual) and large group sizes reduce human labour and building costs. However, they alter the behaviour and the production of animals. Under crowded conditions an animal cannot maintain individual distances and is forced to move around to avoid superiors. In calves, a negative correlation is found between agonistic behaviour and space allowance (Kondo *et al.*, 1989). In cows, increased social aggressiveness occurs under high-density husbandry (Hafez and Bouissou, 1975; Kondo *et al.*, 1989). Under conditions of excessively large group sizes (see Section 2.5), individual animals appear to have difficulty memorizing the social status of all peers, which increases the incidence of aggressive interactions, in both dairy (Hurnik, 1982) and beef cattle (Stricklin *et al.*, 1980).

Most of these conditions are associated with physiological responses indicative of chronic stress. For example, daily gain of heifers and bulls is lower when they have a low space allowance than when they have more space (Andersen *et al.*, 1997; Mogensen *et al.*, 1997). Overcrowding due to insufficient number of headlocks or inadequate manger spaces per cow has a greater impact on behaviour and well-being than does group size *per se* (Albright and Arave, 1997).

5.3 Social Behaviour, Management and Welfare

Cattle are highly adaptable and generally they respond well to modern farming practices. However, this adaptive ability can be overwhelmed. For instance, the intensification of animal housing and management can cause social disturbances resulting in behavioural problems, which in turn may affect productivity and welfare. Social constraints are of lesser importance for cattle reared in open rangelands or at pasture, although social relationships among animals in extensive husbandry may also have implications on productivity. On the other hand, social environment has positive effects on individual adjustments to the environment through social facilitation or learning.

5.3.1 Influence of social partners

During exposure to stressful events, the social group can lower the subject's arousal. For example, heifers are less likely to avoid an unusual noise in the presence of penmates (Boissy and Le Neindre, 1990). In a novel arena, heifers seem less afraid of the situation when they are with social partners than when they are alone (Veissier and Le Neindre, 1992). Animals are not only aware of the presence of partners, but also of their emotional state. When heifers are exposed to a novel environment, they show a lower tendency to feed in the presence of a stressed partner than in the presence of an unstressed one (Boissy *et al.*, 1998). Social influences are likely to be mediated by olfactory signals contained in urine. Heifers are more reluctant to approach a novel object or to go along a corridor when air containing volatile compounds of the urine of a stressed animal has been sprayed (Boissy *et al.*, 1998; Terlouw *et al.*, 1998).

Foraging with experienced social partners decreases food neophobia and facilitates acceptance of novel foods by naive animals (Ralphs *et al.*, 1994). Likewise, cattle learn to avoid a harmful food by observing their conspecifics avoiding this food, or vice versa. Calves that have been previously trained to avoid eating a plant begin to graze this plant when placed with naive cattle that are grazing this plant (Ralphs and Olson, 1990). Social partners can also influence distribution patterns in cattle while grazing in free-ranging conditions. Adult cattle return to the locations where they were reared as calves by their dams (Howery *et al.*, 1998).

5.3.2   Grouping animals

Regrouping and mixing of unfamiliar animals are common practices in beef and dairy husbandry. More than a tenfold increase in agonistic interactions can be observed during the hours following mixing. Many studies report behavioural and physiological consequences of repeated social changes that could reflect social stress in dairy and beef cows (Arave and Albright, 1976; Kondo *et al.*, 1984, 1994; Mench *et al.*, 1990). Introducing strange cows into a stable herd affects mean weight gain, not only for the introduced animals but also for the whole herd (Nakanishi *et al.*, 1991). It is common to regroup cows during lactation according to milk yields and it is said that this usually does not adversely affect production (Arave and Albright, 1981; Konggaard *et al.*, 1982). However, a reduction in milk production after regrouping has been reported by Hasegawa *et al.* (1997) and this can reach 4% during the first 5 days (Jezierski and Podluzny, 1984).

5.3.3   Separation problems (see Chapter 12)

Rearing in isolation

Early isolation can have profound effects on the reactivity and on the subsequent social behaviour of calves. De Wilt (1985) and Webster *et al.* (1985) reported that veal calves reared in individual crates are easily alarmed and Veissier *et al.* (1997) found that the hyper-reactivity is further increased when all physical and visual contact between calves is suppressed. The hyper-reactivity of calves reared in crates (Trunkfield *et al.*, 1991) or the increased response to mixing (Warnick *et al.*, 1977; Waterhouse, 1978; Bouissou, 1985) may be responsible for more intense reactions during transport.

Physiological changes due to isolation during rearing have been reported in dairy calves (Arave *et al.*, 1974). Calves reared in individual crates can have lower weight gains than calves reared in groups (Warnick *et al.*, 1977; Veissier *et al.*, 1994). In contrast, other authors (Purcell and Arave, 1991; Arave *et al.*, 1992) found no detrimental effect of calf isolation on their milk production as cows.

Separation from the dam at weaning

Suckler calves are often separated from their dam for artificial weaning at about 8–9 months of age. This weaning seems highly stressful to calves, as shown by the increase in plasma cortisol level and the disruption of the circadian rhythm of activity (Veissier *et al.*, 1989a, b). At the time of weaning the young receives only a small percentage of its energy from the milk of its dam (Le Neindre *et al.*, 1976). Hence the stressful aspects of this abrupt weaning seem to result from the separation. The calves try to compensate for the lack of their favourite partner by strengthening bonds with peers (Veissier and Le Neindre, 1989). This compensation seems to be effective 3 weeks after weaning since at this time calves prefer other calves to their dam (Veissier *et al.*, 1990b).

The reactivity of calves to external events is altered after this late weaning. During the 2 weeks following weaning, they react very actively to being handled in a crush and have higher cardiac responses to conditioned fear. They are also better at learning a route in a T-maze (Veissier *et al.*, 1989a). It is suggested that weaning can be used by farmers to accustom animals to a new environment and to acquire the proper responses to it. Indeed, the post-weaning period can be used to habituate calves to further human contact (Boivin *et al.*, 1992).

Temporary separation from the group in adulthood

In adult animals, temporary separation from the usual group elicits an immediate increase in locomotion, heart rate and plasma cortisol levels (Adeyemo and Heath, 1982). Stockmen can be prone to accidents especially when they separate an animal from others and when this animal tries to rejoin its group. Holstein heifers are less disturbed by the separation from their group than Aubrac heifers (Boissy and Le Neindre, 1990), and appear better adapted to some husbandry practices, like entering an automatic feeding machine or a milking robot.

5.3.4 Dominance-related problems

Dominance (see Section 1.5) is a potentially adaptive mechanism which can become maladaptive in certain circumstances, such as when the social structure of the group is frequently changed, or in any situation where some resources are limited. Many studies have shown that dominant animals get advantages when food availability is reduced or in the case of food supplementation (Bouissou, 1985; Manson and Appleby, 1990). Position at the food trough is not random as some animals have preferred places (Bouissou, 1985), which can be of importance in the case of unequal mechanical feed distribution. Social relationships also influence positions at the feeder: the greater the difference in rank, the further apart the animals are (Manson and Appleby, 1990). The more intense the competition (reduction in trough length), the stronger the correlation between food intake and dominance value (Friend *et al.*, 1977). Even when food is permanently available and when cows have access to food during the night, low-ranking animals eat and gain less than the others (Bosc *et al.*, 1971; Metz and Mekking, 1978). Surprisingly, most authors have found no correlation (Reinhardt, 1973; Soffié *et al.*, 1976; Friend and Polan, 1978) or only a low correlation between social rank and milk production (Barton *et al.*, 1974). Dividing the feed trough with protective barriers reduces the effect of dominance on eating behaviour (Bouissou, 1970).

Social position also affects the use of lying space as high-ranking animals have priority to choose the best cubicles (Friend and Polan, 1974). Sometimes low-ranking animals cannot enter shelters if a dominant animal is in their way or in front of the door. Resting time of low-ranking animals can also be reduced in loose housing (Bouissou, 1985). Bouissou (1985) found that the adrenal glands of the subordinate cows were significantly hypertrophied compared with glands of the dominant cows. In addition, Kay *et al.* (1977) found a correlation between the number of leukocyte cells in the milk and the social rank. However, no correlation between blood cortisol levels and rank was found by Adeyemo and Heath (1982) or Arave *et al.* (1977). Likewise, no differences in cortisol or neutrophil/lymphocyte ratio were found in groups of steers with varied access to the feeder (Corkum *et al.*, 1994).

Reproduction can also be affected. When several bulls are kept with females and the number of oestrous females is low, the dominant male interrupts the mounting attempts of its subordinates (de Blockey, 1978). The dominant males in groups produce the majority of the calves (Lehrer *et al.*, 1977; Rupp *et al.*, 1977). The fact that the dominant has priority of access to females can be a problem if its fertility is low.

5.3.5 Abnormal behaviour

Aggression towards people (see Chapter 14)
Aggression towards the caretaker by cattle is rare but of great importance due to risk of accidents (Albright and Arave, 1997). Young bulls can engage in social play with a caretaker and this can turn into

dangerous butts. Adult bulls can engage directly in aggressive inter-actions with the caretaker. Dairy bulls might be even more aggressive than beef bulls, probably due to early rearing in social isolation, which prevents the development of a proper social behaviour (Price and Wallach, 1990). Cows can also be aggressive, e.g. when they are with their offspring (at calving and, less frequently, later on).

Abnormal sexual behaviour

Excessive sexual behaviour is sometimes displayed by female and male cattle. Nymphomania is observed in cows with ovarian cysts producing oestrogens. They can be treated by removal of the cyst or by administer-ing gonadotrophin or gonadotrophic releasing hormone (Albright and Arave, 1997). In groups of intact or castrated males which have no contacts with females, excessive mounting (buller-steer syndrome) can be observed, with a particular bull (called 'rider' or 'bullee') always mounting the same animal ('buller') (for a review see Blackshaw et al., 1997). This syndrome is of economic importance because of injuries to the bull being ridden, decreased weight gain and even death. At pres-ent, separating the buller from the group is the only effective treatment.

Abnormal behaviours due to the social environment

Cattle reared in intensive conditions can display behaviours that are considered to be abnormal, such as cross-sucking, biting objects and tongue rolling (see Albright and Arave, 1997). Cross-sucking (of mouth, ears, scrotum and prepuce of penmates) often occurs after a milk meal in calves fed from buckets (Lidfors, 1993). Prepuce sucking can lead to urine drinking and abscesses. The taste of milk is known to trigger sucking (de Passillé et al., 1992) and non-nutritive sucking is reduced when calves are fed milk through a nipple (Hammel et al., 1988). Cross-sucking can be prevented by blocking the calves at the feeding gate for a while after milk meals, rearing calves in individual crates or tethering them for 8 weeks before putting them in groups (Wiepkema, 1987), or providing non-nutritive teats (de Passillé and Rushen, 1997). Cross-sucking directed to the udder can occur in adult cows and affect the health of the udder. In this case, a nose-ring weaner (with sharp points) is used to prevent cross-sucking.

Social isolation can increase oral non-nutritive behaviours. Isolated calves spend more time licking objects than calves having contact with their neighbours (Waterhouse, 1978). Calves reared in individual crates also spent more time in tongue rolling compared with calves in groups (Veissier et al., 1998).

Modern husbandry practices create important constraints on the environment of the animals including those on their social environment. These constraints are of lesser importance in suckling herds where the calves remain with their dams for months and normal social development can occur. However, even in this environment, the calves are actively weaned and gathered into groups of animals of the same age and sex. The social environment of dairy cattle is very different from the pasture or rangeland animals. In particular, the young are isolated from their dams very early, weaned and reared in groups of calves of the same sex.

Under intensive conditions, the role of dominance relationships is exacerbated and they seem to play a major role in regulating the group's life whereas affiliative relationships appear of lesser importance. However, under more natural conditions (rangelands) the relative importance of these two types of relationships is reversed. Environmental conditions can modulate the expression of social relationships and, depending on the circumstances, other mechanisms prevail.

Social relationships should not be considered as having only negative influences, but can also be used as a tool to improve the adaptation of animals to their environment through social facilitation, imitation, transmission of information, leadership, social learning, etc. Thus, training some key animals in a herd to a particular surrounding or procedure is likely to benefit the whole herd. The possibility of modulating the impact of social pressure by means of affiliative relationships created during ontogeny, together with the stress-reducing effect of peers, could also provide useful means for the alleviation of problems due to social tension or modifications of the social environment.

Cattle have adapted differently to specific management techniques and environments that can be considered as ecological niches into which different types of animals have evolved (e.g. beef and dairy breeds). These adaptations do not imply that there is no limit to the adaptive ability of the animals. We have to define environments so that the animals can best cope and maximize their health and welfare.

References

Adeyemo, O. and Heath, E. (1982) Social behaviour and adrenal cortical activity in heifers. *Applied Animal Ethology* 8, 99–108.

Albright, J.L. and Arave, C.W. (1997) *The Behaviour of Cattle.* CAB International, Wallingford, UK, 306 pp.

Andersen, H.R., Jensen, L.R., Munksgaard, L. and Ingvartsen, K.L. (1997) Influence of floor space allowance and access to feed trough on the production of calves and young bulls and on the carcass and meat quality of young bulls. *Acta Agriculturae Scandinavica, Section A, Animal Science* 47, 48–56.

Arave, C.W. and Albright, J.L. (1976) Social rank and physiological traits of dairy cows as influenced by changing group membership. *Journal of Dairy Science* 59, 974–981.

Arave, C.W. and Albright, J.L. (1981) Cattle behavior. *Journal of Dairy Science* 64, 1318–1329.

Arave, C.W., Albright, J.L. and Sinclair, C.L. (1974) Behavior, milking yield, and leucocytes of dairy cows in reduced space and isolation. *Journal of Dairy Science* 57, 1497–1501.

Arave, C.W., Mickelsen, C.H., Lamb, R.C., Svejda, A.J. and Canfield, R.V. (1977) Effects of dominance rank changes, age and body weight on plasma corticoids of mature dairy cattle. *Journal of Dairy Science* 60, 244–248.

Arave, C.W., Albright, J.L., Armstrong, D.V., Foster, W.W. and Larson, L.L. (1992) Effects of isolation of calves on growth, behavior, and first lactation milk yield of Holstein cows. *Journal of Dairy Science* 75, 3408–3415.

Baker, A.E.M. and Seidel, G.E. (1984/85) Why do cows mount other cows? *Applied Animal Behaviour Science* 13, 237–241.

Baldwin, B.A. (1977) Ability of goats and calves to distinguish between conspecific urine samples using olfaction. *Applied Animal Ethology* 3, 145–150.

Barton, E.P., Donaldson, S.L., Ross, M. and Albright, J.L. (1974) Social rank and social index as related to age, body weight, and milk production on dairy cows. *Proceedings of the Indiana Academy of Science* 83, 473.

Beilharz, R.G. and Mylrea, P.J. (1963) Social position and behavior of dairy heifers in yards. *Animal Behaviour* 11, 522–527.

Berteaux, D. and Micol, T. (1992) Population studies and reproduction of the feral cattle (*Bos taurus*) of Amsterdam Island, Indian Ocean. *Journal of Zoology, London* 228, 265–276.

Bilton, L. (1957) The Chillingham herd of wild cattle. *Transactions of the Natural History Society of Northumberland, Durham and Newcastle-upon-Tyne* 12, 137–160.

Blackshaw, J.K., Blackshaw, A.W. and McGlone, J.J. (1997) Buller steer syndrome review. *Applied Animal Behaviour Science* 54, 97–108.

Boissy, A. and Bouissou, M.F. (1994) Effects of androgen treatment on behavioral and physiological responses of heifers to fear-eliciting situations. *Hormones and Behavior* 28, 66–83.

Boissy, A. and Le Neindre, P. (1990) Social influences on the reactivity of heifers: implications for learning abilities in operant conditioning. *Applied Animal Behaviour Science* 25, 149–165.

Boissy, A., Terlouw, C. and Le Neindre, P. (1998) Presence of cues from stressed conspecifics increases reactivity to aversive events in cattle: evidence for the existence of alarm substances in urine. *Physiology and Behavior* 63, 489–495.

Boivin, X., Le Neindre, P. and Chupin, J.M. (1992) Establishment of cattle–human relationships. *Applied Animal Behaviour Science* 32, 325–335.

Bosc, M.J., Bouissou, M.F. and Signoret, J.P. (1971) Conséquences de la hiérarchie sociale sur le comportement alimentaire des bovins domestiques. In: *Comptes Rendus du 93e Congrès National des Sociétés Savantes, Tours*, 1968, pp. 511–515.

Bouissou, M.F. (1970) Rôle du contact physique dans la manifestation des relations hiérarchiques chez les bovins. Conséquences pratiques. *Annales de Zootechnie* 19, 279–285.

Bouissou, M.-F. (1971) Effet de l'absence d'informations optiques et de contact physique sur la manifestation des relations hiérarchiques chez les bovins domestiques. *Annals de Biologie Animale, Biochimie et Biophysique* 11, 191–198.

Bouissou, M.F. (1974a) Etablissement des relations de dominance–soumission chez les bovins domestiques. I. Nature et évolution des interactions sociales. *Annales de Biologie Animale, Biochimie et Biophysique* 14, 383–410.

Bouissou, M.F. (1974b) Etablissement des relations de dominance–soumission chez les bovins domestiques. II. Rapidité et mode d'établissement. *Annales de Biologie Animale, Biochimie et Biophysique* 14, 757–768.

Bouissou, M.F. (1975) Etablissement des relations de dominance–soumission chez les bovins domestiques. III. Rôle de l'expérience sociale. *Zeitschrift für Tierpsychologie* 38, 419–435.

Bouissou, M.F. (1977) Etude du développement des relations de dominance–subordination chez les bovins, à l'aide d'épreuves de compétition alimentaire. *Biology of Behaviour* 2, 213–221.

Bouissou, M.F. (1978) Effects of injections of testosterone propionate on dominance relationships in a group of cows. *Hormones and Behavior* 11, 388–400.

Bouissou, M.F. (1985) Contribution à l'étude des relations interindividuelles chez les bovins domestiques femelles (*Bos taurus* L.). Thèse de Doctorat d'Etat, Université Paris VI, France, 366 pp.

Bouissou, M.F. (1990) Effects of estrogen treatment on dominance relationships in cows. *Hormones and Behavior* 24, 376–387.

Bouissou, M.F. and Andrieu, S. (1977) Etablissement des relations de dominance–soumission chez les bovins domestiques. IV. Etablissement des relations chez les jeunes. *Biology of Behaviour* 2, 97–107.

Bouissou, M.F. and Andrieu, S. (1978) Etablissement des relations préférentielles chez les bovins domestiques. *Behaviour* 64, 148–157.

Bouissou, M.F. and Gaudioso, V. (1982) Effect of early androgen treatment on subsequent social behavior in heifers. *Hormones and Behavior* 16, 132–146.

Bouissou, M.F. and Hövels, J. (1976a) Effet d'un contact précoce sur quelques aspects du comportement social des bovins domestiques. *Biology of Behaviour* 1, 17–36.

Bouissou, M.F. and Hövels, J.H. (1976b) Effet des conditions d'élevage sur le comportement des génisses dans une situation de compétition alimentaire. *Annales de Zootechnie* 25, 213–219.

Canali, E., Verga, M., Montagna, M. and Baldi, A. (1986) Social interactions and induced behavioural reactions in milk-fed female calves. *Applied Animal Behaviour Science* 16, 207–215.

Clutton-Brock, T.H., Greenwood, P.J. and Powell, R.P. (1976) Ranks and relationships in Highland ponies and Highland cows. *Zeitschrift für Tierpsychologie* 41, 202–216.

Corkum, M.J., Bate, L.A., Tennessen, T. and Lirette, A. (1994) Consequences of reduction of number of individual feeders on feeding behaviour and stress level of feedlot steers. *Applied Animal Behaviour Science* 41, 27–35.

Cummins, K.A. and Myers, L.J. (1991) Olfactory and visual cues, individual recognition, and social aggression in lactating dairy cows. *Journal of Dairy Science* 74 (suppl. 1), 301.

Daycard, L. (1990) Structure sociale de la population de bovins sauvages de l'île d'Amsterdam, sud de l'Océan Indien. *Revue d'Ecologie (La Terre et la Vie)* 45, 35–53.

de Blockey, M.A. (1978) Serving capacity and social dominance of bulls in relation to fertility. In: *Proceedings of the 1st World Congress of Ethology Applied to Zootechnie, Madrid*, pp. 503–506.

De Passillé, A.M. and Rushen, J. (1997) Motivational and physiological analysis of the causes and consequences of non-nutritive sucking by calves. *Applied Animal Behaviour Science* 53, 15–31.

De Passillé, A.M., Metz, J.H.M., Mekking, P. and Wiepkema, P.R. (1992) Does drinking milk stimulate sucking in young calves? *Applied Animal Behaviour Science* 34, 23–36.

De Wilt, J.G. (1985) *Behaviour and Welfare of Veal Calves in Relation to Husbandry Systems.* IMAG, Wageningen, the Netherlands, 137 pp.

Dickson, D.P., Barr, G.R. and Wieckert, D.A. (1967) Social relationships of dairy cows in a feed lot. *Behaviour* 29, 195–203.

Dickson, D.P., Barr, G.R., Johnson L.P. and Wiekert, D.A. (1970) Social dominance and temperament of Holstein cows. *Journal of Dairy Science* 53, 904.

Dietrich, J.P., Snyder, W.W., Meadows, C.E. and Albright, J.L. (1965) Rank order in dairy cows. *Animal Zoology* 5, 713 (abstract).

Edwards, S.A. (1983) The behaviour of dairy cows and their newborn calves in individual or group housing. *Applied Animal Ethology* 10, 191–198.

Entsu, S., Dohi, H. and Yamada, A. (1992) Visual acuity of cattle determined by the method of discrimination learning. *Applied Animal Behaviour Science* 34, 1–10.

Estes, R.D. (1974) Social organization of the African Bovidae. In: Geist, V. and Walther, E. (eds) *The Behavior of Ungulates and its Relation to Management.* IUNC Publications new series 24, Morges, Switzerland, pp. 166–205.

Ewbank, R. (1967) Behavior of twin cattle. *Journal of Dairy Science* 50, 1510–1515.

Fraser A.F. (1957) The state of fight or flight in the bull. *British Journal of Animal Behaviour* 5, 48–49.

Fraser, A.F. and Broom, D.M. (1997) *Farm Animal Behaviour and Welfare.* CAB International, Wallingford, UK, 437 pp.

Friend, T.H. and Polan, C.E. (1974) Social rank, feeding behavior and free stall utilization by dairy cattle. *Journal of Dairy Science* 57, 1214–1220.

Friend, T.H. and Polan, C.E. (1978) Competitive order as a measure of social dominance in dairy cattle. *Applied Animal Ethology* 4, 61–70.

Friend, T.H., Polan, C.E. and McGilliard, M.L. (1977) Free stall and feed bunk requirements relative to behavior, production and individual feed intake in dairy cows. *Journal of Dairy Science* 60, 108–116.

George, J.M. and Barger, I.A. (1974) Observations on bovine parturition. *Proceedings of the Australian Society of Animal Production* 10, 314–317.

Gilbert, B.J. Jr and Arave, C.W. (1986) Ability of cattle to distinguish among different wave lengths of light. *Journal of Dairy Science* 69, 825–832.

Guhl, A.M. and Atkeson, F.W. (1959) Social organization in a herd of dairy cows. *Transactions of the Kansas Academy of Science* 62, 80–87.

Hafez, E.S.E. (1974) *Reproduction in Farm Animals.* Lea and Febiger, Philadelphia, pp. 241–254.

Hafez, E.S.E. and Bouissou, M.F. (1975) The behaviour of cattle. In: Hafez, E.S.E. (ed.) *The Behaviour of Domestic Animals.* Baillière Tindall, London, pp. 203–245.

Hall, S.J.G. (1986) Chillingham cattle: dominance and affinities and access to supplementary food. *Ethology* 71, 201–215.

Hall, S.J.G. and Moore, G.F. (1986) Feral cattle of Swona, Orkney Islands. *Mammal Review* 16, 89–96.

Hammel, K.L., Metz, J.H.M. and Mekking, P. (1988) Sucking behaviour of dairy calves fed milk ad libitum by bucket or teat. *Applied Animal Behaviour Science* 20, 275–285.

Hasegawa, N., Nishiwaki, A., Sugawara, K. and Ito, I. (1997) The effects of social exchange between two groups of lactating primiparous heifers on milk production, dominance order, behavior and adrenocortical response. *Applied Animal Behaviour Science* 51, 15–27.

Hinch, G.N., Lynch, J.J. and Thwaites, C.J. (1982a) Patterns and frequency of social interactions in young grazing bulls and steers. *Applied Animal Ethology* 9, 15–30.

Hinch, G.N., Thwaites, C.J., Lynch, J.J. and Pearson, A.J. (1982b) Spatial relationships within a herd of young sterile bulls and steers. *Applied Animal Ethology* 8, 27–44.

Howery, L.D., Provenza, F.D., Banner, R.E. and Scott, C.B. (1996) Differences in home range and habitat use among individuals in a cattle herd. *Applied Animal Behaviour Science* 49, 305–320.

Howery, L.D., Provenza, F.D., Banner, R.E. and Scott, C.B. (1998) Social and environmental factors influence cattle distribution on rangeland. *Applied Animal Behaviour Science* 55, 231–244.

Hudson, S.J. and Mullord, M.M. (1977) Investigations of maternal bonding in dairy cattle. *Applied Animal Ethology* 3, 271–276.

Hurnik, J.F. (1982) Social stress: an often overlooked problem in dairy cattle. *Hoard's Dairyman* 127, 739.

Jezierski, T.A. and Podluzny, M. (1984) A quantitative analysis of social behaviour of different crossbreds of dairy cattle kept in loose housing and its relationship to productivity. *Applied Animal Behaviour Science* 13, 31–40.

Kabuga, J.D. (1992a) Social interactions in N'dama cows during periods of idling and supplementary feeding post-grazing. *Applied Animal Behaviour Science* 34, 11–22.

Kabuga, J.D. (1992b) Social relationships in N'dama cattle during supplementary feeding. *Applied Animal Behaviour Science* 34, 285–290.

Kabuga, J.D., Gari-Kwaku, J. and Annor, S.Y. (1991) Social status and its relationships to maintenance behaviour in a herd of N'dama and West African Shorthorn cattle. *Applied Animal Behaviour Science* 31, 169–181.

Kay, S.J., Collis, K.A., Anderson, J.C. and Grant, A.J. (1977) The effect of inter-group movement of dairy cows on bulk-milk somatic cell numbers. *Journal of Dairy Research* 44, 589–593.

Kiley, M. (1972) The vocalisations of ungulates, their causation and function. *Zeitschrift für Tierpsychologie* 31, 171–222.

Kiley-Worthington, M. (1976) The tail movements of ungulates, canids and felids with particular reference to their causation and function as displays. *Behaviour* 56, 69–115.

Kilgour, R. and Campin, D.N. (1973) The behaviour of entire bulls of different ages at pasture. *Proceedings of the New Zealand Society of Animal Production* 33, 125–138.

Kilgour, R. and Scott, T.H. (1959) Leadership in a herd of dairy cows. *Proceedings of the New Zealand Society of Animal Production* 19, 36–43.

Kilgour, R., Skarsholt, B.H., Smith, J.F., Bremmer, J.K. and Morrison, M.C.L. (1977) Observations on the behaviour and factors influencing the sexually-active group in cattle. *Proceedings of the New Zealand Society of Animal Production* 37, 128–135.

Kimura, D. and Ihobe, H. (1985) Feral cattle (*Bos taurus*) on Kuchinoshima Island, south-western Japan: their stable ranging and unstable grouping. *Journal of Ethology* 3, 39–47.

Klemm, W.R., Sherry, C.J., Sis, R.F., Schake, L.M. and Waxman, A.B. (1984) Evidence of a role for the vomeronasal organ in social hierarchy in feedlot cattle. *Applied Animal Behaviour Science* 12, 53–62.

Kondo, S., Kawakami, N., Kohama H. and Nishino, S. (1984) Changes in activity, spatial pattern and social behavior in calves after grouping. *Applied Animal Ethology* 11, 217–228.

Kondo, S., Sekine, J., Okubo, M. and Asahida, Y. (1989) The effect of group size and space allowance on the agonistic and spacing behavior of cattle. *Applied Animal Behaviour Science* 24, 127–135.

Kondo, S., Yasue, T., Ogawa, K., Nakatuji, H., Okubo, M. and Asahida, Y. (1994) The relationships of group sizes, space allowance and paddock topography to social behavior in grazing cattle after grouping with intro-ducing animals. *Japanese Journal of Livestock Management* 30, 63–68.

Konggaard, S.P., Krohn, C.C. and Agergaad, E. (1982) Investigations concerning feed intake and social behaviour among group fed cows under loose hous-ing conditions. VI. Effects of different grouping criteria in dairy cows. *Beretning Jra Statens Husdyrbrugsforsog* 553, 35pp.

Lazo, A. (1994) Social segregation and the maintenance of social stability in a feral cattle population. *Animal Behaviour* 48, 1133–1141.

Lehrer, A.R., Brown, M.B., Schindler, H., Holzer, Z. and Larsen, B. (1977) Pater-nity tests in multisired beef herds by blood grouping. *Acta Veterinaria Scandinavicae* 18, 433–441.

Le Neindre, P. (1984) La relation mère–jeune chez les bovins: influence de l'environnement social et de la race. Thèse de Doctorat d'Etat, Université de Rennes, France, 274 pp.

Le Neindre, P. (1989) Influence of rearing conditions and breed on social relationships of mother and young. *Applied Animal Behaviour Science* 23, 117–127.

Le Neindre, P. and D'Hour, P. (1988) Effects of a postpartum separation on maternal responses in primiparous and multiparous cows. *Animal Behaviour* 37, 166–168.

Le Neindre, P. and Garel, J.P. (1976) Existence d'une période sensible pour l'établissement du comportement maternel de la vache après la mise-bas. *Biology of Behaviour* 1, 217–221.

Le Neindre, P. and Sourd, C. (1984) Influence of rearing conditions on subsequent social behaviour of Friesian and Salers heifers from birth to six months of age. *Applied Animal Behaviour Science* 12, 43–52.

Le Neindre, P., Petit, M. and Muller, A. (1976) Quantités d'herbe et de lait consommées par des veaux au pis. *Annales de Zootechnie* 25, 521–531.

Lent, P.C. (1974) Mother–infant relationships in ungulates. In: Geist, V. and Walther, F. (eds) *The Behavior of Ungulates and its Relation to Management*. IUNC Publications new series 24, Morges, Switzerland, pp. 14–55.

Lésel, R. (1969) Etude d'un troupeau de bovins sauvages vivant sur l'île d'Amsterdam. *Revue d'Elevage et de Médecine Vétérinaire des Pays Tropicaux* 22, 107–125.

Leyhausen, P. and Heinemann, I. (1975) 'Leadership' in a small herd of dairy cows. *Applied Animal Ethology* 1, 206 (abstract).

Lidfors, L.M. (1993) Cross-sucking in group-housed dairy calves before and after weaning off milk. *Applied Animal Behaviour Science* 38, 15–24.

Lott, D.F. and Minta, S.C. (1983) Random individual association and social group instability in American bison (*Bison bison*). *Zeitschrift für Tierpsychologie* 61, 153–172.

Lucifero, M., Janella, G.G. and Secchiari, P. (1977) *Origini, evoluzione, miglioramento e prospettive delle razza bovina Maremmana*. Edagriocole, Bologna.

Mansard, C. and Bouissou, M.F. (1980) Effect of olfactory bulbs removal on the establishment of the dominance–submission relationships in domestic cattle. *Biology of Behaviour* 5, 169–178.

Manson, F.J. and Appleby, M.C. (1990) Spacing of dairy cows at a food trough. *Applied Animal Behaviour Science* 26, 69–81.

McPhee, C.P., McBridge, G. and James, J.W. (1964) Social behaviour of domestic animals. III. Steers in small yards. *Animal Production* 6, 9–15.

Meese, G.B. and Ewbank, R. (1973) Exploratory behaviour and leadership in domestic pig. *British Veterinary Journal* 129, 251–259.

Mench, J.A., Swanson, J.C. and Stricklin, W.R. (1990) Social stress and dominance among group members after mixing beef cows. *Canadian Journal of Animal Science* 70, 345–354.

Metz, J.H.M. and Mekking, P. (1978) Adaptation in the feeding pattern of cattle according to the social environment. In: *Proceedings of the Zodiacal Symposium on Adaptation, Wageningen, the Netherlands*, pp. 36–42.

Mogensen, L., Nielsen, L.H., Hindhede, J., Sorensen, J.T. and Krohn, C.C. (1997) Effect of space allowance in deep bedding systems on resting behaviour, production and health of dairy heifers. *Acta Agriculturae Scandinavica, Section A., Animal Science* 47, 178–186.

Nakanishi, Y., Mutoh, Y., Umetsu, R., Masuda, Y. and Goto, I. (1991) Changes in social and spacing behaviour of Japanese Black Cattle after introducing a strange cow into a stable herd. *Journal of the Faculty of Agriculture, Kyushu University* 36, 1–11.

Owens, J.L., Edey, T.N., Bindon, B.M. and Piper, L.R. (1985) Parturient behaviour and calf survival in a herd selected for twinning. *Applied Animal Behaviour Science* 13, 321–333.

Poindron, P. and Le Neindre, P. (1975) Comparaison des relations mère–jeune observées lors de la tétée chez la brebis (*Ovis aries*) et chez la vache (*Bos taurus*). *Annales de Biologie Animale, Biochimie et Biophysique* 15, 495–501.

Price, E.O. and Wallach, S.J.R. (1990) Physical isolation of hand-reared Hereford bulls increases their aggressiveness towards humans. *Applied Animal Behaviour Science* 27, 263–267.

Price, E.O., Thos, J. and Anderson, G.B. (1981) Maternal responses of confined beef cattle to single versus twin cattle. *Applied Animal Ethology* 53, 934–939.

Purcell, D. and Arave, C.W. (1991) Isolation vs. group rearing in monozygous twin heifer calves. *Applied Animal Behaviour Science* 31, 147–156.

Ralphs, M.H. and Olson, J.D. (1990) Adverse influence of social facilitation and learning context in training cattle to avoid eating larkspur. *Applied Animal Ethology* 68, 1944–1952.

Ralphs, M.H., Graham, D. and James, L.F. (1994) Social facilitation influences cattle to graze locoweed. *Journal of Range Management* 47, 123–126.

Rankine, G. and Donaldson, L.E. (1968) Animal behaviour and calf mortalities in a North Queensland breeding herd. *Proceedings of the Australian Society of Animal Production* 7, 138–143.

Rathore, A.K. (1982) Order of cow entry at milking and its relationships with milk yield and consistency of the order. *Applied Animal Ethology* 8, 45–52.

Reinhardt, V. (1973) Beiträge zur sozialen Rangordnung und Melkordnung bei Kühen. *Zeitschrift für Tierpsychologie* 32, 281–292.

Reinhardt, V. (1981) Cohesive relationships in a cattle herd (*Bos indicus*). *Behaviour* 77, 121–151.

Reinhardt, V. and Reinhardt, A. (1975) Analysis of the social rank system in a herd of 41 heifers and cows. *Applied Animal Ethology* 1, 206–207 (abstract).

Reinhardt, V. and Reinhardt, A. (1982a) Mock fighting in cattle. *Behaviour* 81, 1–13.

Reinhardt, V. and Reinhardt, A. (1982b) Social behaviour and social bonds between juvenile and sub-adult *Bos indicus* calves. *Applied Animal Ethology* 9, 92–93 (abstract).

Reinhardt, V., Mutiso, F.M. and Reinhardt, A. (1978) Social behaviour and social relationships between female and male prepubertal bovine calves (*Bos indicus*). *Applied Animal Ethology* 4, 43–54.

Riol, J.A., Sanchez, J.M., Eguren, V.G. and Gaudioso, V.R. (1989) Colour perception in fighting cattle. *Applied Animal Behaviour Science* 23, 199–206.

Rupp, G.P., Ball, L., Shoop, M.C. and Chenoweth, P.J. (1977) Reproductive efficiency of bulls in natural service: effects of male to female ratio and single *vs* multiple-sire breeding groups. *Journal of the American Veterinary Medical Association* 171, 639–642.

Rutter, S.M., Jackson, D.A., Johnson, C.L. and Forbes, J.M. (1987) Automatically recorded competitive feeding behaviour as a measure of social dominance in dairy cows. *Applied Animal Behaviour Science* 17, 41–50.

Sambraus, H.H. (1969) Das soziale Lecken des Rindes. *Zeitschrift für Tierpsychologie* 26, 805–810.

Sambraus, H.H., Fries, B. and Osterkorn, K. (1978) Das Sozialgeschehen in einer Herde hornloser Hochleistungsrinder. *Zeitschrift für Tierzüchtung und Züchtungsbiologie* 95, 81–88.

Sato, S. (1984) Social licking pattern and its relationships to social dominance and live weight gain in weaned calves. *Applied Animal Behaviour Science* 12, 25–32.

Sato, S., Wood-Gush, D.G.M. and Wetherill, G. (1987) Observations on creche behaviour in suckler calves. *Behavioural Processes* 15, 333–343.

Sato, S., Sato, S. and Maeda, A. (1991) Social licking patterns in cattle (*Bos taurus*): influence of environmental and social factors. *Applied Animal Behaviour Science* 32, 3–12.

Sato, S., Tarumizu, K. and Hatae, K. (1993) The influence of social factors on allogrooming in cows. *Applied Animal Behaviour Science* 38, 235–244.

Schein, M.W. and Fohrman, M.H. (1955) Social dominance relationships in a herd of dairy cattle. *British Journal of Animal Behaviour* 3, 45–55.

Scheurmann, E. (1974) Untersuchungen über Aktivität und Reheverhalten bei neugeborenen Kälbern. *Zuchthygiene* 9, 58–68.

Schloeth, R. (1956) Quelques moyens d'intercommunication des taureaux de Camargue. *Revue d'Ecologie (La Terre et la Vie)* 2, 83–93.

Schloeth, R. (1958) Le cycle annuel et le comportement social du taureau de Camargue. *Mammalia* 22, 121–139.

Schloeth, R. (1961) Das Sozialleben des Camargue Rindes. *Zeitschrift für Tierpsychologie* 18, 574–627.

Selman, I.E., McEwan, A.D. and Fisher, E.W. (1970a) Studies on natural suckling in cattle during the first eight hours post partum. I. Behaviour studies (dams). *Animal Behaviour* 18, 276–283.

Selman, I.E., McEwan, A.D. and Fisher, E.W. (1970b) Studies on natural suckling in cattle during the first eight hours post partum. II. Behaviour studies (calves). *Animal Behaviour* 18, 284–289.

Sinclair, A.R.E. (1977) *The African Buffalo. A Study of Resource Limitation of Populations.* University of Chicago Press, Chicago.

Soffié, M. and Zayan, R. (1978) Responsiveness to 'social' releasers in cattle. II. Relation between social status and responsiveness and possible effect of previous familiarization with the test conditions. *Behavioural Processes* 3, 241–258.

Soffié, M., Thinès, G. and De Marneffe, G. (1976) Relation between milking order and dominance value in a group of dairy cows. *Applied Animal Ethology* 2, 271–276.

Soffié, M., Thinès, G. and Falter, U. (1980) Colour discrimination in heifers. *Mammalia* 44, 97–121.

Stricklin, W.R. (1983) Matrilinear social dominance and spatial relationships among Angus and Hereford cows. *Journal of Animal Science* 57, 1397–1405.

Stricklin, W.R., Graves, H.B., Wilson, L.L. and Singh, R.K. (1980) Social organization among young beef cattle in confinement. *Applied Animal Ethology* 6, 211–219.

Stricklin, W.R., Kautz-Scanavy, C.C. and Greger, D.L. (1985) Determination of dominance–subordinance relationships among beef heifers in a dominance tube. *Applied Animal Behaviour Science* 14, 111–116.

Syme, G.J. and Syme, L.A. (1979) *Social Structure in Farm Animals.* Elsevier Science Publications, 200 pp.

Terlouw, E.M.C., Boissy, A. and Blinet, P. (1998) Behavioural responses of cattle to the odours of blood and urine from conspecifics and to the odour of faeces from carnivores. *Applied Animal Behaviour Science* 57, 9–21.

Trunkfield, H.R., Broom, D.M., Maatje, K., Wierenga, H.K., Lambooy, E. and Kooijman, J. (1991) The effects of housing on responses of veal calves to handling and transport. In: Metz, J.H.M. and Groenestein, C.M. (eds) *New Trends in Veal Calf Production.* EEAP Publications 52, Wageningen, the Netherlands, pp. 40–48.

Veissier, I. and Le Neindre, P. (1989) Weaning in calves: its effects on social organization. *Applied Animal Behaviour Science* 24, 43–54.

Veissier, I. and Le Neindre, P. (1992) Reactivity of Aubrac heifers exposed to a novel environment alone or in groups of four. *Applied Animal Behaviour Science* 33, 11–15.

Veissier, I., Le Neindre, P. and Trillat, G. (1989a) Adaptability of calves during weaning. *Biology of Behaviour* 14, 66–87.

Veissier, I., Le Neindre, P. and Trillat, G. (1989b) The use of circadian behaviour to measure adaptation of calves to changes in their environment. *Animal Behaviour Science* 22, 1–12.

Veissier, I., Lamy, D. and Le Neindre, P. (1990a) Social behaviour in domestic beef cattle when yearling calves are left with the cows for the next calving. *Applied Animal Behaviour Science* 27, 193–200.

Veissier, I., Le Neindre, P. and Garel, J.P. (1990b) Decrease in cow–calf attachment after weaning. *Behavioural Processes* 21, 95–105.

Veissier, I., Gesmier, V., Le Neindre, P., Gautier, J.Y. and Bertrand, G. (1994) The effects of rearing in individual crates on subsequent social behaviour of veal calves. *Applied Animal Behaviour Science* 41, 199–210.

Veissier, I., Chazal, P., Pradel, P. and Le Neindre, P. (1997) Providing social contacts and objects for nibbling moderates reactivity and oral behaviors in veal calves. *Applied Animal Ethology* 75, 356–365.

Veissier, I., Ramirez de la Fe, A.R. and Pradel, P. (1998) Non-nutritive oral activities and stress responses of veal calves in relation to feeding and housing conditions. *Applied Animal Behaviour Science* 57, 35–49.

Vitale, A.F., Tenucci, M., Papini, M. and Lovari, S. (1986) Social behaviour of the calves of semi-wild Maremma cattle, *Bos primegenius taurus. Applied Animal Behaviour Science* 16, 217–231.

Wagnon, K.A. (1963) Behavior of beef cows on a California range. *California Agricultural Experimental Station Bulletin* 799, 38–44.

Warnick, V.D., Arave, C.W. and Mickelsen, C.H. (1977) Effects of group, individual and isolated rearing of calves on weight gain and behavior. *Journal of Dairy Science* 60, 947–953.

Waterhouse, A. (1978) The effects of pen conditions on the development of calf behaviour. *Applied Animal Ethology* 4, 285–294.

Webster, A.J.F., Saville, C., Church, B.M., Granasakthy, A. and Moss, R. (1985) The effects of different rearing systems on the development of calf behaviour. *British Veterinary Journal* 141, 249–264.

Whitehead, G.K. (1953) *The Ancient White Cattle of Britain and Their Descendants.* Faber and Faber, London.

Wiepkema, P.R. (1987) Developmental aspects of motivated behavior in domestic animals. *Journal of Animal Science* 65, 1220–1227.

Wierenga, H.K., Hopster, H., Engel, B. and Buist, W. (1991) Measuring social dominance in dairy cows. Thesis, Agricultural University, Wageningen, the Netherlands, pp. 39–52.

Wood, M.T., (1977) Social grooming in two herds of monozygotic twin dairy cows. *Animal Behaviour* 25, 635–642.

The Social Behaviour of Pigs

6

Harold W. Gonyou

Prairie Swine Centre, PO Box 21057, Saskatoon, Saskatchewan S7H 5N9, Canada

(Editors' comments: Studies on wild, feral and free-ranging pigs reveal that the most common social grouping is that of several sows and their juvenile offspring, living within a home range. Within this group avoidance behaviour is used to maintain the social organization. Males exist in small groups of young boars, or as solitary males when older, except during the breeding season when they join the sow and offspring groups. Another social group is the sow and newborn litter, which exists for approximately 10 days following parturition.

The large litters are maintained with their mothers for several weeks under commercial conditions. A unique aspect of managing pigs is the extensive fostering which occurs among litters at this time, resulting in changing social groupings. Pigs are typically regrouped either at weaning or at the beginning of the grow/finish phase of production, or both. It is interesting that, in a species that normally exists in stable social groupings of mature females, one of the most common socially mediated practices is the individual penning of gestating animals. Some of the greatest challenges in swine management arise from attempts to keep sows or sows and piglets in groups.)

6.1 Basic Social Characteristics

Our information on the basic or natural social behaviour of pigs is derived primarily from studies on wild boar, feral and free-ranging pigs. Wild boar have been studied in their free-ranging state (Fradrich, 1974; Mauget, 1981), production and zoo environments (Schnebel and

©CAB *International* 2001. *Social Behaviour in Farm Animals*
(eds L.J. Keeling and H.W. Gonyou)

LIVERPOOL JOHN MOORES UNIVERSITY
LEARNING & INFORMATION SERVICES

Griswold, 1983; Barrette, 1986; Blasetti *et al.*, 1988). Populations of feral pigs, not under the control of humans for several generations, have been studied in the United States (Singer *et al.*, 1981; Graves, 1984) and New Zealand (Martin, 1975). Domestic pigs have been placed in free-ranging environments for behaviour studies in Scotland (Newberry and Wood-Gush, 1986; Stolba and Wood-Gush, 1989), Sweden (Jensen and Wood-Gush, 1984; Jensen, 1986) and Denmark (Petersen *et al.*, 1989, 1990).

6.1.1 Composition and structure of social groups

The primary social grouping of pigs consists of two to four sows, their most recent litters and juvenile offspring of previous litters (Mauget, 1981; Graves, 1984). It is generally believed that the sows are closely related, either mother–daughter or sibling (full or half) groups. This assumption is based on the observations that non-member sows are rarely allowed to incorporate into a group (Stolba and Wood-Gush, 1989). Early associations, particularly among females, often persist into adulthood (Graves, 1984). The number of sows in a group is likely dependent upon the availability of resources (see Section 2.5). Larger groups will exist if food is plentiful, but smaller groups are observed during seasons of sparse and widely dispersed resources (Mauget, 1981; Graves, 1984).

Within the sow and offspring groups, sows will be dominant to all other members, and maintain a linear hierarchy within their class (Mauget, 1981). Similarly, juveniles also maintain a well-defined social order among themselves. Littermates interact primarily with each other and with their dam (Newberry and Wood-Gush, 1986; Petersen *et al.*, 1989). In larger social groups, two or three litters will form preferences for each other, although they will interact with all other litters to some extent (Newberry and Wood-Gush, 1986).

During mating season, the sow and offspring group is joined by a male to form the breeding group (Fradrich, 1974; Graves, 1984; see Section 1.6). The boar assumes dominance (Mauget, 1981; Schnebel and Griswold, 1983), while non-breeding females and juveniles remain on the periphery of the group at this time (Mauget, 1981; Blasetti *et al.*, 1988).

Juvenile boars leave the sow and offspring group at approximately 7–8 months of age (Fradrich, 1974; Graves, 1984). Small groups of two or three young boars may exist, particularly during the non-breeding season (Graves, 1984). However, as boars mature they lead an increasingly solitary life. It is rare to find boars older than 3 years of age in a group with other boars (Mauget, 1981).

A temporary social group is formed when a sow leaves the sow and offspring group to give birth (Jensen, 1986; see Section 3.4). During this period the sow and newborn group exists in three stages: sow and litter in or near the nest for 2–3 days; the sow foraging away from the litter and nest between days 3 and 6; and the sow with following litter from day 6 to the time of rejoining the primary sow and offspring group (Stangel and Jensen, 1991).

6.1.2 Use of space

Sow and offspring groups maintain home ranges of 100–500 ha (Wood and Brenneman, 1980; Mauget, 1981). The size of the home range will vary with the availability of resources. A central feature of the home range is the communal nest where all sows and offspring will sleep except during the farrowing season (Jensen, 1986; Stolba and Wood-Gush, 1989). The pigs also maintain a distinct dunging zone, 5–15 m away from the sleeping site (Stolba and Wood-Gush, 1989).

Most sows will build their farrowing nest on the periphery of the group's home range, at least 100 m away from the communal nest, and there is evidence that greater separation is favourable to piglet survival (Jensen, 1989). The farrowing sow and newborns maintain a much smaller home range, approximately 1 ha in size (Mauget, 1981). During the initial 2 weeks following their return to the communal nest, nursing sows maintain a small home range of only 20 ha.

Singer *et al.* (1981) and Wood and Brenneman (1980) estimated that the size of home ranges for boars was similar to that of sow and offspring groups. However, Martin (1975) reported that the maximum distances between repeated captures of boars were approximately six times greater than those for sows. The size of the boars' home ranges may vary with resources and reproductive seasons. It is also possible that the greater distances observed for boars represent dispersion rather than the extremes of a home range.

6.1.3 Communication

Dominant boars mark the environment with odours more than do subordinate animals, with the metacarpal glands on the front legs being most commonly used in this behaviour (Mayer and Brisbin, 1986). It is hypothesized that the secretions of these glands are related to dominance and reproduction (Mayer and Brisbin, 1986), but not to territorial defence (Fradrich, 1974). Preputial secretions are mixed with urine and may play a role in the 'covering' of urine from sows and other boars (Mayer and Brisbin, 1986). Salivary pheromones are released in frothy saliva of boars, produced in response to sexual stimuli. This saliva may be deposited on the female during courtship or on trees and bushes marked by the boars' tusks (Mayer and Brisbin, 1986; Stolba and Wood-Gush, 1989). Salivary and preputial pheromones, consisting of androgens, are involved in eliciting the standing response in oestrous sows (Signoret, 1970), the induction of oestrus in gilts (Pearce *et al.*, 1988), and may play a role in synchronizing mating in groups of sows following their being joined by a boar (Mauget, 1981; Rowlinson and Bryant, 1982). Females, in turn, will use urine to signal oestrus (Fradrich, 1974).

Auditory stimuli are used extensively by pigs. Wild and feral pigs communicate by means of grunts, squeals, snarls and snorts, as well as by champing of jaws and the clacking of teeth (Fradrich, 1974; Graves, 1984). The vocal repertoire of adult boars includes a 'roar' exhibited during social encounters (Mayer and Brisbin, 1986), and a 'mating song' during courtship (Fradrich, 1974). Piglets use open and closed mouth grunts and squeals to maintain contact between littermates and their dam (Fradrich, 1974; Fraser, 1975). Sows use a series of grunts, varying in frequency, tone and magnitude, to indicate the stages of nursing to the piglets (Whittemore and Fraser, 1974; Algers, 1993). Piglets respond with vocalizations as well (Jensen and Algers, 1984). A nursing bout, including those shortly after birth, is often preceded by the vocalization of a piglet near the head of the sow (Petersen *et al.*, 1990). Sows respond to the alarm call of an overlain piglet (Cronin and Cropley, 1991), and piglet vocalizations are indicative of the degree of distress experienced (Weary and Fraser, 1995). The role of communication within a social group is discussed more extensively in Chapter 2.3.

6.1.4 Cohesion and dispersion

Although sows within a social group are likely to be closely related, it is not clear how new groups form. A new group could form if a sow and her juvenile female offspring leave, or if several female offspring disperse together without an adult. Artificial systems have been managed to retain a juvenile daughter within the sow group in order to mimic the natural pattern (Stolba and Wood-Gush, 1983). A new boar joins the group for each mating season and so dispersal of young females is not necessary to avoid inbreeding. However, the high reproductive capacity of pigs would necessitate the formation of new groups following a year of abundant resources.

Mauget (1981) and Blasetti *et al.* (1988) have reported a temporary dispersal of juvenile males and females during the mating season. However, these offspring remain nearby and return once the breeding male departs. Fradrich (1974) and Graves (1984) reported that dispersal of juveniles occurs at approximately 7–8 months of age. Males leaving at this time may form small all-male groups but eventually become solitary (see Section 1.6).

6.1.5 Inter-group interactions

Solitary males and sow and offspring groups overlap in their use of space, but do not interact except during the mating season. Sow and offspring groups may also share common space in their home ranges, but will not merge to form a single unit. Stolba and Wood-Gush (1989) observed that two groups maintained a distance of at least 50 m between them while foraging. Sows added to a free-range system are not allowed into the common nest for several months. During seasons of heavily concentrated food resources, such as the acorn or mast season, sow and litter groups exist in close proximity, but continue to maintain the integrity of their social group (Graves, 1984).

6.1.6 Intra-group interactions (see Section 2.2)

Male–male

Adult males develop a thick shield of skin on their neck and shoulders that apparently serves as protection during male–male aggression. It is not clear if aggression is involved in the break-up of small male groups into solitary units, or if the aggression is usually among isolated males. Under zoo conditions, males have been observed to 'wrestle'; an interaction in which two animals are erect upon their hind legs, supported by each other as they fight (Barrette, 1986). In general, aggression in pigs will not be head-on, but rather involves lateral attacks (Fradrich, 1974). Visual displays, such as an arched back, head down and eyes averted, are used in non-aggressive interactions (Schnebel and Griswold, 1983).

Female–female

Within sow and offspring groups, sows form a hierarchy, which is maintained by subordinate animals avoiding dominants, rather than dominant sows attacking those of lower status (Jensen, 1980, 1982). Jensen and Wood-Gush (1984) reported that the interactions involved in this 'avoidance' order are similar in free-range and confined conditions, with the exception that 'aiming' was only observed in the spacious situation (Table 6.1). However, dominant sows will displace subordinates from choice feeding sites (Graves, 1984). In captive situations, most of these displacements involve rank neighbours (Schnebel and Griswold, 1983). Stolba and Wood-Gush (1989) reported

Table 6.1.   Agonistic interaction patterns observed among pigs. (Adapted from Jensen, 1982, and Jensen and Wood-Gush, 1984.)

Behaviour	Description
Inverse parallel pressing	Pressing of shoulders against each other, facing opposite directions
Parallel pressing	Pressing of shoulders against each other, facing same direction
Head-to-body knock	Hitting with the snout against the body of the receiver
Head-to-head knock	Hitting with the snout against the head of the receiver
Nose-to-nose	The nose approaches the snout or head of the receiver
Nose-to-body	The nose approaches the body of the receiver
Anal-genital nosing	The nose approaches the anogenital area of the receiver
Head tilt	The head is lowered and turned away from another animal
Aiming	An upward-directed thrust of the snout, slightly directed at the receiver, from a distance of 2–3 m
Retreat	Takes several steps away from the other animal

that aggression among sows within a group was greatest during and following parturition. Social grooming also occurs within these groups.

Parent–offspring (see Chapter 3)

The focal point of interactions between the sow and her litter during the first few days after farrowing is the nest. The isolation provided by the nest allows the sow and piglets to form a bond. In litter-bearing species, proximity to the nest is the primary means of distinguishing between offspring and alien young immediately following birth. Nursing occurs within the nest and involves teat seeking by the piglets immediately after birth. Piglets move from the rear of the sow towards her head, usually maintaining contact with the sow as they move. The sow does little in terms of reorientation to facilitate teat location. Subtle means of directing the piglet include vocalization, odours from the mammary and birth fluids, and hair patterns on the sow (Rhode Parfet and Gonyou, 1991). Piglets often move near the head of the sow, engage in nose-to-nose contact, vocalize and then begin suckling (Petersen *et al.*, 1990). During the initial few hours following birth, nursing shifts from being virtually continuous, to becoming episodic, occurring at approximately 1-h intervals (Lewis and Hurnik, 1985). During each episode of nursing a complex pattern of vocalization occurs which serves to call the piglets and indicate that milk flow is imminent (Fraser, 1980; see Section 3.7). The piglets, in turn, indicate their presence and motivation to suckle by massaging the udder of the sow. The extent of this pre- and post-suckling massage may be a means of communicating nutritional status to the sow (Algers, 1993).

The same pattern of nursing continues after the sow and piglets leave the farrowing nest.

Recognition between the sow and her piglets involves olfactory and vocal cues (Jensen and Redbo, 1987). Nose-to-nose contacts between a sow and her piglets are frequent, but decrease over the first 6 days (Stangel and Jensen, 1991). The majority of these contacts are associated with nursing (Blackshaw and Hagelso, 1990). Other contact of the piglets with the sow is also extensive. Piglets climb on to the sow while she is lying and will bite her (Whatson and Bertram, 1983). Piglets are able to recognize the faeces of their mother within 7 days of age (Horrell and Eaton, 1984). Sows and piglets respond to the isolation calls of each other once they begin to leave the farrowing nest (McBride, 1963; Fraser, 1975).

During the first 10 days after farrowing, sows and piglets will always lie within 15 m of each other, regardless of whether the sow is foraging away from the nest. By 6 days of age, piglets begin to follow their mother when she leaves the nest and eventually the farrowing nest is abandoned between days 7 and 10 post-farrowing (Jensen, 1986; Stangel and Jensen, 1991). By 10 days post-farrowing, when the sow and litter rejoin the main group, a strong bond between the mother and young exists. After returning to the main sow and offspring group, piglets remain close to their mother, and interact more intensively with her than with other sows in the herd, until weaning (Newberry and Wood-Gush, 1986).

If weaning is defined as the time at which all nursing of a litter ceases, then piglets are naturally weaned at approximately 17 weeks of age (Jensen and Recen, 1989). However, weaning may also be considered a gradual process in which the independence of the offspring from the sow gradually increases. Weaning in this sense may begin as early as 4 weeks after birth as the sow begins to reduce her nursing efforts (Jensen, 1988). This process is characterized by the sow terminating a greater proportion of, and initiating fewer, nursings (Jensen, 1988; Jensen and Recen, 1989).

Among juveniles

Other than their mother, sibling piglets have only themselves with which to interact during the first week of life. Play among piglets begins at 3–5 and peaks at 21–25 days of age (Blackshaw *et al.*, 1997). The bonds that develop during this time are quite strong and are maintained after the litter joins the main sow and offspring social group. For several weeks thereafter, the majority of interactions a piglet is involved in will be with siblings (Newberry and Wood-Gush, 1986; Petersen *et al.*, 1989).

The interactions with non-siblings that do occur at this point rarely involve aggressive behaviour (Newberry and Wood-Gush, 1986;

Petersen *et al.*, 1989). In groups which involve more than two litters, each litter forms a preferential attachment to another which is closest to them in age (Petersen *et al.*, 1989). In very large groups, two or three subgroups of two or more litters may exist. However, piglets will interact with piglets from any other litter within the sow and offspring group, even if they are not preferred. Occasionally a piglet from one litter joins a second litter, even though piglets in a free-ranging state do not normally suckle from a sow other than their mother (Jensen, 1986). On the rare occasions that this does occur, the piglet also interacts primarily with its 'adopted' siblings rather than with its own (Newberry and Wood-Gush, 1986).

6.2   Social Groupings Under Commercial Conditions

6.2.1   Social groupings

Sow and litter

Of the social groupings present in wild and feral pigs, only the sow and newborn group regularly exists on a commercial farm (see Fig. 6.1). Sows are usually moved to a farrowing facility 2–5 days prior to the expected farrowing date, slightly earlier than feral sows would leave their sow group. However, rather than being well separated from other sows during parturition, the females are no more than a few metres from each other. In most systems piglets remain in the farrowing environment until weaning, rather than leaving the farrowing site as they would in extensive or feral systems.

Much of the interaction among newborn piglets in commercial systems is likely to be similar to that in feral or wild pigs, but has been more intensively studied in confinement. Piglets engage in considerable fighting over teats during the first few hours after birth (Hartsock *et al.*, 1977; de Passille and Rushen, 1989). From this period of competition for teats, the teat order emerges wherein each piglet has a preferred teat to which it returns in successive sucklings (McBride, 1963; Hemsworth *et al.*, 1976; de Passille *et al.*, 1988). Fostering of piglets among sows is commonly practised in order to create litters of equal number and size. Unrelated piglets fostered into a litter are not differentiated from other pigs in that litter when pigs are regrouped at weaning (Stookey and Gonyou, 1998). Similarly, in extensive situations, the occasional piglet which does suckle from an alien sow associates with her piglets in other social situations as well (Newberry and Wood-Gush, 1986).

Although the majority of sows and litters remain in individual sow farrowing accommodation until weaning, some are either farrowed in groups or are grouped together during lactation. Group farrowing

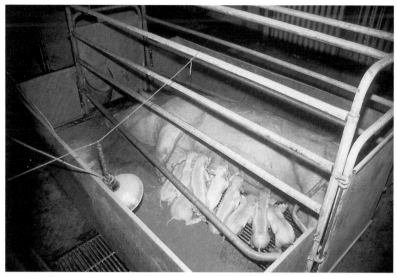

Fig. 6.1. The sow and newborn litter group. A social group common to both free-ranging and commercial conditions.

facilities usually consist of several individual farrowing pens which sows may access from a common area (Arey and Sancha, 1996; Marchant *et al.*, 1999). Such a facility attempts to mimic the individual nests of free-ranging sows, but does not provide the spatial separation available in the extensive habitat. Consequently, problems that contribute to the high mortality seen in free-ranging litters that are farrowed close together (Jensen, 1989) should be expected. One problem is the invasion of a nest by an alien sow, which has been reported at levels of 6.4% (Boe, 1994). Cross-suckling may also be high, particularly if more than one sow farrows within a nest (van Putten and van de Burgwal, 1990).

Sows and litters that have farrowed separately may be grouped during lactation. Often this grouping is accomplished between 10 and 14 days of age, similar to the time that the sow and newborn groups join the main sow and offspring group in free-ranging conditions. Aggression among piglets is less if they are regrouped during lactation than if it occurs post-weaning (Olsson and Samuelsson, 1993). Grouping during lactation may be accomplished by removing the dividers between farrowing crates, allowing only the piglets to move freely, or by moving sows and litters to group pens. If sows remain in their farrowing crate, piglets intermingle with each other but rarely suckle from an alien sow. However, if sows are moved from their farrowing location, either to another farrowing crate or to a group pen, inter-suckling is common (Wattanakul *et al.*, 1997; Pedersen *et al.*, 1998). Cross-suckling and the presence of additional piglets disrupt suckling, resulting in a higher proportion of false nursings and more suckling attempts (Wattanakul *et al.*, 1996).

In free-ranging conditions the sow spends more time away from the piglets and suckling frequency declines after a few days of age. In typical farrowing crates the sow cannot escape the piglets and the decline in suckling is less evident. When sows are allowed to leave the farrowing crate they spend less time with the piglets, nurse less often, are less likely to terminate suckling bouts, and the piglets gain less weight compared with conventional management (Rantzer, 1993). Access to a sow by an alien litter in a group lactation situation will result in earlier weaning. Weaning will also proceed more quickly if the sow is able to escape from the piglets (Pedersen *et al.*, 1998). The parent–offspring conflict (see Section 3.2), whereby parents attempt to reduce investment over time while offspring attempt to maintain it, is biased in conventional production situations in favour of the offspring. Alternatively, if the sow is able to leave the piglets, weaning proceeds quickly and favours the parent (Boe, 1991).

Juvenile groups

Once weaned, pigs proceed through what are commonly referred to as the nursery, grower and finisher phases of production, with body weight ranges of 10–25 kg, 25–60 kg and 60–120 kg, respectively. Many production systems maintain these three distinct phases by providing different housing for the pigs during each stage and perhaps regrouping the animals as they are moved. However, we are seeing an increase in systems that combine two or all three phases into one facility, thereby reducing the costs of moving animals as well as the social disturbance that is involved. The extreme situation, in which litters remain together in the same pen from birth to market is uncommon in commercial operations, but may have some advantages in terms of avoiding social conflict. Regardless of the production system, pigs are grouped by age, often with a variation of only 4–7 days within a pen, throughout these phases.

During the wean-to-market period pigs are fed several different diets in order to match their protein intake to their lean growth potential. In recent years there has been a move to feed males and females different diets as their lean growth potential differs. Most large operations, and some smaller ones, will house the sexes in different pens to accomplish this feeding practice.

Pigs are often regrouped when they are placed in nursery, grower or finisher facilities and the ensuing aggression can be quite dramatic (Fig. 6.2). The most intense aggression occurs during the first 1–2 h, after which it steadily decreases to a very low level by 24–48 h post-grouping (Meese and Ewbank, 1973; McGlone, 1986; Arey and Franklin, 1995). A lack of familiarity, rather than degree of relatedness,

Fig. 6.2. Head-to-head knock. A form of aggression between recently grouped grow/finish pigs.

is the basis for this aggression, as littermates separated shortly after birth are as aggressive towards each other as they are to unrelated and unfamiliar pigs when grouped together after weaning (Stookey and Gonyou, 1998). Within a given group size, the amount of fighting increases with the number of unfamiliar pigs within the pen (Arey and Franklin, 1995). Pigs that have never been penned together, but which have experienced some degree of contact through pen dividers, are less aggressive towards each other than totally unfamiliar pigs (Fraser, 1974). However, attempts to create familiarity through the use of common odours applied to pigs prior to regrouping have not reduced the level of aggression (Friend *et al.*, 1983; Gonyou, 1997a).

When all pigs within a pen are unfamiliar with each other, the pig that eventually becomes dominant will fight with every other pig in the pen to achieve this status. If several littermates from multiple litters are placed together, one pig from each litter will fight initially. The winner of that encounter will then attack the remaining pigs in the loser's litter (Rundgren and Lofquist, 1989). As a result, littermates tend to achieve similar dominance status within a pen but large litters are no more likely to be dominant than are those with only a few pigs present (Gonyou, 1997b).

Breeding animals

Males and females destined for the breeding herd are usually housed in single-sex groups until their first mating. Rearing females in isolation from other females delays the standing response of the females once they are introduced to boars (Soede and Schouten, 1991). Rearing males in isolation from other males has little effect on subsequent sexual behaviour (Tonn et al., 1985). In the case of gilts, the presence of stimuli from boars will induce earlier puberty (Hemsworth et al., 1988; Paterson et al., 1989). Odour appears to be the principal stimulus involved (Pearce et al., 1988).

The breeding phase of pig management consists of the period during which the females are observed for oestrus, actually bred, and until they are confirmed as being pregnant. During this period the females may be penned individually or kept in groups. The onset of oestrus is unpredictable in gilts and so they are often housed separately from post-weaning sows. Following weaning, sows return to oestrus in approximately 5–10 days, and are usually housed without regard for age. Housing sows in groups following weaning may facilitate their return to and synchronization of oestrus. However, individual housing in stalls reduces the level of aggression (Mendl et al., 1993), the degree of aggression-induced injuries, and facilitates certain breeding practices such as artificial insemination.

Although it is possible to operate the breeding facility without having a boar present, the most common practice would be to use boars to facilitate recognition of oestrus in the females (Hughes et al., 1985). In situations in which the females are in individual stalls, boars are routinely housed nearby so that their pheromones stimulate the sows and gilts. In many facilities the boars are allowed to walk past the heads of the sows to aid in detecting oestrus. The response of the sow is more critical than the response of the boar in detection of oestrus, and thus the sows are allowed to sniff the passing boar, rather than allowing the boar to attempt to mount.

When breeding females are group housed the boar may be penned with the sows or in an adjacent pen. Contact with the boar will reduce the weaning-to-mating interval (Petchey and English, 1980; Hemsworth et al., 1982). Oestrous females are proceptive and spend significant proportions of their time near a boar. In situations in which boars are penned adjacent to the breeding group, oestrous females may be detected by monitoring their presence near the boar (Bressers et al., 1991).

Under certain conditions sows may be bred during lactation. A significant proportion of females will cycle during lactation provided they are group housed, well fed and exposed to a boar (Rowlinson et al., 1975; Rowlinson and Bryant, 1982). Although lactational breeding has been incorporated into some management systems (Stolba and Wood-Gush, 1983), it is generally not reliable enough for common use.

Breeding males may be kept in small groups but the need for individual penning increases as the animals age. This management practice is similar to the behaviour observed in feral pigs in which boars become more solitary as they age. Regrouping of post-pubertal boars involves a great deal of fighting and is generally avoided.

Gestation

During pregnancy sows are fed restricted amounts of feed to prevent excessive accumulation of body fat. Restricting such an important resource makes the social behaviour of the animals a critical factor in management designs. One approach is to house the sows and gilts individually, usually in stalls but occasionally in tethers (Fig. 6.3). Individual penning allows each animal to be fed separately. When feed bowls are provided the control is nearly absolute, but if feed troughs are used some feed may be stolen from adjacent sows. Although the animals are separated by a partition, social behaviour is not completely prevented. Social encounters among sows in stalls persist over several gestations, with non-agonistic encounters exceeding the frequency observed among sows in groups (Mendl *et al.*, 1993). Social relationships between stalled sows remain unsettled (Jensen, 1984). Aggression does occur through the stall dividers, and it has been suggested that the nature of these dividers is important. Barnett *et al.* (1986) reported that aggression persisted between sows separated by horizontal bars, while that between sows separated by vertical bars suggested social conflict was resolved shortly after penning. Stereotypies are socially

Fig. 6.3. Individual stalls for sows. A controversial means of reducing social problems among gestating sows. These stalls allow sows to turn around but limit interaction.

transmitted in stall systems as an animal adjacent to a stereotyping animal is more likely to stereotype herself (Appleby *et al.*, 1989).

The composition of gestation groups is highly variable. Females may differ in age, size and stage of gestation. Two strategies have emerged when placing animals together. The first strategy is to minimize disturbances to the social group by limiting changes in its composition. Once the group is formed at the beginning of gestation, no other females will be added and the group only disbands at the time of farrowing. These 'static' groups are relatively uniform in terms of stage of lactation. The other strategy is to add recently bred animals and remove those approaching parturition, on a regular basis. 'Dynamic' systems involve post-regrouping aggression on a regular basis, but allow farms to operate with only a few groups rather than many.

6.2.2   Social effects on production

Competition for resources affects production in all phases of pigs' lives, but is most noticeable if those resources are limited (see Section 2.2). Our production methods have generally attempted to reduce the effects of competition by either providing *ad libitum* access to resources or physically preventing contact among animals. In interpreting the results of a study it is important to note the amount of feed provided and the housing conditions of the animals in order to assess the degree of competition encountered.

Suckling pigs

Most sows have more functional teats than they have piglets in their litter. Access to a teat is not limited, but competition for specific teats does occur. During the initial 2 h after birth, piglets fight over teats (Hartsock and Graves, 1976). The result of this competition is that the winning piglets control the higher-producing teats and have more stable suckling patterns, resulting in increased growth (Hartsock *et al.*, 1977; de Passille *et al.*, 1988; de Passille and Rushen, 1989). The anterior teats produce slightly more milk than the central and posterior teats (Dyck *et al.*, 1987), but considerable variation exists among teats within these general locations (Kim *et al.*, 1999). Fraser and Jones (1975) reported that teat order explains less than 5% of the total variation in weight gain. The size of a piglet relative to its littermates is more important that its actual size in determining its growth rate. Small piglets perform better if they are part of a litter of smaller piglets than if they compete with larger piglets (Fraser *et al.*, 1979). Partial correlations within litters indicate that variation in birth weight accounts for 3% of the variation in teat order, but that these two factors combined account for 25% of the variation in weaning weight (Fraser and Jones, 1975).

As the teat order of the recipient litter may already be established when fostering occurs, the new piglets must compete with the original litter to obtain access to high-producing teats. Fostered piglets do not perform as well as original piglets within the litter, particularly if they attempt to control a teat claimed by an original piglet (Horrell and Bennett, 1981). If piglets have a similar level of vigour, then fostered piglets are less likely to survive than are non-fostered (Neal and Irvin, 1991). The deleterious effects of fostering are greatest if only one piglet is fostered, and least if fostering is performed at 1 day of age (Horrell *et al.*, 1985). To reduce the cost of fostering it is recommended that it be done as early as possible, before the teat order is established, and that larger pigs be moved rather than the small piglets, which cannot compete as well.

Growing pigs

Within the nursery, growing and finishing phases of production, pigs are generally fed *ad libitum* in order to reduce social effects on production. In some management systems restricted feeding is practised to control fat deposition in the older animals. When feed is restricted, sources of social competition affect productivity to a greater extent. Feeding space must be severely restricted in order to result in reduced productivity if feed is available *ad libitum* (Walker, 1991). However, when feeding space is restricted as well as feed, competition for space results in highly variable levels of intake and growth rate (Botermans *et al.*, 1997). Post-regrouping aggression reduces subsequent weight gain if feed is restricted, but has little effect relative to non-regrouped pigs if feed is available *ad libitum* (Sherritt *et al.*, 1974). McBride *et al.* (1964) reported that variation in initial body weight accounts for 30% of the variation in social dominance, and approximately 13% of the variation in subsequent growth. However, Blackshaw *et al.* (1994) found no correlation between social status and either initial weight or rate of gain.

Pigs may be sorted upon entry into nursery, grower or finisher facilities in an attempt to create uniform competition within pens. If feed were restricted then such sorting would probably be advantageous. Under conditions of *ad libitum* feeding, sorting by sex or size may have little effect on productivity. Males generally grow faster than females, and the magnitude of this difference is unaffected by the sex ratio of the social group. Males perform as well when they are in an all-male group as they do when females are present (H. Gonyou, unpublished data).

It is generally believed within the industry that uniform grouping results in less social stress within the pen and reduces weight variation at marketing. There is reason to question both of these assumptions.

Weight variation at marketing is similar in groups that were either very uniform or heterogeneous when grouped together at the beginning of the grower phase (Tindsley and Lean, 1984). The growth rate of pigs in heterogeneous groups has been reported to be equal to or to exceed that of pigs in uniform groups (Gonyou et al., 1986; Francis et al., 1996). Aggression within pens following regrouping is either less in heterogeneous pens (Rushen, 1987; Francis et al., 1996) or unaffected by the variation in weight among pigs (Jensen and Yngvesson, 1998). It may be that social groups of pigs are most stable when clear differentiation by weight is possible, and thus groups of low initial weight variation increase that variation to a certain level.

Gestating sows

The major challenge in gestation systems is to control feed intake when sows are housed in groups. Group housing systems vary in the degree of physical control over competition for feed. In some systems sows are fed as individuals within a feeding stall or stalls. When all sows are fed simultaneously in feeding stalls, competition shifts from feed to the stall itself. Dominant animals will claim the preferred stalls – those that are fed first. When sows are fed sequentially from a single or a few stalls, as in electronic feeding stations, competition shifts to accessing the stall early in the daily feeding cycle. Dominant animals eat first, and often return to the feeder to remove any feed left by another animal (Hunter et al., 1988). The resulting aggression can result in injuries to the animals, and the exclusion of timid animals from the feeding area. Other systems, such as trickle feeding and floor feeding, provide less protection to the sows as they eat and competition increases. To maintain similar levels of intake among the sows, it is necessary to have animals of similar size and temperament together in the group. Group size is usually less than ten in such systems.

Some systems attempt to control competition among gestating sows by providing a modification of the ad libitum feeding situation. The use of high-fibre diets allows the animals to eat larger volumes and achieve some degree of satiety (Robert et al., 1993). An alternative is to intermittently provide ad libitum feed. Animals are given access to ad libitum feeders every second or third day.

In systems in which competition for feed is not well controlled, dominant sows become fat and subordinate animals become thin. The thin sows will not be able to maintain a high level of milk production during lactation and will fail to re-breed. The fat sows are also likely to have reproductive problems.

Competition also appears to be related to the sex ratio of litters produced by group-housed sows. In systems in which sows compete for feed during breeding and shortly thereafter, dominant animals produce a higher proportion of males than do subordinates (Meikle *et al.*, 1993, 1998). However, when access to feed is better controlled or groups are not formed until after pregnancy is confirmed, dominance status has no effect on litter sex ratio (Mendl *et al.*, 1985).

Social facilitation

When a hungry pig is placed with a satiated animal, the satiated pig will begin eating again. Similarly, when a pig can see another pig eating in a neighbouring pen, it is likely to begin eating as well (Hsia and Wood-Gush, 1983; Hutson, 1995). Thus, social facilitation results in a temporary increase in eating, which results in simultaneous or synchronized eating. However, synchronized eating does not necessarily result in increased intake over the duration of a trial (Gonyou *et al.*, 1999). An exception would appear to be suckling piglets, which can be induced to suckle more often and gain more weight through the playback of sow and piglet nursing grunts (Stone, 1974). Social facilitation may also be useful in inducing feed consumption following weaning.

6.2.3   Effects of group size and space allowance on social behaviour

The natural social groupings of pigs are relatively small, with a few adults and their offspring. Pigs are often kept in much larger groups in production systems. Small groups of pigs generally have a very stable social hierarchy with linear relationships. In larger groups the proportion of intransitive relationships increases (Moore *et al.*, 1996). Individually penned pigs and groups of three perform very well, but there is a dramatic reduction in productivity when five pigs are penned together (Gonyou *et al.*, 1992; Gonyou and Stricklin, 1998). It would appear that the less complex social system involved in very small groups has advantages in terms of productivity. The number of agonistic encounters each pig is involved in increases with group size up to six pigs per pen, and decreases with subsequent increases in group size (Moore *et al.*, 1996). Two possible explanations for this pattern in aggression are that animals form distinct subgroups which avoid each other, or the animals develop a tolerance of other individuals in large groups. Subgrouping has been noted in a number of studies, at least in terms of lying position. Sows added to an existing group form a distinct subgroup for up to 21 days post-regrouping (Moore *et al.*, 1993). Grower pigs will also form subgroups when added to a pen of larger pigs (Moore *et al.*, 1994). However, it is not clear if these subgroups remain distinct during activity periods and only exist during lying. If the subgroups remain distinct during activity periods, it would suggest territorial or at least well-defined home ranges within the pen. In many group systems pigs must share common feeding and drinking areas, which would increase aggression by forcing subgroups to intermingle. An alternative explanation to the reduced aggression in larger groups is the development of tolerance for less familiar pigs. The animals may still lie in subgroups, although this occurs less frequently in large groups (Penny *et al.*, 1997), but are able to interact freely when active. We have observed that pigs in groups of 40 will investigate much of the pen during activity bouts, suggesting that they have no hesitation in meeting less familiar pigs. Group sizes in excess of 100 pigs per pen are becoming more common in commercial practice.

Small group sizes have an advantage over large groups in that pigs can be effectively sorted by weight, age or other criteria. As indicated above, sorting grower pigs that are provided with feed *ad libitum* may not be beneficial. However, when feed is restricted, as with gestating sows, uniformity is desirable unless access to the feed is well controlled. Larger groups of sows may be used on electronic sow feeders because of the control provided over intake, but smaller groups are recommended for most other group housing systems.

The quantity (amount) and quality (configuration) of space affect social behaviour by allowing animals to avoid and escape from each other. Providing large amounts of space allows the animals to move about the pen to obtain resources without entering the personal space of other animals. If a fight does occur, an animal may signal submission by fleeing (Kay *et al.*, 1999). Within the feeding area, adequate trough space and dividers between spaces reduce the incidence of aggression (Baxter, 1991). Strategically placed partitions also allow animals to avoid being within sight of another animal, even if they are very close together. Equipping a pen with 'hide' areas at the time of regrouping reduces subsequent aggression and improves growth for the immediate post-regrouping period (McGlone and Curtis, 1985). Dynamic groupings of sows can benefit by providing well-defined areas within the pen which new groups can claim as their own during the period when integration into the main group is occurring (van Putten, 1990).

Space allowance can be expressed as an amount per animal or as the reciprocal, the number of animals per unit of space. In terms of feeders, we have generally used the latter. The number of animals that can eat from a single feeding space depends upon the total duration of eating. Small pigs take longer to eat than large pigs (Hyun *et al.*, 1997), and pigs eating dry feed require more time than those eating from a wet/dry feeder (Gonyou and Lou, 2000). Pigs are able to adapt their eating behaviour as the number of pigs eating from a feeder increases. Under some conditions, groups as large as 30 pigs can maintain intake levels when eating from a single-space feeder but must increase their eating speed to do so (Walker, 1991).

Reducing space allowance, either in terms of floor area or by increasing the number of pigs on a feeder, eventually results in a reduction in daily feed intake and growth rate (Walker, 1991; Gonyou and Stricklin, 1998). It is not clear if reduced intake results in reduced growth, or if a reduction in growth potential results in a reduced appetite (Chapple, 1993). Attempting to compensate for crowded conditions by providing a more nutrient-dense diet does not result in improved growth (Edmonds *et al.*, 1988). Rather, the pigs continue to limit their energy and protein intake. These results suggest that the direct cause for the reduction in intake is a stress-induced metabolic change rather than environmental restrictions on access to feed.

Floor space allowance has traditionally been expressed as area per animal. However, the space requirements of an animal change with its size. Space allowance should be expressed in relation to body weight, but a linear relationship would underestimate the requirement for small pigs compared with large ones. The use of an allometric expression, a form of estimating the surface area of the animal, has proved useful in expressing space allowance over a wide range of weights (Gonyou and Stricklin, 1998). Calculating space allowance in this fashion, a constant k is multiplied by body weight$^{0.667}$. It would appear that pigs reach maximum growth when k is approximately 0.035, with area expressed in square metres and body weight in kilograms (Edwards *et al.*, 1988; Gonyou and Stricklin, 1998).

6.3 Social Behaviour, Management and Welfare

6.3.1 Grouping animals

With the exception of the sow and her week-old litter rejoining the main sow group, pigs rarely allow newcomers into their social group under natural conditions. Under production conditions, pigs fight vigorously to exclude unfamiliar animals from their group. This fighting can affect productivity, particularly if resources such as feed are limited. Even when resources are quite adequate, production is reduced for 2–4 weeks (McGlone and Curtis, 1985). The greatest reduction in growth is among those pigs that received the greatest amount of injuries to their ears and shoulders (Gonyou *et al.*, 1988). Regrouping young piglets results in less aggression than among older animals (Jensen, 1994). Nursing piglets can be grouped with little resultant aggression by removing the partitions between farrowing crates. However, by the time of weaning in most commercial operations, aggression at regrouping is intense.

Odour is a major means of identifying familiar and unfamiliar pigs. Odours have been used in a variety of ways to attempt to reduce aggression following regrouping. One approach has been to mask the odour of the unfamiliar pig (Friend *et al.*, 1983). In this case the odour was applied at the time that the pigs were regrouped. In general, such attempts have failed to reduce the aggression, and may actually increase the aggression among familiar pigs. If an odour is masked by this method, it would appear that it is the familiar odour, and littermates may attack each other as if they were unfamiliar.

A second means of using odour is to create a common familiar odour on all pigs. This can be attempted by placing the same artificial odour on all pigs, or by exposing pigs to the manure of other pigs for several days prior to regrouping. Unfortunately, neither method has proved to be effective in reducing aggression (Gonyou, 1997a).

A third use of odour is the application of pig-derived compounds which may act as pheromones. Urine obtained from pigs that have recently fought is effective in reducing aggression when applied to unfamiliar pigs prior to regrouping. Urine from pigs treated with ACTH, and sprays containing androstenone are also reported to reduce aggression (McGlone, 1985; McGlone *et al.*, 1987).

Several pharmacological compounds have been used to reduce regrouping aggression. By administering these compounds at the time of regrouping, the animals become inactive and so aggression is greatly reduced for the subsequent few hours (Symoens and van den Brande, 1969; Bjork *et al.*, 1988). However, once the pigs recover from the drug, aggression begins and may be as severe as normal aggression (Blackshaw, 1981b). In general, these agents do not result in an improvement in weight gain (Gonyou *et al.*, 1988).

The composition of the group being assembled can affect regrouping aggression. A large difference in weight among the pigs may reduce the resulting aggression (Rushen, 1987), although not all studies support this finding (Jensen and Yngvesson, 1998). A dynamic grouping system in grower/finisher pigs, in which small pigs are added to a pen comprised of older pigs, reduces aggression compared with mixing pigs within age groups (Moore *et al.*, 1994). However, dynamic systems for gestating sows may result in more aggression. The presence of a large boar in a pen of market pigs will reduce aggression (Grandin and Bruning, 1992), but does not appear to be effective when sows are regrouped (Luescher *et al.*, 1990).

Litters have been combined using varying numbers of animals from each litter. Equal numbers of pigs from several litters, whether it be one, two or four pigs per litter forming groups of eight, do not appear to affect aggression, or at least post-regrouping growth (Friend *et al.*, 1983; Blackshaw *et al.*, 1987). When unequal numbers of pigs from two litters are grouped together, post-regrouping growth is similar between the minority and majority litters (Gonyou, 1997b). However, when individual sows are added to an established group the resulting aggression can produce serious injuries. Again, the importance of availability of resources, primarily feed, should be considered.

Pigs may be scored according to certain behaviour characteristics such as resistance to handling or propensity to fight. When a group is formed of pigs having a high propensity to fight, aggression is more severe than when non-aggressive pigs are grouped together, and even less if aggressive and non-aggressive pigs are grouped in the same pen (Erhard *et al.*, 1996). Similarly, if pigs that are resistant to restraint are penned together, there is a high level of aggression. Productivity is best if different behaviour types are combined in the same pen (Hessing *et al.*, 1994).

6.3.2 Separation problems

Although removal from a social group does not involve the aggression associated with joining a group, problems do occur when pigs are removed from the social environment. In commercial production weaning is accomplished by removing the sow from the piglets. When weaning is abrupt the piglets no longer have the sow as a focus for feeding. The initial result is a virtual cessation of eating for 1–3 days, depending upon the age of the piglet (Metz and Gonyou, 1990). It is possible that social facilitation may be useful in initiating eating among weaned pigs as they continue to synchronize eating for several days thereafter. Without the sow present, newly weaned pigs begin directing nosing behaviour towards the belly of their littermates approximately 4 days following weaning (Blackshaw, 1981a; Metz and Gonyou, 1990). This time is after they have begun eating a normal amount of solid feed, and so does not appear to be due to hunger. However, the behaviour is very like the udder massage directed towards the sow and may have some relationship to suckling behaviour. The behaviour peaks approximately 2–3 weeks after weaning, but persists at a higher level among earlier-weaned pigs through the grow/finish period (Gonyou *et al.*, 1998). Not all pigs engage in belly-nosing, and some are more likely to be nosed than to nose (Blackshaw, 1981a; Gonyou *et al.*, 1998).

Pigs are also removed from their social group to be penned individually. Isolation is very stressful for pigs and it is often necessary to keep at least two animals together to facilitate handling or experimental testing. When sows are isolated and placed in stalls or tethered they attempt to escape. These escape attempts eventually subside, but may be the basis for future stereotypic behaviour under some conditions (Cronin *et al.*, 1984). Temporary removal of a pig from a stable social group may lead to rejection when reintroduced. Among growing pigs, subordinate pigs may be rejected after being absent for only a few days, while dominant animals can return after several weeks without extensive fighting (Ewbank and Meese, 1971). However, if the social group was not stable when the pigs were removed, the dominant pig may also encounter considerable aggression (Otten *et al.*, 1997).

6.3.3 Abnormal behaviour

Some of the most perplexing behaviour problems in pigs are those which cause physical injuries to the recipient and fall under the general term of cannibalism. These include tail- and ear-biting, and flank- and navel-sucking. The incidence of these behaviours can be quite high, exceeding 10% of pigs in several studies (Arey, 1991). These behaviours are generally, but not always, directed at animals that are lower in the social hierarchy (Blackshaw, 1981a). There are many known and suspected causes of these behaviours. The causative factors may be classified as nutritional, due to discomfort, and lack of environmental enrichment. Social causes relate primarily to the issue of discomfort. Early weaning is a significant factor affecting the incidence of belly-nosing among nursery pigs, but also continues as higher levels of chewing on penmates during the grow/finish period (Gonyou *et al.*, 1998). Overcrowding is generally recognized as a causative factor of tail-biting. Group size has also been implicated in tail-biting, but evidence for this causative link is limited and the association may be unfounded.

Stereotypies are one of the most intensively studied abnormal behaviours of gestating sows investigated in recent years. Although feed restriction appears to be the primary cause of these behaviours, environmental factors may also play a role (Lawrence and Terlouw, 1993). Sows in stalls or tethered adjacent to stereotyping animals are more likely to develop similar behaviour (Appleby *et al.*, 1989). Although social facilitation may be involved, the association could be due to increased arousal caused by the activity of the neighbour (Lawrence and Terlouw, 1993).

References

Algers, B. (1993) Nursing in pigs: communicating needs and distributing resources. *Journal of Animal Science* 71, 2826–2831.

Appleby, M.C., Lawrence, A.B. and Illius, A.W. (1989) Influence of neighbours on stereotypic behaviour of tethered sows. *Applied Animal Behaviour Science* 24, 137–146.

Arey, D.S. (1991) Tail-biting in pigs. *Farm Building Progress* 105, 20–23.

Arey, D.S. and Franklin, M.F. (1995) Effects of straw and unfamiliarity on fighting between newly mixed growing pigs. *Applied Animal Behaviour Science* 45, 23–30.

Arey D.S. and Sancha, E.S. (1996) Behaviour and productivity of sows and pig-lets in a family system and in farrowing crates. *Applied Animal Behaviour Science* 50, 135–145.

Barnett, J.L., Hemsworth, P.H., Winfield, C.G. and Hansen, C. (1986) Effects of social environment on welfare status and sexual behaviour of female pigs. I. Effects of group size. *Applied Animal Behaviour Science* 16, 249–257.

Barrette, C. (1986) Fighting behaviour of wild *Sus scrofa*. *Journal of Mammalogy* 67, 177–179.

Baxter, M.R. (1991) The design of the feeding environment for pigs. In: Batterham, E.S. (ed.) *Manipulating Pig Production III*. Australian Pig Science Association, Attwood, Victoria, Australia, pp. 150–157.

Bjork, A., Olsson, N.G. Christensson, E., Martinsson, K. and Olsson, O. (1988) Effects of amperozide on biting behaviour and performance in restricted-fed pigs following regrouping. *Journal of Animal Science* 66, 669–675.

Blackshaw, J.K. (1981a) Some behavioural deviations in weaned domestic pigs: persistent inguinal nose thrusting, and tail and ear biting. *Animal Production* 33, 325–332.

Blackshaw, J.K. (1981b) The effect of pen design and the tranquilising drug, azaperone, on the growth and behaviour of weaned pigs. *Australian Veterinary Journal* 57, 272–276.

Blackshaw, J.K. and Hagelso, A.M. (1990) Getting-up and lying-down behaviours of loose-housed sows and social contacts between sows and piglets during day 1 and day 8 after parturition. *Applied Animal Behaviour Science* 25, 61–70.

Blackshaw, J.K., Bodero, D.A.V. and Blackshaw, A.W. (1987) The effect of group composition on behaviour and performance of weaned pigs. *Applied Animal Behaviour Science* 19, 73–80.

Blackshaw, J.K., Thomas, F.J. and Blackshaw, A.W. (1994) The relationship of dominance, forced and voluntary leadership and growth rate in weaned pigs. *Applied Animal Behaviour Science* 41, 263–268.

Blackshaw, J.K., Swain, A.J., Blackshaw, A.W., Thomas, F.J.M. and Gillies, K.J. (1997) The development of playful behaviour in piglets from birth to weaning in three farrowing environments. *Applied Animal Behaviour Science* 55, 37–49.

Blasetti, A., Boitani, L., Riviello, M.C. and Visalberghi, E. (1988) Activity budgets and use of enclosed space by wild boars (*Sus scrofa*) in captivity. *Zoo Biology* 7, 69–79.

Boe, K. (1991) The process of weaning in pigs: when the sow decides. *Applied Animal Behaviour Science* 30, 47–59.

Boe, K. (1994) Variation in maternal behaviour and production of sows in integrated loose housing systems in Norway. *Applied Animal Behaviour Science* 41, 53–62.

Botermans, J.A.M., Svendsen, J. and Westrom, B. (1997) Competition at feeding of growing–finishing pigs. In: Bottcher, R.W. and Hoff, S.J. (eds) *Livestock Environment*, Vol. V. American Society of Agricultural Engineers, USA, pp. 591–598.

Bressers, J.P.M., Te Brake, J.H.A. and Noordhuizen, J.P.T.M. (1991) Oestrus detection in group-housed sows by analysis of data on visits to the boar. *Applied Animal Behaviour Science* 31, 183–193.

Chapple, R.P. (1993) Effects of stocking arrangement on pig performance. In: Batterham, E.S. (ed.) *Manipulating Pig Production*, Vol. IV. Australian Pig Science Association, Attwood, Victoria, pp. 87–97.

Cronin, G.M. and Cropley, J.A. (1991) The effect of piglet stimuli on the posture changing behaviour of recently farrowed sows. *Applied Animal Behaviour Science* 30, 167–172.

Cronin, G.M., Wiepkema, P.R. and Hofstede, G.J. (1984) The development of stereotypies in tethered sows. In: Unshelm, J., van Putten, G. and Zeeb, K. (eds) *Proceedings of the International Congress on Applied Ethology in Farm Animals*. Kuratorium für Technik und Bauwesen in der Landwirtschaft, Kiel, Germany, pp. 97–100.

de Passille, A.M.B. and Rushen, J. (1989) Suckling and teat disputes by neonatal piglets. *Applied Animal Behaviour Science* 22, 23–38.

de Passille, A.M.B., Rushen, J. and Hartsock, T.G. (1988) Ontogeny of teat fidelity in pigs and its relation to competition at suckling. *Canadian Journal of Animal Science* 68, 325–338.

Dyck, G.W., Swierstra, E.E., Mckay, R.M. and Mount K. (1987) Effect of location of the teat suckled, breed and parity on piglet growth. *Canadian Journal of Animal Science* 67, 929–939.

Edmonds, M.S., Arentson, B.E. and Mente, G.A. (1998) Effect of protein levels and space allocations on performance of growing–finishing pigs. *Journal of Animal Science* 76, 814–821.

Edwards, S.A., Armsby, A.W. and Spechter, H.H. (1988) Effects of floor area allowance on performance of growing pigs kept on fully slatted floors. *Animal Production* 46, 453–459.

Erhard, H.W., Mendl, M. and Ashley, D.D. (1996) How information about individual pigs can help to reduce aggression and injuries after mixing. In: *Proceedings of British Society of Animal Science*. British Society of Animal Science, Scarborough.

Ewbank, R. and Meese, G.B. (1971) Aggressive behaviour in groups of domesticated pigs on removal and return of individuals. *Animal Production* 13, 685–695.

Fradrich, H. (1974) A comparison of behaviour in the Suidae. In: *The Behaviour of Ungulates and its Relation to Management*. IUCN Publications, New Series No. 6, pp. 133–143.

Francis, D.A., Christison, G.I. and Cymbaluk, N.F. (1996) Uniform or heterogeneous weight groups as factors in mixing weanling pigs. *Canadian Journal of Animal Science* 76, 171–176.

Fraser, D. (1974) The behaviour of growing pigs during experimental social encounters. *Journal of Agricultural Science, Cambridge* 82, 147–163.

Fraser, D. (1975) Vocalizations of isolated piglets. II Some environmental factors. *Applied Animal Ethology* 2, 19–24.

Fraser, D. (1980) A review of the behavioural mechanism of milk ejection of the domestic pig. *Applied Animal Ethology* 6, 247–255.

Fraser, D. and Jones, R.M. (1975) The 'teat order' of suckling pigs. 1 Relation to birth weight and subsequent growth. *Journal of Agricultural Science, Cambridge* 84, 387–391.

Fraser, D., Thompson, B.K., Ferguson, D.K. and Darroch, R.L. (1979) The 'teat order' of suckling pigs. 3. Relation to competition within litters. *Journal of Agricultural Science, Cambridge* 92, 257–261.

Friend, T.H., Knabe, D.A. and Tanksley, T.D. Jr (1983) Behaviour and performance of pigs grouped by three different methods at weaning. *Journal of Animal Science* 57, 1406–1411.

Gonyou, H.W. (1997a) Can odours be used to reduce aggression in pigs? In: *1997 Annual Research Report*. Prairie Swine Centre, Saskatoon, pp. 59–62.

Gonyou, H.W. (1997b) Behaviour and productivity of pigs in groups composed of disproportionate numbers of littermates. *Canadian Journal of Animal Science* 77, 205–209.

Gonyou, H.W. and Lou, Z. (2000) Effects of eating space and availability of water in feeders on productivity and eating behaviour of grower/finisher pigs. *Journal of Animal Science* 78, 865–870.

Gonyou, H.W. and Stricklin, W.R. (1998) Effects of floor area allowance and group size on the productivity of growing/finishing pigs. *Journal of Animal Science* 76, 1326–1330.

Gonyou, H.W., Rohde, K.A. and Echeverri, A.C. (1986) Effects of sorting pigs by weight on behaviour and productivity after mixing. *Journal of Animal Science* 63, 163–164.

Gonyou, H.W., Rohde Parfet, K.A., Anderson, D.B. and Olson, R.D. (1988) Effects of amperozide and azaperone on aggression and productivity of growing–finishing pigs. *Journal of Animal Science* 66, 2856–2864.

Gonyou, H.W., Chapple, R.P. and Frank, G.R. (1992) Productivity, time budgets and social aspects of eating in pigs penned in groups of five or individually. *Applied Animal Behaviour Science* 34, 291–301.

Gonyou, H.W., Beltranena, E., Whittington, D.L. and Patience, J.F. (1998) The behaviour of pigs weaned at 12 and 21 days of age from weaning to market. *Canadian Journal of Animal Science* 78, 517–523.

Gonyou, H.W., Peterson, C. and Getson, K. (1999) Effects of feeder and penning design on social facilitation of eating in pigs in adjoining pens. In: *Proceedings of the British Society of Animal Science*. British Society of Animal Science, Scarborough, 182 pp.

Grandin, T. and Bruning, J. (1992) Boar presence reduces fighting in mixed slaughter-weight pigs. *Applied Animal Behaviour Science* 33, 272–276.

Graves, H.B. (1984) Behaviour and ecology of wild and feral swine (*Sus scrofa*). *Journal of Animal Science* 58, 482–492.

Hartsock, T.G. and Graves, H.B. (1976) Neonatal behaviour and nutrition-related mortality in domestic swine. *Journal of Animal Science* 42, 235–241.

Hartsock, T.G., Graves, H.B. and Baumgardt, B.R. (1977) Agonistic behaviour and the nursing order in suckling piglets: relationships with survival, growth and body composition. *Journal of Animal Science* 44, 320–330.

Hemsworth, P.H., Winfield, C.G. and Mullaney, P.D. (1976) A study of the development of the teat order in piglets. *Applied Animal Ethology* 2, 225–233.

Hemsworth, P.H., Salden, N.T.C.J. and Hoogerbrugge, A. (1982) The influence of the post-weaning social environment on the weaning to mating interval of the sow. *Animal Production* 35, 41–48.

Hemsworth, P.H., Hansen, C., Winfield, C.G. and Barnett, J.L. (1988) Effects on puberty attainment in gilts of continuous or limited exposure to boars. *Australian Journal of Experimental Agriculture* 28, 469–472.

Hessing, M.J.C., Schouten, W.G.P., Wiepkema, P.R., and Tielen, M.J.M. (1994) Implications of individual behavioural characteristics on performance in pigs. *Livestock Production Science* 40, 187–196.

Horrell, I. and Bennett, J. (1981) Disruption of teat preferences and retardation of growth following cross-fostering of 1-week-old pigs. *Animal Production* 33, 99–106.

Horrell, I., Hodgson, J. and Lumb, S. (1985) The effect on growth of fostering piglets under various conditions. *Proceedings of the British Society of Animal Production*, Scarborough, April 1985.

Horrell, R.I. and Eaton, M. (1984) Recognition of maternal environment in piglets: effects of age and some discrete complex stimuli. *The Quarterly Journal of Experimental Psychology* 36B, 119–130.

Hsia, L.C. and Wood-Gush, D.G.M. (1983) A note on social facilitation and competition in the feeding behaviour of pigs. *Animal Production* 37, 149–152.

Hughes, P.E., Hemsworth, P.H. and Hansen, C. (1985) The effects of supplementary olfactory and auditory stimuli on the stimulus value and mating success of the young boar. *Applied Animal Behaviour Science* 14, 245–252.

Hunter, E.J., Broom, D.M., Edwards, S.A. and Sibly, R.M. (1988) Social hierarchy and feeder access in a group of 20 sows using a computer controlled feeder. *Animal Production* 47, 139–148.

Hutson, G.D. (1995) Effect of enclosure of the feeding space on feeding behaviour of growing pigs. In: Batterham, E.S. (ed.) *Manipulating Pig Production*, Vol. V. Australian Pig Science Association, Attwood, Victoria, Australia.

Hyun, Y., Ellis, M., McKeith, F.K. and Wilson, E.R. (1997) Feed intake pattern of group-housed growing–finishing pigs monitored using a computerized feed intake recording system. *Journal of Animal Science* 75, 1443–1451.

Jensen, P. (1980) An ethogram of social interaction patterns in group-housed dry sows. *Applied Animal Ethology* 6, 341–350.

Jensen, P. (1982) An analysis of agonistic interaction patterns in group-housed dry sows – aggression regulation through an 'avoidance order'. *Applied Animal Ethology* 9, 47–61.

Jensen, P. (1984) Effects of confinement on social interaction patterns in dry sows. *Applied Animal Behaviour Science* 12, 93–101.

Jensen, P. (1986) Observation on the maternal behaviour of free-ranging domestic pigs. *Applied Animal Behaviour Science* 16, 131–142.

Jensen, P. (1988) Maternal behaviour and mother–young interactions during lactation in free-ranging domestic pigs. *Applied Animal Behaviour Science* 20, 297–308.

Jensen, P. (1989) Nest site choice and nest building of free-ranging domestic pigs due to farrow. *Applied Animal Behaviour Science* 22, 13–21.

Jensen, P. (1994) Fighting between unacquainted pigs – effects of age and of individual reaction pattern. *Applied Animal Behaviour Science* 41, 37–52.

Jensen, P. and Algers, B. (1984) An ethogram of piglet vocalizations during suckling. *Applied Animal Ethology* 11, 237–248.

Jensen, P. and Recen, B. (1989) When to wean – observations from free-ranging domestic pigs. *Applied Animal Behaviour Science* 23, 49–60.

Jensen, P. and Redbo I. (1987) Behaviour during nest leaving in free-ranging domestic pigs. *Applied Animal Behaviour Science* 18, 355–362.

Jensen, P. and Wood-Gush, G.M. (1984) Social interactions in a group of free-ranging sows. *Applied Animal Behaviour Science* 12, 327–337.

Jensen, P. and Yngvesson, J. (1998) Aggression between unacquainted pigs – sequential assessment and effects of familiarity and weight. *Applied Animal Behaviour Science* 58, 49–61.

Kay, R.M., Burfoot, A., Spoolder, H.A.M. and Docking, C.M. (1999) The effect of flight distance on aggression and skin damage of newly weaned sows at mixing. In: *Proceedings of the British Society of Animal Science*. British Society of Animal Science, Scarborough.

Kim, S.W., Hurley, W.L. and Easter, R.A. (1999) Mammary gland order and piglet growth during lactation. In: *Proceedings of the British Society of Animal Science*. British Society of Animal Science, Scarborough, 190 pp.

Lawrence, A.B. and Terlouw, C.E.M. (1993) A review of behavioral factors involved in the development and continued performance of stereotypic behaviors in pigs. *Journal of Animal Science* 71, 2815–2825.

Lewis, N. and Hurnik, J.F. (1985) The development of nursing behaviour in swine. *Applied Animal Behaviour Science* 14, 225–232.

Luescher, U.A., Friendship, R.M. and McKeown, D.B. (1990) Evaluation of methods to reduce fighting among regrouped gilts. *Canadian Journal of Animal Science* 70, 366–370.

Marchant, J.N., Forde, R.M., Corning, S. and Broom, D.M. (1999) The effects of farrowing system design on gilt behaviour. In: *Proceedings of British Society of Animal Science*. British Society of Animal Science, Scarborough, 185 pp.

Martin, J.T. (1975) Movement of feral pigs in North Canterbury, New Zealand. *Journal of Mammalogy* 56, 914–915.

Mauget, R. (1981) Behavioural and reproductive strategies in wild forms of *Sus scrofa* (European wild boar and feral pigs). In: Sybesma, W. (ed.) *The Welfare of Pigs*. Martinus Nijhoff, Brussels, pp. 3–13.

Mayer, J.J. and Brisbin, I.L. Jr (1986) A note on the scent-marking behavior of two captive-reared feral boars. *Applied Animal Behaviour Science* 16, 85–90.

McBride, G. (1963) The 'teat order' and communication in young pigs. *Animal Behaviour* 11, 53–56.

McBride, G., James, J.W. and Hodgens, N. (1964) Social behaviour of domestic animals. IV. Growing pigs. *Animal Production* 6, 129–139.

McGlone, J.J. (1985) Olfactory cues and pig agonistic behavior: evidence for a submissive pheromone. *Physiology and Behavior* 34, 195–198.

McGlone, J.J. (1986) Influence of resources on pig aggression and dominance. *Behavioural Processes* 12, 135–144.

McGlone, J.J. and Curtis, S.E. (1985) Behavior and performance of weanling pigs in pens equipped with hide areas. *Journal of Animal Science* 60, 20.

McGlone, J.J., Curtis, S.E. and Banks, E.M. (1987) Evidence for aggression-modulating pheromones in prepubertal pigs. *Behavioral and Neural Biology* 47, 27–39.

Meese, G.B. and Ewbank, R. (1973) The establishment and nature of the dominance hierarchy in the domesticated pig. *Animal Behaviour* 21, 326–334.

Meikle, D.B., Drickamer, L.C., Vessey, S.H., Rosenthal, T.L. and Fitzgerald, K.S. (1993) Maternal dominance rank and secondary sex ratio in domestic swine. *Animal Behaviour* 46, 79–85.

Meikle, D.B., Vessey, S.H. and Drickamer, L.C. (1998) Mechanisms of sex-ration adjustment in domestic swine: reply to James. *Animal Behaviour* 55, 770–772.

Mendl, M.T., Broom, D.M. and Zanella, A.J. (1993) The effects of three types of dry sow housing on sow welfare. In: Collins, E. and Boon, C. (eds) *Livestock Environment*, Vol. IV. American Society of Agricultural Engineers, USA, pp. 461–467.

Mendl, M., Zanella, A.J., Broom, D.M. and Whittmore, C.T. (1995) Maternal social status and birth sex ratio in domestic pigs: an analysis of mechanisms. *Animal Behaviour* 50, 1361–1370.

Mendl, M., Broom, D.M. and Zanella, A.J. (1998) Multiple mechanisms may affect birth sex ratio in domestic pigs. *Animal Behaviour* 55, 773–776.

Metz, J.H.M. and Gonyou, H.W. (1990) Effect of age and housing conditions on the behavioural and haemolytic reaction of piglets to weaning. *Applied Animal Behaviour Science* 27, 229–309.

Moore, A.S., Gonyou, H.W. and Ghent, A.W. (1993) Integration of newly introduced and resident sows following grouping. *Applied Animal Behaviour Science* 38, 257–267.

Moore, A.S., Gonyou, H.W., Stookey, J.M. and McLaren, D.G. (1994) Effect of group composition and pen size on behaviour, productivity and immune response of growing pigs. *Applied Animal Behaviour Science* 40, 13–30.

Moore, C.M., Zhou, J.Z., Stricklin, W.R. and Gonyou, H.W. (1996) The influence of group size and floor area space on social organization of growing–finishing pigs. In: Duncan, I.J.H., Widowski, T.M. and Haley, D.B. (eds) *Proceedings of the 30th International Congress of the International Society of Applied Ethology*. The Colonel K.L. Campbell Centre for the Study of Animal Welfare, Guelph, Ontario, p. 34.

Neal, S.M. and Irvin, K.M. (1991) The effects of crossfostering pigs on survival and growth. *Journal of Animal Science* 69, 41–46.

Newberry, R.C. and Wood-Gush, D.G.M. (1986) Social relationships of piglets in a semi-natural environment. *Animal Behaviour* 34, 1311–1318.

Olsson, A.-Ch. and Samuelsson, O.V. (1993) Grouping studies of lactating and newly weaned sows. In: Collins, E. and Boon, C. (eds) *Livestock Environment*, Vol. IV. American Society of Agricultural Engineers, USA, pp. 475–482.

Otten, W., Puppe, B., Stabenow, B., Kanitz, E., Schon, P.C., Brussow, K.P. and Nurnberg, G. (1997) Agonistic interactions and physiological reactions of top- and bottom-ranking pigs confronted with a familiar and an unfamiliar group: preliminary results. *Applied Animal Behaviour Science* 55, 79–90.

Paterson, A.M., Hughes, P.E. and Pearce, G.P. (1989) The effect of season, frequency and duration of contact with boars on the attainment of puberty in gilts. *Animal Reproduction Science* 21, 115–124.

Pearce, G.P., Hughes, P.E. and Booth, W.D. (1988) The involvement of boar submaxillary salivary gland secretions in boar-induced precocious puberty attainment in the gilt. *Animal Reproduction Science* 16, 125–134.

Pedersen, L.J., Studnitz, M., Jensen, K.H. and Giersing, A.M. (1998) Suckling behaviour of piglets in relation to accessibility to the sow and the presence of foreign litters. *Applied Animal Behaviour Science* 58, 267–279.

Penny, P.C., Stewart, A.H. and English, P.R. (1997) The behaviour of high and low performing pigs and location preferences in large groups of pigs housed on deep bedded straw. *Proceedings of British Society of Animal Science*. British Society of Animal Science, Scarborough, p. 112.

Petchey, A.M. and English, P.R. (1980) A note on the effects of boar presence on the performance of sows and their litters when penned as groups in late lactation. *Animal Production* 31, 107–109.

Petersen, H.V., Vestergaard, K. and Jensen, P. (1989) Integration of piglets into social groups of free-ranging domestic pigs. *Applied Animal Behaviour Science* 23, 223–236.

Petersen, V., Recen, B. and Vestergaard, K. (1990) Behaviour of sows and piglets during farrowing under free-range conditions. *Applied Animal Behaviour Science* 26, 169–179.

Rantzer, D. (1993) Weaning of pigs in a sow-controlled and a conventional housing system for lactating sows. In: Collins, E. and Boon, C. (eds) *Livestock Environment*, Vol. IV. American Society of Agricultural Engineers, USA, pp. 468–474.

Robert, S., Matte, J.J., Farmer, C., Girard, C.L. and Martineau, G.P. (1993) High-fiber diets for sows: effects on stereotypies and adjunctive drinking. *Applied Animal Behaviour Science* 37, 297–309.

Rohde Parfet, K.A. and Gonyou, H.W. (1991) Attraction of newborn piglets to auditory, visual, olfactory and tactile stimuli. *Journal of Animal Science* 69, 125–133.

Rowlinson, P. and Bryant, M.J. (1982) Lactational oestrus in the sow. 2. The influence of group-housing, boar presence and feeding level upon the occurrence of oestrus in lactating sows. *Animal Production* 34, 283–290.

Rowlinson, P., Boughton, H.G. and Bryant, M.J. (1975) Mating of sows during lactation: observations from a commercial unit. *Animal Production* 21, 233–241.

Rundgren, M. and Lofquist, I. (1989) Effects on performance and behaviour of mixing 20-kg pigs fed individually. *Animal Production* 49, 311–315.

Rushen, J. (1987) A difference in weight reduces fighting when unacquainted newly weaned pigs first meet. *Canadian Journal of Animal Science* 67, 951–960.

Schnebel, E.M. and Griswold, J.G. (1983) Agonistic interactions during competition for different resources in captive European wild pigs (*Sus scrofa*). *Applied Animal Ethology* 10, 291–300.

Sherritt, G.W., Graves, H.B., Gobble, J.L. and Hazlett, V.E. (1974) Effects of mixing pigs during the growing–finishing period. *Journal of Animal Science* 39, 834–837.

Signoret, J.P. (1970) Sexual behaviour patterns in female domestic pigs (*Sus scrofa* L.) reared in isolation from males. *Animal Behaviour* 18, 165–168.

Singer, F.J., Otto, D.K., Tipton, A.R. and Hable, C.P. (1981) Home ranges, movements, and habitat use of European wild boar in Tennessee. *Journal of Wildlife Management* 45, 343–353.

Soede, N.M. and Schouten, W.G.P. (1991) Effect of social conditions during rearing on mating behaviour of gilts. *Applied Animal Behaviour Science* 30, 373–379.

Stangel, G. and Jensen, P. (1991) Behaviour of semi-naturally kept sows and piglets (except suckling) during 10 days postpartum. *Applied Animal Behaviour Science* 31, 211–227.

Stolba, A. and Wood-Gush, D.G.M. (1983) The identification of behavioural key features and their incorporation into a housing design for pigs. *Annales de Recherches Vétérinaires* 15, 287–298.

Stolba, A., and Wood-Gush, D.G.M. (1989) The behaviour of pigs in a semi-natural environment. *Animal Production* 48, 419–425.

Stone, C.C. (1974) Auditory stimuli as a means to increase weaning weight of swine. MSc thesis, The Southern Illinois University, Carbondale, Illinois.

Stookey, J.M. and Gonyou, H.W. (1998) Recognition in swine: recognition through familiarity or genetic relatedness? *Applied Animal Behaviour Science* 55, 291–305.

Symoens, J. and van den Brande, M. (1969) Prevention and cure of aggressiveness in pigs using the sedative azaperone. *The Veterinary Record* 85, 64–67.

Tindsley, W.E.C. and Lean I.J. (1984) Effects of weight range at allocation on production and behaviour in fattening pig groups. *Applied Animal Behaviour Science* 12, 79–92.

Tonn, S.R., Davis, D.L. and Craig, J.V. (1985) Mating behavior, boar-to-boar behavior during rearing and soundness of boars penned individually or in groups from 6 to 27 weeks of age. *Journal of Animal Science* 61, 287–296.

van Putten, G. (1990) Schweinehaltung-modern und tiergerecht. *Deutsche Tierarztliche Wochenschrist* 97, 137–192.

van Putten, G. and van de Burgwal, J.A. (1990) Vulva biting in group-housed sows: preliminary report. *Applied Animal Behaviour Science* 26, 181–186.

Walker, N. (1991) The effects on performance and behaviour of number of growing pigs per mono-place feeder. *Animal Feed Science and Technology* 35, 3–13.

Wattanakul, W., Stewart, A.H., Edwards, S.A. and English, P.R. (1996) The effect of cross-suckling and presence of additional piglets on suckling behaviour and performance. In: *Proceedings of the British Society of Animal Science*. British Society of Animal Science, Scarborough, 202 pp.

Wattanakul, W., Stewart, A.H., Edwards, S.A. and English, P.R. (1997) Effects of grouping piglets and changing sow location on suckling behaviour and performance. *Applied Animal Behaviour Science* 55, 21–35.

Weary, D.M. and Fraser, D. (1995) Calling by domestic piglets: reliable signals of need? *Animal Behaviour* 50, 1047–1055.

Whatson, T.S. and Bertram, J.M. (1983) Some observations on mother–infant interactions in the pig (*Sus scrofa*). *Applied Animal Ethology* 9, 253–261.

Whittemore, C.T. and Fraser, D. (1974) The nursing and suckling behaviour of pigs. II Vocalization of the sow in relation to suckling behaviour and milk ejection. *British Veterinary Journal* 130, 346–356.

Wood, G.W. and Brenneman, R.E. (1980) Feral hog movements and habitat use in coastal South Carolina. *Journal of Wildlife Management* 44, 420–427.

The Social Behaviour of Domestic Birds

Joy Mench[1] and Linda J. Keeling[2]

[1]Department of Animal Science, University of California, Davis, CA 95616, USA; [2]Department of Animal Environment and Health, Swedish University of Agricultural Sciences, PO Box 234, Skara, Sweden

(Editors' comments: A variety of bird species have been domesticated, and these represent a wide range of social behaviour. Natural group sizes range from solitary to large aggregations. Mating systems include both polygamous and monogamous species. Although most domestic birds are precocial, the pigeon is altricial and requires substantial parental care.

Among the most commonly farmed birds, the number of animals kept on individual farms is often astounding. Flocks of tens or even hundreds of thousands exist on commercial farms. Group sizes within these operations vary from three to six hens in a cage, to groups of over 10,000 birds. Such group sizes require the birds to adjust their social organization compared with the smaller groups under natural conditions. The authors address the possibilities available for birds in such large flocks. Birds can try to maintain a social hierarchy based on individual recognition, maintain subgroups with such an organization, or resort to a system in which physical characteristics are used to estimate social status without ever establishing a definitive order.)

7.1 Basic Social Structure

Species of birds from many orders have been domesticated and selected for egg, meat or feather production (Crawford, 1990), or for companionship, fighting ability or ornamental purposes (Mason, 1984). The most common domesticated birds are chickens and turkeys, but other galliforms (guinea-fowl, quail, grouse, pheasant and partridge), waterfowl (ducks and geese) and pigeons (or squab) are also used for

©CAB *International* 2001. *Social Behaviour in Farm Animals*
(eds L.J. Keeling and H.W. Gonyou)

food production. The most recently domesticated birds are probably the ratites (ostrich, emu and rhea), now increasingly farmed mainly for their feathers and hide.

The ancestors of the species that have been domesticated display a variety of different forms of social organization (see Section 1.6). Some, like jungle fowl and turkeys, are polygamous, while others (like geese, bobwhite quail and emu) are monogamous and pair-bond for at least a single breeding season. Some, like ostriches, may be found in large aggregations for at least part of the year, while others, like pheasants, are essentially solitary. Incubation, guarding and care of the offspring may be done primarily by the male (emus), by the male and female together (ostriches, bobwhite quail, geese) or by the female alone (fowl and mallard ducks). While the offspring of domesticated species are typically precocial and can develop without parental care in captivity, the young of pigeons are altricial and require intensive parental care until fledging.

Although the social behaviour of all the wild ancestors of domesticated species has been studied to at least some extent, there is surprisingly little information available about the social behaviour of many of their domesticated relatives under commercial conditions. We will therefore concentrate on the better studied species: chickens, turkeys and quail. Information about the social behaviour of the ancestors of some other domesticated species can be found in other reviews (e.g. ostriches: Bertram, 1992; Deeming and Bubier, 1999; waterfowl: McKinney, 1975; Reiter, 1997).

7.1.1 Composition and structure of social groups

Chickens were domesticated in Thailand and the surrounding region about 8000 years ago from the red jungle fowl, *Gallus gallus gallus* (West and Zhou, 1989; Fumihito *et al.*, 1996). Red jungle fowl are wary and difficult to study in their native habitat, but long-term observations have been carried out on free-ranging flocks living at the San Diego zoo (Collias and Collias, 1996). The flocks were comprised of between four and 30 adults, both males and females, although small all-male flocks are also sometimes observed in the wild (Johnson, 1963). Jungle fowl are a harem polygynous species, and within each flock at San Diego there was a dominant male who defended the flock's territorial boundaries, and within whose territory the flock members always roosted. Dominant males were generally tolerant of young subordinate males in the flock, but drove older subordinate males to the periphery of the territory. The females in each flock formed their own dominance hierarchy. Social organization in feral fowl appears to be strikingly similar to that described for the San Diego jungle fowl (McBride *et al.*, 1969; Wood-Gush and Duncan, 1976).

Turkeys were domesticated in Mexico from the smallest of the four native American species of turkeys, *Meleagris gallopavo gallopavo*. This species was introduced by Europeans to North America, where the birds cross-bred with indigenous wild turkey subspecies to create the bronze turkey that was the foundation stock for the modern commercial turkey (Schorger, 1966). Social organization in wild turkeys is similar to that of jungle fowl, although the mating system of turkeys varies by subspecies and habitat (Schorger, 1966; Watts and Stokes, 1971; Latham, 1976). During the winter, turkeys may live in either mixed-sex family groups or all-male and all-female flocks. Male flocks are comprised of groups of siblings that remain together throughout their lifetime, while female flocks are comprised of females integrated from different broods. Males and females each form a dominance hierarchy. As the breeding season approaches, male sibling groups compete with one another for dominance, and then court hens at a lekking ground where females have congregated (see Section 7.1.6). The dominant male in the dominant sibling group secures most matings during the height of the breeding season, but other males may sometimes mate later. In some habitats, turkeys show the harem polygynous mating system characteristic of jungle fowl, with a (usually older) territorial male who defends a harem of about four to six females (Schorger, 1966).

Two species of quail are used for food production: the Japanese quail (*Coturnix japonica*), used for both meat and egg production; and the larger common bobwhite (*Colinus virginianus*) from North America, which is used mainly for meat or is released for hunting. The process of domestication of quail is less well documented than that of chickens or turkeys, but quail have been maintained in captivity in Japan since at least the 12th century (Kovach, 1975; Crawford, 1990). Wild *Coturnix* form large flocks during the non-breeding season (Crawford, 1990). During the breeding season they establish or migrate to breeding grounds (Kovach, 1975; Wakasugi, 1984), where the males establish territories. Females may mate with the resident male in a territory, or males may compete for females at a crowing ground. Both polygamous mating and single-season pair bonding have been observed (Mills *et al.*, 1997).

Bobwhite quail (Johnsgard, 1973) occupy coveys averaging 12–15 birds of mixed ages and sexes during the winter. The composition of these coveys changes in the spring when males and females begin to pair for breeding. Unmated males may set up 'whistling territories' close to the nesting areas of mated pairs, from which they expel younger males, or unmated birds may continue to occupy coveys during the breeding season. Coveys re-form again in the autumn, as males from adjacent whistling territories, other unmated individuals and pairs that have not been successful in raising a brood join the brood that has been raised at a particular nesting site.

7.1.2   Use of space

Many factors, including the distribution of food and water and the distribution of conspecifics and predators, influence how wild and feral birds distribute themselves. The availability of cover (McBride *et al.*, 1969; Johnsgard, 1973; Duncan *et al.*, 1978; Wood-Gush *et al.*, 1978) and roosting sites (Schorger, 1966; Collias and Collias, 1996) is also important.

The amount of space used by an individual or a flock is thus strongly dependent on resource availability, which can vary both daily and seasonally. The daily range used by turkeys depends primarily on food availability, but turkeys can cover many kilometres (8–16) while foraging each day (Schorger, 1966), and their ranges may encompass 32 km in the winter when food is scarce (Latham, 1976). Male turkeys may also range over as many as 10,000 ha during the breeding season (Schorger, 1966). Although bobwhite quail are comparatively sedentary and do not move far from their roosting area when food is plentiful (Johnsgard, 1973), the winter ranges of coveys can occupy as much as 80 km, and bobwhite may also move considerable distances in the autumn. However, mated pairs rarely move more than 1.5 km from the winter range to establish their nesting range and broods generally move less than 370 m from the nest.

In contrast, adult jungle fowl at the San Diego zoo showed extreme locality fixation, with individuals rarely moving more than 50 m from their home roost (Collias and Collias, 1996). However, San Diego has a year-round temperate climate, and food was probably always readily available to the birds in this setting. Johnson (1963) reports, although anecdotally, that wild jungle fowl flocks in Thailand may move into the rainforest, which is 8–32 km away from their usual location in the bamboo forest, during the rainy season. Longer-distance migrations are also seen seasonally in some species. Some Japanese quail populations in Asia migrate as far as 1000 km to breeding grounds in the north (Wakasugi, 1984).

As mentioned above, the males of several domesticated species of birds show territorial behaviour, at least during the breeding season. The size of these territories varies widely from one male to another. Jungle fowl territories at the San Diego zoo averaged about 50–75 m across, but actually varied a great deal both in size and in the numbers of flock members that they contained (Collias and Collias, 1996). Jungle fowl males at the San Diego zoo occupied their territory continuously until they died or were deposed (Collias and Collias, 1996). Patterns of movement in female birds have been less well characterized. Feral fowl hens appear to occupy overlapping home ranges within male territories (McBride *et al.*, 1969).

7.1.3 Communication

Most bird species have excellent colour vision and acute hearing (Waldvogel, 1990), and communication (see Section 2.3) within and between flocks thus takes place primarily via signals provided by postures, displays and vocalizations. Postures and displays are used to signal threat and submission (Kruijt, 1964; Hale *et al.*, 1969; Wood-Gush, 1971) for example, and particularly elaborate displays are given by all of the ancestors of domestic species during courtship (see Section 7.1.6 below on male–female interactions).

The vocal repertoire of most of the species that have undergone domestication has been studied to at least some extent. Although the numbers and types of vocalizations given by these species are limited as compared with most avian species that have a learned vocal repertoire (like songbirds or parrots), it is still impressive. At least 15 different calls have been identified in Japanese quail (Guyomarc'h, 1967; Potash, 1970) and at least 31 in domestic fowl and jungle fowl (Wood-Gush, 1971; Collias, 1987). Turkeys and bobwhites also appear to have a relatively large vocal repertoire (Schorger, 1966; Hale *et al.*, 1969; Johnsgard, 1973).

Perhaps the most striking vocalizations are the ones that males use in territorial advertisement: the crow call of Japanese quail and fowl and the whistle of bobwhites. These calls can carry great distances and are a very effective means of territorial defence, minimizing the need for direct confrontation between males on neighbouring territories (Collias and Collias, 1996). The crow call of jungle fowl and domestic fowl males is also individually acoustically distinctive (Siegel *et al.*, 1965; Miller, 1978), unlike any of the other calls so far characterized in fowl. Crow frequency characteristics are correlated with comb length (Furlow *et al.*, 1998) and males use crows to assess the dominance status of other males (Leonard and Horn, 1995).

We will not attempt to describe each of the vocalizations given by the different species, but they fall into the general (although sometimes overlapping) categories of warning and predator alarm calls; reinstatement and/or contact calls; territorial calls; laying and nesting calls; mating calls; threat calls; submissive calls; distress, alarm or fear calls; contentment calls; and food calls. Since the stimuli that elicit many of these calls have not been investigated thoroughly, and since there is no consistent nomenclature for call types, the communication functions of many of these calls have not been defined with certainty.

In addition to vocalizations and displays, morphological features associated with the head and neck are important in some species for both communication and social recognition. There has been considerable work in this area using domestic fowl (Guhl and Ortman, 1953; Jones and Mench, 1991; Bradshaw and Dawkins, 1993). In fowl, comb size and hue in both males and females are influenced by sex hormone levels and are indicators of social status (Guhl and Ortman, 1953). In quail, the head and neck area is important for male–female recognition, since that area contains a large number of sexually dimorphic feathers (Domjan and Nash, 1988). The necks of turkeys are featherless, so colour becomes important. In turkeys, neck and head coloration varies from white to red to blue depending on the animal's state (Schorger, 1966; Hale et al., 1969). Male turkeys have a pendant snood above the beak that is normally flaccid or retracted, as well as an area of spongy tissue on the breast. Both of these become enlarged during aggression and courtship (Hale et al., 1969).

7.1.4 Cohesion and dispersion

As with spacing behaviour, many factors influence cohesion and dispersion (see Section 1.4) in flocks of birds. One factor maintaining group cohesion in some species is the presence of attractive roosting sites (McBride et al., 1969). Large jungle fowl roosts can contain as many as 30 adults, and birds in a flock rarely move far from their roosting areas (Collias and Collias, 1996). Roosting serves a thermoregulatory function in bobwhite quail, and Johnsgard (1973) suggests that bobwhites disperse when covey size becomes so large that it impairs the thermoregulatory efficiency of their roosting pattern.

Relationships between individuals also contribute to flock cohesion. Fowl females maintain proximity to dominant males (Graves et al., 1985; Collias and Collias, 1996). Females of some species also tend to maintain relatively close spacing relationships with one another during the non-breeding season, and it has been suggested that female–female relationships are a major contributor to flock cohesion in jungle fowl (Sullivan, 1991). In turkeys, on the other hand, cohesion is promoted by the maintenance of relationships and proximity among sibling males (Watts and Stokes, 1971).

Although movements between groups can occur at any time, dispersal often occurs at the beginning of the breeding season. In bobwhite quail the males (probably yearlings) are expelled from the territories (Johnsgard, 1973). Whole groups can even sometimes break apart. In bobwhite quail new coveys are formed after the breeding season (Johnsgard, 1973).

7.1.5   Inter-group interactions

Inter-group interactions are also influenced by resource availability. For example, during winter, when food is scarce, two or three flocks of turkeys may join into a loosely associated group in an area where food is plentiful (Latham, 1976). Since jungle fowl usually stay within their territories year-round inter-group interactions are less common than for the other species, although individuals may occasionally move to join a new flock (Collias and Collias, 1996).

Inter-group interactions probably occur most frequently at the beginning of the breeding season, as breeding aggregations or breeding pairs are formed by members of different groups, and as males from different groups compete with one another for territories or females. The lek mating system of turkeys (see below) is an example of one such competitive inter-group interaction during the breeding season.

7.1.6   Intra-group interactions

Male–male

Interactions among adult male birds are primarily competitive, particularly during the breeding season. When there is more than one male in an established flock the males form a dominance hierarchy (Hale *et al.*, 1969; Wood-Gush, 1971; Collias and Collias, 1996; see Section 1.5). Dominant males are typically relatively tolerant of subordinate males outside the breeding season, but the presence of a dominant male can lead to the suppression or modification of territorial and reproductive behaviours in subordinate males. For example, dominant jungle fowl and fowl roosters crow at a higher rate than subordinates (Salomon *et al.*, 1966; Leonard and Horn, 1995; Collias and Collias, 1996). Dominant males often attack subordinates that crow (Leonard and Horn, 1995), and crowing in subordinates is also suppressed in situations in which the dominant rooster can be seen or heard but in which he has no direct contact with the subordinate males (Mench and Ottinger, 1991).

In contrast, the lek mating system of turkeys is an example of a cooperative interaction among adult males (Watts and Stokes, 1971). Sibling males cooperate in displaying to females on a lekking ground where other sibling groups are also displaying. Although the dominant male in the sibling group secures most of the matings during the height of the breeding season, lifetime reproductive success (fitness) of all of the brothers in the sibling group is presumably increased by mutual display.

Female–female

Like males, the females of domesticated bird species typically form dominance hierarchies. Unlike males, however, dominance relationships between adult females are often reported to be extremely stable (Schjelderupp-Ebbe, 1922; Hale *et al.*, 1969), and may persist for years (Schjelderupp-Ebbe, 1922) even after individuals' competitive abilities with strangers have declined (Lee *et al.*, 1982). Overt aggression in established groups is rarer among females than among males (Wechsler and Schmid, 1998). Establishing and maintaining a social order requires that individuals recognize one another, or at least recognize certain signals of status (Pagel and Dawkins, 1997). In fowl hens, status is affected by individual physical characteristics like age, breed, comb size/colour and body weight; these latter characteristics can in turn be affected by state of moult and health, including parasite load (Guhl and Ortman, 1953; Cloutier *et al.*, 1996; Zuk *et al.*, 1998). These characteristics can have additive influences on status outcomes when dominance relationships are established (Cloutier *et al.*, 1996). One factor of particular importance in predicting dominance in a new dyad is each hen's recent experience of victory or defeat in an encounter with an unfamiliar hen (Martin *et al.*, 1997). Hens are also influenced by interactions occurring among other hens. Hens who observe a dominant flockmate being defeated by a stranger do not initiate attacks against that stranger, but they will initiate attacks against a stranger if they see their dominant flockmate defeat that stranger (Hogue *et al.*, 1996). Steroid hormones also appear to have important effects on female dominance status. Hens injected repeatedly with androgens demonstrate a persistent elevation of dominance rank (Allee *et al.*, 1955), while hens injected with oestrogen either maintain or lose status (Allee and Collias, 1940; Guhl, 1968).

Affiliative behaviour in females has been less well studied than has competitive behaviour, but it may be a significant factor contributing to flock cohesion (Sullivan, 1991). Fowl hens show preferences for particular hens in their flock, preferences that appear to be unrelated to the dominance status of the birds preferred (Mench, 1996). Many behaviours of hens are socially facilitated or synchronized (Mench *et al.*, 1986; Webster and Hurnik, 1994; Duncan *et al.*, 1998), and hens also learn by observing one another's behaviour (Johnson *et al.*, 1986; Nicol and Pope, 1999).

Male–female

The most conspicuous male–female interactions in domesticated birds occur during mating. Male courtship displays are generally elaborate, involving vocalizations and noises, postures, spreading of the feathers in such a way that the male appears larger and that also emphasizes male plumage characteristics, and sometimes colour changes or enlargement of structures like the snood of male turkeys (Hale *et al.*, 1969; Wood-Gush, 1971; Kovach, 1975). Females also engage in behaviours that initiate courtship and also during the courtship sequence, primarily changes in posture or proximity to particular males (e.g. Fischer, 1975; Graves *et al.*, 1985).

There has been considerable work on mate selection by females, particularly in jungle fowl and with respect to the relationship between mate selection and fitness characteristics. Plumage colour appears to have little effect on the jungle fowl female's choice of a mate, although comb size does (Zuk *et al.*, 1990b; Ligon and Zwartjes, 1995). Males infected with nematode parasites have smaller combs (Zuk *et al.*, 1990a), so this suggests that females are assessing fitness using comb size as a cue. However, jungle fowl females do not select mates based solely on comb size, but seem to use a suite of labile morphological characteristics, including comb colour, eye colour and spur length, when selecting mates (Zuk *et al.*, 1992, 1995). Similarly, turkey females prefer males with longer snoods and wider skullcaps, features that are correlated with a lower parasite load in wild males; snood length may also be an indicator of the male's energy reserves (Buchholz, 1995). In contrast, domesticated fowl hens appear to use aspects of the mating display (Leonard and Zanette, 1998) as well as morphological features (Graves *et al.*, 1985) in selecting mates.

In some species males and females may also interact during the incubation and rearing of the young. In fowl, the hen selects the nest site, incubates the eggs and rears the young. Bobwhite males and females, on the other hand, build their nest together, and both incubate the eggs. The male also remains with the female after the chicks hatch to help in defending the brood (Johnsgard, 1973).

In fowl, single-sex dominance hierarchies are formed early in life and aggressive interactions between adult males and females are rare (Rushen, 1982), except under unusual circumstances (see Section 7.3.3 below on dominance-related problems). However, female quail do direct aggression towards males (Gerken and Mills, 1993).

Parent–offspring (see Chapter 3)

Social interactions occur for the first time in birds even before hatching (Rogers, 1995), both among eggs that are in close contact with one another and between the developing embryos and the incubating parent. Embryonic calls influence the behaviour of the hen by stimulating her to turn the eggs or to return to the nest to resume incubation. Embryos respond to particular behaviours and vocalizations of the hen with calls that further influence her behaviour (Tuculescu and Griswold, 1983). Exposure to maternal calls during embryonic development may be important for the development of post-hatch species-specific maternal call recognition (Gottleib, 1976).

Although precocial birds (see Section 3.3) are self-sufficient immediately after hatching, parents serve an important protective function and also teach the chicks about edible and inedible foods during the first few weeks of life (Schorger, 1966; Nicol and Pope, 1996). Precocial chicks imprint on their parents during the first few days of life (Bateson, 1966; Lickliter et al., 1993; Rogers, 1995). Imprinted chicks maintain close proximity to the imprinting object, which in nature would normally be the parent. However, under laboratory conditions chicks will imprint upon a variety of different objects, although there is a predisposition for them to imprint upon hen-like objects (Johnson and Horn, 1988). Visual characteristics of the parents are particularly important in facilitating imprinting (see review in Rogers, 1995), and particularly cues associated with the head and neck region (Johnson and Horn, 1988), but olfactory (Vallortigara and Andrew, 1994) and auditory cues probably also play a role. Chicks can distinguish their own hen's maternal call from those of other hens (Kent, 1987). When the parents and the chicks begin to move further away from the nesting area, different broods may join together to form 'crèches'.

Siblings

Hatching synchrony in precocial birds is influenced by social interactions among siblings in the nest. Quail and domestic fowl embryos make low-frequency sounds that retard the development of more advanced embryos in the brood and 'clicking' vocalizations that accelerate the development of less advanced embryos (Vince, 1970) and facilitate hatching synchrony (Vince, 1964). Fowl chicks and ducklings also produce other vocalizations while in the egg and during hatching (Guyomarc'h, 1974; Gottleib, 1976). In chicks, some of these calls (Gottleib and Vandenbergh, 1968; Tuculescu and Griswold, 1983) are similar to those given by hatchlings when they are in close contact, suggesting that they may serve a communicative function during pre-hatch development.

In addition to following the parents upon whom they have become imprinted, chicks recognize and follow their broodmates (Salzen and Cornell, 1969; Vallortigara and Andrew, 1991). This recognition of siblings (sexual imprinting) influences the development of later sexual preferences, since individuals prefer sexual partners who are slightly different in appearance from their siblings, ensuring outbreeding (Bateson, 1983; Bolhuis, 1991; Mills et al., 1997).

The post-hatching development of other social behaviours, and primarily social dominance, has been studied in most detail in jungle fowl and domestic fowl chicks (Kruijt, 1964; Dawson and Siegel, 1967), which show remarkably similar patterns of development. Chickens give aggressive pecks when they are as young as 2 weeks of age, although submissive behaviours occur infrequently before 4 weeks. Separate male and female dominance hierarchies are formed between 6 and 10 weeks of age, with the peak period of hierarchy formation occurring approximately 1 week earlier in males than in females (Guhl, 1958; Rushen, 1982). Pecking and threatening are unidirectional from the outset, and chicks initiating agonistic encounters at earlier ages have higher initial status. Caponization (castration) delays peck order formation by approximately 4 weeks, probably because it reduces aggression (Guhl, 1958). Early maturation and endogenous androgen levels therefore appear to be important factors influencing the development of dominance in chicks. Levels of aggression and the peak period for hierarchy formation during development are also affected by genetics, since broiler chickens show earlier and lower levels of aggression than laying stocks (Mench, 1988).

Turkeys develop more slowly than chickens, and the onset of aggression and dominance hierarchy formation is correspondingly later (Hale *et al.*, 1969). Aggression begins to be apparent in turkeys at about 3 months of age, and increases to a peak at 5 months of age when hierarchies are finally well established. Both males and females form hierarchies, although males fight more vigorously than females.

7.2 Social Behaviour Under Commercial Conditions

Large-scale commercial poultry production practices for turkeys, quail and chickens are relatively uniform throughout the world (North and Bell, 1990; Appleby *et al.*, 1992). However, many people still keep poultry in 'backyard flocks', and the husbandry conditions and social groupings for these flocks will obviously vary considerably.

7.2.1 Social groupings

Laying flocks
By far the greatest number of laying flocks are of commercial laying hens (*Gallus gallus domesticus*) and most of these are kept in small groups of three to ten birds in cages with space allowances of 350–600 cm^2 per bird (Fig. 7.1). There are several different designs of cages, but they are usually made of wire and arranged in rows and tiers within a building. Most management is automated. The number of birds on a typical farm varies, but tends to be large, several thousand to several million birds. The birds are usually reared at one location and then transported at 16–18 weeks of age as pullets to the laying farm. The hens start laying eggs at about 20 weeks of age and are kept until

Fig. 7.1. Caged laying hens.

72 weeks, or to the point where the production is reduced to an unprofitable level. In some countries the birds are moulted and kept for a second or even third production cycle. The only other poultry species kept primarily for egg production is the Japanese quail. Quail are also kept in cages but the group size is usually larger, between 60 and 80 birds per cage with a stocking density of 120–160 birds m^{-2} (Gerken and Mills, 1993).

Alternatively, laying hens are kept in large groups in litter floor systems. These are often traditional systems, with a low stocking density (< 7 birds m^{-2}), and in some countries are combined with an outdoor run, so-called free-range egg production. In recent years there has been increased interest, especially in Europe, in alternative housing systems for laying hens (Kuit *et al.*, 1989; Sherwin, 1994; Tauson, 1998; Huber-Eicher and Audigé, 1999) and quail (Wechsler and Schmid, 1998). These alternative systems are usually designed to combine high stocking densities with the availability of resources onsidered important for bird welfare, such as litter, nest boxes and perches.

Meat-type flocks

Many billions of chickens are reared for meat each year in the world. In the 1950s, the poultry industry began an intense selection programme to develop a fast-growing strain of chicken that would provide higher yields of meat in a shorter time than was possible when spent (end-of-lay) hens were used as the primary source of chicken meat. The result was the modern broiler chicken, which grows so quickly that it can be marketed when only 6–7 weeks of age, while still juvenile.

Broiler chickens are raised in large groups of 10,000–70,000 on litter in either semi-enclosed or environmentally closed houses (Fig. 7.2). The typical stocking density varies from 35 to 50 kg liveweight m^{-2}. Since males grow more quickly than females, broilers are usually raised in single-sex groups to ensure that body weights are more uniform when the birds are processed. Females may be processed at an even younger age (3 weeks) and sold as Cornish game hens, while males may sometimes be kept for as long as 10–12 weeks and then sold as roasters. Although attempts have been made to develop cage-rearing systems for broilers there are problems with carcass quality in these systems, so they have not been widely adopted.

Meat-type quail and turkeys are housed under conditions very similar to those of broiler chickens, in groups of several thousand birds on litter floors, although cage rearing systems, particularly for late-stage growing, are also under development for quail (Gerken and Mills, 1993; Shanaway, 1994). Approximately 70–100 quail are housed per square metre on the floor in mixed-sex groups. Bobwhite are given more space, particularly if they require flight practice because they are to be released for hunting (Skewes and Wilson, 1990). Turkeys used to be kept on range for at least part of the year but are now typically housed intensively.

Breeder flocks

Natural mating is still the rule for the production of chicken and quail young, although all turkey offspring are now produced by artificial insemination. Breeder chickens are housed in bird flocks of several

Fig. 7.2. Broiler chickens. Photograph courtesy of Carolyn Stull.

thousands in semi-enclosed or enclosed housing on litter or wire. The male–female ratio in breeder flocks is about one male to 8–15 females, depending on the strain of chicken, kept at a space allowance of about 0.2–0.3 m² per bird. There is more opportunity for the birds to use the three-dimensional space in breeding houses than in broiler houses, since breeding flocks are provided with nest boxes and often also a slatted elevated area that can be used for perching (Fig. 7.3). Since strains of broiler chickens have been selected for rapid weight gain, birds to be kept for breeding are kept on a restricted diet so they are not too heavy as adults. There may even be separate sex feeding in breeding flocks to prevent males from becoming overweight and thus showing reduced mating activity. Breeding quail can be housed either in floor pens or in small breeding groups in cages (Skewes and Wilson, 1990; Shanaway, 1994). For Japanese quail, one male is housed for each two to three females in a cage, but a few more females can be housed per male in floor pens. Caged breeder bobwhites are generally kept either in pairs or trios (one male and two females).

The poultry industry moved to artificial insemination for turkeys some time ago because of problems with fertility, primarily because males had difficulty mating due to their size and conformation. Males (toms) and females are housed on litter in separate buildings, the males in small (10–15 bird) stud flocks and the females in bird flocks of several thousands. Space allowances are about 2.5–3.5 m² per bird (*Breeder Management Guide*, undated). Attempts to house breeder turkeys and chickens in cages have so far been unsuccessful because the birds develop foot and leg problems that interfere with mating.

Fig. 7.3. Broiler breeders in a typical housing facility. The slatted area with nest boxes can be seen at the side of the house. Photograph courtesy of Joseph Mauldin.

7.2.2   Social effects on production

Almost all domesticated birds used for food production live in groups, so there are many direct and indirect social effects on production. For example, in laying hens there is a complex interrelationship among social rank, aggression, feeding behaviour and egg production. Hens selected for early sexual maturity and high egg production are more aggressive than, and dominant to, unselected hens, although these differences tend to disappear somewhat as the birds mature (Lowry and Abplanalp, 1972; Craig *et al.*, 1975). Higher-ranking hens may have better egg production than the lowest ranked bird in a cage (Cunningham and van Tienhoven, 1983), possibly because the higher ranked birds have greater access to the feed. However, this effect may be related to whether or not there is sufficient trough space (Cunningham and van Tienhoven, 1983; Adams and Craig, 1984). Most aggression is seen at the feed trough, implying some competition. Nevertheless, the level of aggression in cages is generally low. The most common reason proposed for this is that the small group size in cages allows the hen to establish a stable dominance hierarchy.

In broiler breeders, social effects on feeding behaviour can have negative consequences for both egg production and fertility. Broiler breeders are feed restricted to control body weight. If dominant birds are successful at out-competing subordinates and thus consume more than their daily allocation of feed, they can become obese and lose reproductive condition. For this reason it is especially important that broiler breeder flocks have sufficient feed trough space so that all birds can feed at the same time.

Social effects can enhance as well as decrease feeding. Laying hens choose to feed close to each other when given a choice of feeding locations (Meunier-Salaün and Faure, 1984), demonstrating the importance of social attraction. Hens in the same cage and in neighbouring cages synchronize their feeding (Hughes, 1971; Mench *et al.*, 1986; Webster and Hurnik, 1994) and birds show socially facilitated feeding, in particular pecking more at the feed when with a companion than when alone (Keeling and Hurnik, 1996).

Besides feeding behaviour, social effects are also important with respect to the choice of nest sites. In non-cage systems, eggs laid outside the nest boxes are a problem because they are more time consuming to collect and are often downgraded. Since hens choose to lay their egg near other eggs and other birds (Appleby *et al.*, 1984: Sherwin and Nicol, 1993) eggs that are laid on the floor and not collected may attract other birds to lay on the floor. In addition, hens may lay their eggs on the floor due to social effects. Birds may be less likely to approach a nesting site during the initial stage of nest searching if there is a dominant or unfamiliar individual nearby

(Friere *et al.*, 1997), or they may stay for a shorter time at the nest site after laying (Lundberg and Keeling, 1999). Subordinate hens are also displaced more often from the nests (Friere *et al.*, 1998).

Social factors, and in particular social status (see Section 1.5), can also affect patterns of reproduction in breeding flocks (Ottinger and Mench, 1989). Subordinate roosters are delayed in showing sexual behaviour even when they are not in direct competition with other males (Grosse and Craig, 1960). In small to moderate-sized flocks of fowl, high-ranking roosters mate more frequently (Guhl *et al.*, 1945; Guhl and Warren, 1946; Lill and Wood-Gush, 1965) and sire more off-spring (Jones and Mench, 1991) than low-ranking males. Several factors appear to contribute to the lower mating frequency of subordinate males. Dominant cockerels may interfere directly with matings by subordinate males (Guhl and Warren, 1946; Kratzer and Craig, 1980; Pamment *et al.*, 1983). In addition, hens play an active role in selecting mates (Lill and Wood-Gush, 1965); they may space more closely to dominant males (Graves *et al.*, 1985) and actively avoid matings by undesirable males (Rappaport and Soller, 1966). Paradoxically, high-status females are less likely to crouch for males than are low-status females (Guhl *et al.*, 1945). The result in a large breeder unit could be that only a limited number of males mate and that a proportion of females are rarely mated, lowering overall flock fertility. This would represent a particular problem in broiler breeder flocks if the dominant males secure most matings but are also the ones that consume most feed and hence are in the poorest reproductive condition. Social effects on mating behaviour are influenced by factors such as flock size and housing density (Craig *et al.*, 1977; Kratzer and Craig, 1980; Pamment *et al.*, 1983), so some of these effects may be minimized in the large, high-density flocks typical of the poultry industry. But there is no definitive information about this, since it is extremely difficult to study individual patterns of behaviour in such large flocks.

Group composition can also affect production. It is common to house breeding males and females separately during the early part of the growing cycle. In chickens, exposure to females at an appropriate age is important for the development of normal male sexual behaviour (Siegel and Siegel, 1964). Males raised in single-sex groups show less sexual behaviour at 20 weeks of age than males raised in mixed-sex groups, although these differences disappear as the males gain sexual experience (Leonard *et al.*, 1993). Quail males have increased testosterone concentrations when they are housed with females (Delville *et al.*, 1984). Production in females is also influenced by the presence of males. The onset of egg production is earlier in laying hens, and egg production is higher in female turkeys, if they are able to at least see and hear males (Jones and Leighton, 1987; Widowski *et al.*, 1998).

7.2.3   Effects of group size and space allowances on social behaviour

The effects of group size (see Section 2.5) and space allowances on social behaviour have been studied in most detail in laying hens. In many early experiments there was a confounding of group size and space allowance because different numbers of birds were placed in the same size enclosure. Nevertheless from these experiments, in combination with experiments investigating group size and area per bird separately, certain general trends are apparent. With increasing group size there is an increase in mortality, feather and skin damage, and a decrease in egg production (Hughes and Duncan, 1972; Allen and Perry, 1975; Bilčík and Keeling, 1999). Increasing group size also results in larger adrenal and thyroid gland weights (Flickinger, 1961). Increased stocking densities result in increased mortality, decreased egg production (Adams and Craig, 1984) and increased feather damage and cannibalism (Hansen, 1976; Adams *et al.*, 1978; Simonsen *et al.*, 1980).

Although there is generally clear evidence for detrimental effects of larger group sizes and higher stocking densities (Adams and Craig, 1984), there is an apparent curvilinear effect of increasing stocking density on aggression. Agonistic behaviour is highest when birds are kept in cages at an intermediate density of 824 cm^2 per bird as compared with a higher density of 412 cm^2 per bird or lower densities of 1442 and 2884 cm^2 per bird (Al-Rawi and Craig, 1975). This is probably attributable to inhibitory effects on aggression at high densities in cages. For example, agonistic acts between domestic hens are inhibited by the presence of an overriding dominant third party (Craig *et al.*, 1969; Ylander and Craig, 1980). In addition, cages are too low to allow the birds to raise their heads in a threat, and aggression is provoked by an approaching bird rather than by a bird who is in continuous close proximity (Hughes and Wood-Gush, 1977). In observations conducted in a perchery system with stocking densities of 9.9, 13.5, 16.0 and 19.0 birds m^{-2} there were no effects of density on agonistic interactions (Carmichael *et al.*, 1999). This difference between cages and perchery systems suggests that the interaction between group size and density may be important. Such effects were studied in hens given a choice to approach groups of different sizes at different densities in a T-maze. The results showed that, while small group sizes were preferred, there also needed to be sufficient space (Lindberg and Nicol, 1996a).

Basic research on the spatial requirements of laying hens has provided little evidence that birds possess a personal space around them (Lill, 1969; Faure, 1985), and it is more likely that spacing is a balance between attraction and repulsion between birds resulting in different appropriate inter-individual distances under different circumstances (Keeling and Duncan, 1991). However, using this to determine optimal space allocations for commercial groups of laying hens is far from simple (Keeling, 1995) and it is suggested that space should not only be regarded quantitatively, but consideration given to what is in the space (Hughes, 1983). Inter-individual distances have also been recorded in selected lines of quail. While birds selected for many generations for high social reinstatement had the predicted small inter-individual distances initially, this did not hold at older ages (Mills and Faure, 1991), hinting that spatial requirement may not be constant over time. Certainly familiarity between hens affects how they space themselves and this changes over time (Lindberg and Nicol, 1996b).

Basic research on the effect of group size on spatial organization has proved equally complicated. As described previously, jungle fowl usually live in small heterogeneous groups, but domestic fowl kept for commercial egg production are kept in homogeneous groups composed entirely of females. The group size is small if the birds are kept in cages, or may contain several thousand individuals in floor housing systems. Even in breeding flocks, where males and females are kept together, the group size is usually several thousand. At no time are young, juvenile and adult birds kept together in the same flock and, in fact, this is strongly discouraged on disease grounds.

At close distances (Dawkins, 1995), laying hens are able to distinguish between familiar flockmates and unfamiliar birds and usually show agonistic behaviour with unfamiliar birds (Craig et al., 1969). Hens can also discriminate between flockmates that are higher in the hierarchy and flockmates that are lower in the hierarchy (King, 1965; D'Eath and Dawkins, 1996). Dim or coloured lighting can affect this ability to discriminate, which has implications for light management strategies in houses. For example, hens in a choice test discriminated between familiar and unfamiliar hens only in bright white light and seemed to have most difficulty under red lighting conditions (D'Eath and Stone, 1999). There is an often quoted paper (Guhl, 1953) stating that birds can recognize and form stable relationships with at least 100 others, but this experiment is open to alternative interpretations and has not been confirmed. More information on individual recognition is presented in Chapter 14 on social cognition.

Although it is theoretically possible that dominance hierarchies can exist without individual recognition (Wood-Gush, 1971; Barnard and Burk, 1979) it is generally thought to be very unlikely that birds in large flocks could form a hierarchy. Thus, birds in large commercial flocks may be in a constant state of trying to establish a hierarchy but never achieving it. Alternatively, they might restrict their movements to stay with subgroups of birds they know and recognize or, finally, they may use a social mechanism other than hierarchy formation in large groups. There is some evidence to support each of the above arguments. As reported previously, mortality, production and behavioural problems are all worse in large groups of hens, implying that they might not have succeeded in establishing a stable social organization. There is some evidence that hens form subgroups (Oden *et al.*, 2000), although other researchers have not found any evidence for them (Hughes *et al.*, 1974). Whereas hens in small groups are aggressive towards each other when mixed, hens in large groups are not (Hughes *et al.*, 1997). The proposal that laying hens in very large groups may move away from a mechanism based on individual recognition and remembered dominance relationships towards estimating dominance relationships based on visual cues such as body size or comb size has some theoretical basis (Pagel and Dawkins, 1997).

Group size and stocking density seem to have fewer obvious effects on broiler chickens, perhaps because they are quite unaggressive and too young to have fully formed a social hierarchy (Mench, 1988; Preston and Murphy, 1988; Estevez *et al.*, 1997). Movement patterns in broilers decrease with age as stocking density increases, but this is probably due to their increasing problems with mobility due to leg problems rather than to social effects associated with high densities (Newberry and Hall, 1990; Weeks *et al.*, 2000).

7.3 Social Effects on Behaviour, Welfare and Production

7.3.1 Grouping animals

It is not as usual to regroup poultry as it is other farm animal species, although when large groups of birds are disturbed or moved, as when young fowl hens (pullets) are moved from the rearing farm to the laying house, it can happen that unfamiliar birds are mixed. Preference studies show that hens prefer to be near flockmates and to avoid unfamiliar birds (Hughes, 1977; Dawkins, 1982), and they also seem to avoid passing by a cage containing unfamiliar birds (Grigor *et al.*, 1995). There is also physiological evidence from small groups of birds that chickens find it stressful to encounter strangers (Candland *et al.*, 1969; Gross and Siegel, 1983; Anthony *et al.*, 1988). Nevertheless, mixing of pullets does not seem to result in severe or lasting problems. Bradshaw (1992) found that, although hens initially chose to aggregate with familiar rather than unfamiliar hens in a multichoice arena, within only several hours they were aggregating with unfamiliar birds, suggesting that they were quickly becoming familiar. Similarly, Mench and Mayeaux (1997) found that mixed pullets showed only small and transient elevations in corticosterone levels and aggression, and quickly spaced as close to previously unfamiliar birds as they did to familiar birds with which they had been reared. Birds in neighbouring cages may be regarded as familiar, since mixing birds from adjacent cages results in less aggression than mixing birds from more widely separated cages (Zayan et al., 1983).

Other examples of regrouping may occur when males and females that have been reared separately are mixed into adult breeder flocks, or when younger, lighter males are introduced into an established broiler breeder flock to replace the original breeding males as a method of improving fertility. The effects on social behaviour and flock structure of these types of regrouping have not been studied.

Effects similar to those of mixing could also be triggered by changing the appearance of birds such that they appear unfamiliar. As mentioned previously, features of the head are important for individual recognition and communication and serve as indicators of social status. Although commercial manipulations that alter the appearance of the heads of chickens and turkeys are becoming less common, dubbing (trimming of the combs and wattles) and desnooding are still sometimes practised. Aggression is higher in flocks of dubbed hens than in undubbed flocks (Siegel and Hurst, 1962).

7.3.2 Separation problems

In commercial poultry production, the young are almost never kept together with the mother bird either during the pre- or post-hatch period. However, all domestic birds are social species and most will show some form of social reinstatement behaviour if they are separated from conspecifics (Mills and Faure, 1991). There is some evidence for gender differences, with females showing stronger social reinstatement behaviour than males (Cailotto et al., 1989; Vallortigara, 1992). Removing one chick from a pair results in increased corticosterone levels in both sexes, but female chicks vocalize more than male chicks (Jones and Williams, 1992). There is evidence that chicks that are reared without a hen are more aggressive than chicks reared by a hen (Fält, 1978), suggesting that there may be separation problems that have not yet been fully recognized or investigated.

7.3.3 Dominance-related problems

In many cases, agonistic interactions in established flocks are subtle and not easily observed, although there can be pecking, chasing and even fighting. This is particularly noticeable between males, but can occur between females as well, and if the agonistic interaction is severe or prolonged wounds can result. Turkey and quail males are particularly aggressive towards one another (Gerken and Mills, 1993; Sherwin and Kelland, 1998), and the severe pecks at the head often result in death. It is for this reason that turkeys are usually kept under very low light intensities (Manser, 1996), although providing pecking substrates and intermittent lighting can also reduce injurious pecking (Sherwin *et al.*, 1999). Despite systematic experiments to reduce aggression among male quail using visual barriers, introducing males at different ages and other techniques, severe aggression persisted and it was concluded that it was not possible to keep more than one quail male per breeding group (Wechsler and Schmid, 1998).

In broiler breeder flocks, aggression between young males is exacerbated by feed restriction (Mench, 1988; Shea-Moore *et al.*, 1990). Older males also direct intense aggression towards, and even sometimes kill, females. It has been suggested that broiler breeder males show deficient courtship behaviour, which results in the hens crouching less and so males resort to aggressive forced copulation (Millman and Duncan, 2000).

In laying hens kept in large groups, the only noticeable dominance-related problem is that occasionally some birds seem to become so low ranking that they are pecked and continually harassed by other birds when they try to feed. These birds eventually go out of production and moult. Keeping a small number of cockerels in large commercial layer flocks can reduce agonistic behaviour between hens (Oden *et al.*, 2000).

7.3.4   Abnormal behaviour

The most commonly occurring behavioural problems among domestic fowl are feather pecking and cannibalism. Feather pecking is the pecking and pulling at the feathers of another bird, while cannibalism is the pecking and tearing of the skin and underlying tissues of another bird (Keeling, 1994). Most research has been carried out on feather pecking in commercial strains of laying hens (Vestergaard, 1994; Huber-Eicher and Wechsler, 1997; Blokhuis and Wiepkema, 1998) but it has also been studied in turkeys (Sherwin and Kelland, 1998; Sherwin *et al.*, 1999) and bantams (Savory and Mann, 1999). Having feathers pulled out has been shown to be painful (Gentle and Hunter, 1991), and the poorer plumage condition that results significantly reduces feed efficiency due to greater heat loss (Emmans and Charles, 1977). Feather pecking can range from allopreening and gentle feather pecking, behaviours that result in little if any feather damage and are thought to be normal, to severe feather pecking and pulling that is thought to be abnormal (Leonard *et al.*, 1993; Keeling, 1994; Savory, 1995) (Fig. 7.4).

There also seem to be different types of cannibalism. One type involves skin damage that occurs as a result of feather pecking, for example when a large flight feather is pulled out and the skin is damaged. In the other type the pecks are directed towards the bird's cloaca (Keeling, 1994; Savory, 1995; Yngvesson, 1997). Cloacal cannibalism (vent pecking), which is generally thought to be unrelated to feather pecking (Allen and Perry, 1975; Gunnarsson *et al.*, 1999), often results in the death of the cannibalized bird. Both feather pecking and cannibalism are welfare and economic problems.

Many factors influence the incidence of feather pecking and cannibalism, including light intensity, position in the building, housing system, strain, group size and stocking density (Hughes and Duncan, 1972). Food deficiencies are also recognized as a trigger (Wahlström *et al.*, 1998). Although it is often argued that feather pecking and cannibalism are multifactorial, two hypotheses have been proposed to explain the causation of feather pecking. Vestergaard and Lisborg (1993) suggest that the primary cause is an abnormal development of the perceptual mechanism responsible for the detection of dust for dust bathing. Blokhuis (1986) and Huber-Eicher and Weschler (1997), on the other hand, argue that these behaviours result from a redirection of the ground pecking normally associated with foraging and/or exploratory behaviour. While the motivational system may differ in these two hypotheses, the practical implications are the same, i.e. that birds should be

Fig. 7.4. Feather pecking is a major economic and welfare problem in commercial laying hens.

reared with litter. But, while some studies have shown that early access to sand (Johnsen *et al.*, 1998), wood shavings (Blokhuis and Van der Haar, 1989) and straw (Huber-Eicher and Wechsler, 1998) reduce feather pecking under experimental conditions, the beneficial effects of rearing with litter are not as clear under commercial conditions (Gunnarsson *et al.*, 1999). However, it might be that the type and way in which the litter is presented to the birds is important (Johnsen *et al.*, 1998). Surprisingly, rearing density has no effect on later feather-pecking behaviour (Hansen and Braastad, 1994), and neither does whether the chicks are reared with or without a hen (Roden and Wechsler, 1998).

Factors that result in higher levels of feather pecking also result in higher levels of cloacal cannibalism, but flocks do not necessarily experience both problems at the same time. There is very little systematic research on cannibalism.

The usual method for reducing feather damage and cannibalism is beak trimming, a practice that is banned in some countries. Since there is work showing that feather pecking is heritable (Craig and Muir, 1993; Keeling and Wilhelmsson, 1997; Kjaer and Sørensen, 1997), it is hoped that the behaviour will be reduced in the future by genetic selection. This may also be the case for cannibalism (Craig and Muir, 1993).

The only other social behaviour problem in poultry could be that of hysteria. This is when a flock of birds panic and in the resulting crush some birds are suffocated. There seems to be a greater problem with hysteria in some housing systems, and is one of the reasons why so-called Pennsylvania systems, which involve colony housing of birds on wire floors, are rarely used. In one study, the incidence of hysteria among hens in colony cages was decreased by providing an enriched environment (Hansen, 1976).

References

Adams, A.W. and Craig, J.V. (1984) Effect of crowding and cage shape on productivity and profitability of caged layers: a survey. *Poultry Science* 54, 238–242.

Adams, A.W., Craig, J.W. and Bhagwat, A.L. (1978) Effects of flock size, age at housing and mating experience on two strains of egg-type chickens in colony cages. *Poultry Science* 57, 48–53.

Allee, W.C. and Collias, N.E. (1940) The influence of estradiol on the social organization of flocks of hens. *Endocrinology* 27, 87–94.

Allee, W.C., Foreman, D. and Banks, E.M. (1955) Effects of an androgen on dominance and subordinance in six common breeds of *Gallus gallus*. *Physiological Zoology* 38, 89–115.

Allen, J. and Perry, G.C. (1975) Feather pecking and cannibalism in a caged layer flock. *British Poultry Science* 16, 441–451.

Al-Rawi, B. and Craig, J.V. (1975) Agonistic behaviour of caged chickens related to group size and area per bird. *Applied Animal Behaviour Science* 2, 69–80.

Anthony, N.B., Kantanbaf, M.N. and Siegel, P.B. (1988) Responses to social disruption in two lines of White Leghorn chickens. *Applied Animal Behaviour Science* 21, 243–250.

Appleby, M.C., McRae, H.E., Duncan, I.J.H. and Bisazza, A. (1984) Choice of social conditions by laying hens. *British Poultry Science* 25, 111–117.

Appleby, M.C., Hughes, B.O. and Elson, H.A. (eds) (1992) *Poultry Production Systems: Behaviour, Management and Welfare.* CAB International, Wallingford, UK.

Barnard, C.J. and Burk, T. (1979) Dominance hierarchies and the evolution of 'individual recognition'. *Journal of Theoretical Biology* 81, 65–73.

Bateson, P.P.G. (1966) The characteristics and context of imprinting. *Biological Review* 41, 177–220.

Bateson, P.P.G. (1983) Optimal outbreeding. In: Bateson, P. (ed.) *Mate Choice.* Cambridge University Press, Cambridge, pp. 257–277.

Bertram, B.C.R. (1992) *The Ostrich Communal Nesting System.* Princeton University Press, Princeton, New Jersey.

Bilčík, B. and Keeling, L.J. (1999) Changes in feather pecking in relation to feather pecking and aggressive behaviour in laying hens. *British Poultry Science* 40, 444–451.

Blokhuis, H.J. (1986) Feather pecking in poultry: its relation with ground pecking. *Applied Animal Behaviour Science* 16, 63–67.

Blokhuis, H.J. and Van der Haar, J.W. (1989) Effects of floor type during rearing and of beak trimming on ground pecking and feather pecking in laying hens. *Applied Animal Behaviour Science* 22, 359–369.

Blokhuis, H.J. and Wiepkema, P.R. (1998) Studies of feather pecking in poultry. *Veterinary Quarterly* 20, 6–9.

Bolhuis, J.J. (1991) Mechanisms of avian imprinting: a review. *Biological Reviews* 66, 303–345.

Bradshaw, R.H. (1992) Conspecific discrimination and social preference in the laying hen. *Applied Animal Behaviour Science* 33, 69–75.

Bradshaw, R.H. and Dawkins, M.S. (1993) Slides of conspecifics as representatives of real animals in laying hens (*Gallus domesticus*). *Behavioural Processes* 28, 165–172.

Breeder Management Guide (undated) Nicholas Turkey Breeding Farms, Sonoma, California.

Buchholz, R. (1995) Female choice, parasite load and male ornamentation in wild turkeys. *Animal Behaviour* 50, 929–943.

Cailotto, M., Vallortigara, G. and Zanflorin, M. (1989) Sex differences in the response to social stimuli in young chicks. *Ethology, Ecology and Evolution* 1, 323–328.

Candland, D.K , Taylor, D.B., Dresdale, L., Leiphart, J.M. and Solow, S.P. (1969) Heart rate, aggression and dominance in the domestic chicken. *Journal of Comparative and Physiological Psychology* 69, 70–76.

Carmichael, N.L., Walker, A.W. and Hughes, B.O. (1999) Laying hens in large flocks in a perchery system: influence of stocking density on location, use of resources and behaviour. *British Poultry Science* 40, 165–176.

Cloutier, S., Beaugrand, J.P. and Lague, P.C. (1996) The role of individual differences and patterns of resolution in the formation of dominance orders in domestic hen triads. *Behavioural Processes* 38, 227–239.

Collias, N.E. (1987) The vocal repertoire of the red junglefowl: a spectrographic classification and the code of communication. *Condor* 89, 510–524.

Collias, N.E. and Collias, E.C. (1996) Social organization of red junglefowl, *Gallus gallus*, population related to evolution theory. *Animal Behaviour* 51, 1337–1354.

Craig, J.V. and Muir, W.M. (1993) Selection for reduced beak-inflicted injuries among caged hens. *Poultry Science* 72, 411–420.

Craig, J.V., Biswas, D.K. and Guhl, A.M. (1969) Agonistic behaviour influenced by strangeness, crowding and heredity in female domestic fowl (*Gallus domesticus*). *Animal Behaviour* 17, 498–506.

Craig, J.V., Jan, M.L., Polley, C.R., Bhagwat, A.L. and Dayton, A.D. (1975) Changes in relative aggressiveness and social dominance associated with selection for early egg production in chickens. *Poultry Science* 54, 1647–1658.

Craig, J.V., Al-Rawi, B. and Kratzer, D.D. (1977) Social status and sex ratio effects on mating frequency of cockerels. *Poultry Science* 56, 767–772.

Crawford, R.D. (1990) Origin and history of poultry species. In: Crawford, R.D. (ed.) *Poultry Breeding and Genetics*. Elsevier, Amsterdam, pp. 1–41.

Cunningham, D.L. and van Tienhoven, A. (1983) Relationship between production factors and dominance in White Leghorn hens in a study on social rank and cage design. *Applied Animal Ethology* 11, 33–44.

Dawkins, M.S. (1982) Elusive concept of group size in domestic hens. *Applied Animal Ethology* 8, 365–375.

Dawkins, M.S. (1995) How do hens view other hens? The use of lateral and binocular vision in social recognition. *Behaviour* 132, 591–606.

Dawson, J.S. and Siegel, P.B. (1967) Behavior patterns of chickens to ten weeks of age. *Poultry Science* 46, 615–622.

D'Eath, R.B. and Dawkins, M.S. (1996) Laying hens do not discriminate between video images of conspecifics. *Animal Behaviour* 52, 903–912.

D'Eath, R.B. and Stone, R.J. (1999) Chickens use visual cues in social discrimination: an experiment with coloured lighting. *Applied Animal Behaviour Science* 62, 233–242.

Deeming, D.C. and Bubier, N.E. (1999) Behaviour in natural and captive environments. In: Deeming, D.C. (ed.) *The Ostrich. Biology, Production and Health*. CAB International, Wallingford, UK.

Delville, Y., Sulon, J., Hendrick, J.C. and Balthazart, J. (1984) Effect of the presence of females on the pituitary–testicular activity in male Japanese quail (*Coturnix coturnix japonica*). *General and Comparative Endocrinology* 55, 295–305.

Domjan, M. and Nash, S. (1988) Stimulus control of social behaviour in male Japanese quail, *Coturnix coturnix japonica*. *Animal Behaviour* 36, 1006–1015.

Duncan, I.J.H., Savory, C.J. and Wood-Gush, D.G.M. (1978) Observations on the reproductive behaviour of domestic fowl in the wild. *Applied Animal Ethology* 4, 29–42.

Duncan, I.J.H., Widowski, T.M., Malleau, A.E., Lindberg, A.C. and Petherick, J.C. (1998) External factors and causation of dustbathing in domestic hens. *Behavioural Processes* 43, 219–228.

Emmans, G.C. and Charles, D.R. (1977) Climatic environment and poultry feeding in practice. In: Haresign, W., Swan, H. and Lewis, D. (eds) *Nutrition and the Climatic Environment*. Butterworths, London, pp. 31–49.

Estevez, I., Newberry, R.C. and de Reyna, L.A. (1997) Broiler chickens: a tolerant social system? *Etologia* 5, 19–29.

Fält, B. (1978) Differences in aggressiveness between brooded and non-brooded domestic chicks. *Applied Animal Ethology* 4, 211–221.

Faure, J.M. (1985) Social space in small groups of laying hens. In: Zayan, R. (ed.) *Social Space for Domestic Animals*. CEC Seminar, Martinus Nijhoff, Brussels/Amsterdam, pp. 99–112.

Fischer, G.J. (1975) The behaviour of chickens. In: Hafez, E.S.E. (ed.) The *Behaviour of Domestic Animals*, 3rd edn. Baillière Tindall, London, pp. 454–489.

Flickinger, G.L. (1961) Effect of grouping on adrenals and gonads of chickens. *General and Comparative Endocrinology* 1, 332–340.

Friere, R., Appleby, M.C. and Hughes, B.O. (1997) Assessment of pre-laying motivation in the domestic hen using social interaction. *Animal Behaviour* 54, 313–319.

Friere, R., Appleby, M.C. and Hughes, B.O. (1998) Effects of social interactions on pre-laying behaviour in hens. *Applied Animal Behaviour Science* 56, 47–57.

Fumihito, A., Miyake, T., Takada, M., Shingu, R., Endo, T., Gojobori, T., Kondo, N. and Ohno, S. (1996) Monophyletic origin and unique dispersal patterns of domestic fowls. *Proceedings of the National Academy of Sciences USA* 93, 6792–6795.

Furlow, B., Kimball, R.T. and Marshall, M.C. (1998) Are rooster crows honest signals of fighting ability? *The Auk* 115, 763–766.

Gentle, M.J. and Hunter, L.N. (1991) Physiological and behavioural responses associated with feather removal in *Gallus gallus* var. *domesticus*. *Research in Veterinary Science* 50, 95–101.

Gerken, M. and Mills, A. (1993) Welfare of domestic quail. In: Savory, C.J. and Hughes, B.O. (eds) *4th European Symposium on Poultry Welfare*. Universities Federation for Animal Welfare, Potters Bar, pp. 58–176.

Gottleib, G. (1976) Consequences of prenatal development: behavioral embryology. *Psychological Review* 83, 215–234.

Gottleib, G. and Vandenbergh, J.G. (1968) Ontogeny of vocalization in duck and chick embryos. *Journal of Experimental Zoology* 168, 307–326.

Graves H.B., Hable, C.P. and Jenkins, T.H. (1985) Sexual selection in *Gallus*: effects of morphology and dominance on female spatial behavior. *Behavioural Processes* 11, 189–197.

Grigor, P.N., Hughes, B.O. and Appleby, M.C. (1995) Social inhibition of movement in domestic hens. *Animal Behaviour* 49, 1381–1388.

Gross, W.B. and Siegel, H.S. (1983) Evaluation of the heterophil/lymphocyte ratio as a measure of stress in chickens. *Avian Diseases* 27, 972–979.

Grosse, A.E. and Craig, J.V. (1960) Sexual maturity of males representing twelve strains of six breeds of chickens. *Poultry Science* 39, 164–172.

Guhl, A.M. (1953) Social behaviour of the domestic fowl. *Kansas Agricultural Experimental Station Technical Bulletin* 73, 1–48.

Guhl, A.M. (1958) The development of social organization in the domestic chick. *Animal Behaviour* 6, 92–111.

Guhl, A.M. (1968) Social inertia and social stability in chickens. *Animal Behaviour* 16, 219–232.

Guhl, A.M. and Ortman, L.L. (1953) Visual patterns in the recognition of individuals among chickens. *Condor* 55, 287–298.

Guhl, A.M. and Warren, D.C. (1946) Number of offspring sired by cockerels related to social dominance in chickens. *Poultry Science* 25, 460–472.

Guhl, A.M., Collias, N.E. and Allee, W.C. (1945) Mating behavior and the social hierarchy in small flocks of White Leghorns. *Physiological Zoology* 18, 365–390.

Gunnarsson, S., Keeling, L.J. and Svedberg, J. (1999) Effect of rearing factors on the prevalence of mislaid eggs, cloacal cannibalism and feather pecking in commercial flocks of loose housed laying hens. *British Poultry Science* 40, 12–18.

Guyomarc'h, J.-C. (1967) Contribution à l'étude des cris de contact chez la caille japonaise (*Coturnix c. japonica*). *Comptes Rendus Congrès Nationale Société Sav.* 3, 354–360.

Guyomarc'h, J.-C. (1974) L'empreinte auditive prénatale chez le poussin domestique. *Revue de Comportement Animal* 6, 79–94.

Hale, E.B., Schleidt, W.M. and Schein, M.W. (1969) The behaviour of turkeys. In: Hafez, E.S.E. (ed.) *The Behaviour of Domestic Animals*, 2nd edn. Williams and Wilkins, Baltimore, pp. 22–44.

Hansen, I. and Braastad, B.O. (1994) Effect of rearing density on pecking behaviour and plumage condition of laying hens in two types of aviary. *Applied Animal Behaviour Science* 40, 263–272.

Hansen, R.S. (1976) Nervousness and hysteria in mature female chickens. *Poultry Science* 55, 531–543.

Hogue, M.-E., Beaugrand, J.P. and Lague, P.C. (1996) Coherent use of information by hens observing their former dominant defeating or being defeated by a stranger. *Behavioral Processes* 38, 241–252.

Huber-Eicher, B. and Audigé, L. (1999) Analysis of risk factors for the occurrence of feather pecking in laying hen growers. *British Poultry Science* 40, 599–604.

Huber-Eicher, B. and Wechsler, B. (1997) Feather pecking in domestic chicks: its relation to dustbathing and foraging. *Animal Behaviour* 54, 757–768.

Huber-Eicher, B. and Wechsler, B. (1998) The effect of quality and availability of foraging materials on feather pecking in laying chicks. *Animal Behaviour* 55, 861–873.

Hughes, B.O. (1971) Allelomimetic feeding in the domestic fowl. *British Poultry Science* 12, 359–366.

Hughes, B.O. (1977) Selection of group size by individual laying hens. *British Poultry Science* 18, 9–18.

Hughes, B.O. (1983) Space requirements in poultry. In: Baxter, S.H., Baxter, M.R. and MacCormack, J.A. (eds) *Farm Animal Welfare and Housing*. Aberdeen. CEC Seminar, Martinus Nijhoff Publishers, pp. 121–128.

Hughes, B.O. and Duncan, I.J.H. (1972) The influence of strain and environmental factors upon feather pecking and cannibalism in fowls. *British
Poultry Science* 13, 525–547.

Hughes, B.O. and Wood-Gush, D.G.M. (1977) Agonistic behaviour in domestic
hens: the influence of housing method and group size. *Animal Behaviour*
25, 1056–1062.

Hughes, B.O., Wood-Gush, D.G.M. and Morely Jones, R. (1974) Spatial
organisation in flocks of domestic fowl. *Animal Behaviour* 22, 438–445.

Hughes, B.O., Carmichel, N.L., Walker, A.W. and Grigor, P.N. (1997) Low
incidence of aggression in large flocks of laying hens. *Applied Animal
Behaviour Science* 54, 215–234.

Johnsen, P.F., Vestergaard, K.S. and Norgaard Nielsen, G. (1998) The influence
of early rearing on the development of feather pecking and cannibalism in
domestic fowl. *Applied Animal Behaviour Science* 60, 25–41.

Johnsgard, P.A. (1973) *Grouse and Quails of North America.* University of
Nebraska, Lincoln.

Johnson, M.H. and Horn, G. (1988) Development of filial preferences in
dark-reared chicks. *Animal Behaviour* 36, 675–683.

Johnson, R.A. (1963) Habitat preference and behavior of breeding jungle fowl in
central western Thailand. *Wilson Bulletin* 75, 270–272.

Johnson, S.B., Hamm, R.J. and Leahey, T.H. (1986) Observational learning in
Gallus gallus domesticus with and without a conspecific model. *Bulletin of
the Psychonomic Society* 24, 237–239.

Jones, M.C. and Leighton, A.T. Jr (1987) Research note: effect of presence or
absence of the opposite sex on egg production and semen quality of
breeder turkeys. *Poultry Science* 66, 2056–2059.

Jones, M.E.J. and Mench, J.A. (1991) Behavioral correlates of male mating
success in a multisire flock as determined by DNA fingerprinting. *Poultry
Science* 70, 1493–1498.

Jones, R.B. and Williams, J.B. (1992) Responses of pair-housed male and
female domestic chicks to the removal of a companion. *Applied Animal
Behaviour Science* 32, 375–380.

Keeling, L.J. (1994) Feather pecking – who in the group does it, how often
and under what circumstances? In: *Proceedings of 9th European Poultry
Conference.* Glasgow, UK, pp. 288–289.

Keeling, L.J. (1995) Spacing behaviour and an ethological approach to assessing
optimal space allocations for laying hens. *Applied Animal Behaviour
Science* 44, 171–186.

Keeling, L.J. and Duncan, I.J.H. (1991) Social spacing in domestic fowl under
semi-natural conditions. The effect of behavioural activity and activity
transitions. *Applied Animal Behaviour Science* 32, 205–217.

Keeling, L.J. and Hurnik, J.F. (1996) Social facilitation acts more on the
appetitive than the consummatory phase of feeding behaviour. *Animal
Behaviour* 52, 11–15.

Keeling, L.J. and Wilhelmsson, M. (1997) Selection based on direct observa-
tions of feather pecking behaviour in adult laying hens. In: Koene, P. and
Blokhuis, H.J. (eds) *Proceedings of 5th European Symposium on Poultry
Welfare.* Wageningen, the Netherlands, pp. 77–79.

Kent, J.P. (1987) Experiments on the relationship between the hen and the chick (*Gallus gallus*): the role of the auditory mode in recognition and the effects of maternal separation. *Behaviour* 102, 1–14.

King, M.G. (1965) Peck frequency and minimum approach distance in domestic fowl. *Journal of General Psychology* 106, 35–38.

Kjaer, J.B. and Sørensen, P. (1997) Feather pecking behaviour in White Leghorns, a genetic study. *British Poultry Science* 38, 333–341.

Kovach, J.K. (1975) The behavior of quail. In: Hafez, E.S.E. (ed.) *The Behaviour of Domestic Animals*, 3rd edn. Baillière Tindall, London, pp. 437–444.

Kratzer, D.D. and Craig, J.V. (1980) Mating behavior of cockerels: effects of social status, group size and group density. *Applied Animal Ethology* 6, 49–62.

Kruijt, J.P. (1964) Ontogeny of social behaviour in the Burmese Red Junglefowl (*Gallus gallus spadecius bonaterre*). *Behaviour* Suppl. XII, 1–201.

Kuit, A.R., Ehlhardt, D.A. and Blokhuis, H.J. (eds) (1989) *Alternative Improved Housing Systems for Poultry*. CEC Seminar, Beekbergen, the Netherlands, 163 pp.

Latham, R.M. (1976) *Complete Book of the Wild Turkey*. Stackpole Books, Harrisburg, Pennsylvania.

Lee, Y.P., Craig, J.V. and Dayton, A.D. (1982) The social rank index as a measure of social status and its association with egg production in White Leghorn pullets. *Applied Animal Ethology* 8, 377–390.

Leonard, M.L. and Horn, A.G. (1995) Crowing in relation to status in roosters. *Animal Behaviour* 49, 1283–1290.

Leonard, M.L. and Zanette, L. (1998) Female mate choice and male behaviour in domestic fowl. *Animal Behaviour* 56, 1099–1105.

Leonard, M.L., Horn, A.G. and Fairfull, R.W. (1993) Correlates and consequences of allopecking in White Leghorn chickens. *Applied Animal Behaviour Science* 43, 17–26.

Lickliter, R., Dyer, A.B. and McBride, T. (1993) Perceptual consequences of early social experience in precocial birds. *Behavioural Processes* 30, 185–200.

Ligon, J.D. and Zwartjes, P.W. (1995) Ornate plumage of male red junglefowl does not influence mate choice by females. *Animal Behaviour* 49, 117–125.

Lill, A. (1969) Spatial organisation in small flocks of domestic fowl. *Behaviour* 32, 256–290.

Lill, A. and Wood-Gush, D.G.M. (1965) Potential ethological isolating mechanisms and assortative mating in the domestic fowl. *Behavior* 25, 16–44.

Lindberg, A.C. and Nicol, C.J. (1996a) Space and density effects on group size preferences in laying hens. *British Poultry Science* 37, 709–721.

Lindberg, A.C. and Nicol, C.J. (1996b) Effects of social and environmental familiarity on group preferences and spacing behaviour in laying hens. *Applied Animal Behaviour Science* 49, 109–123.

Lowry, D.C. and Abplanalp, H. (1972) Social dominance, given limited access to common food, between hens selected and unselected for increasing egg production. *British Poultry Science* 13, 365–373.

Lundberg, A. and Keeling, L.J. (1999) The impact of social factors on nesting in laying hens (*Gallus gallus domesticus*). *Applied Animal Behaviour Science* 64, 57–69.

Manser, C.E. (1996) Effects of lighting on the welfare of domestic poultry: a review. *Animal Welfare* 5, 341–360.

Martin, F., Beaugrand, J.P. and Lagu, P.C. (1997) The role of the hen's weight and recent experience on dyadic conflict outcome. *Behavioral Processes* 41, 139–140.

Mason, I.L. (1984) *Evolution of Domestic Animals*. Longman, London.

McBride, G., Parer, I.P. and Foenander, F. (1969) The social organization and behavior of the feral domestic fowl. *Animal Behavior Monographs* 2, 125–181.

McKinney, F. (1975) The behaviour of ducks. In: Hafez, E.S.E. (ed.) *The Behaviour of Domestic Animals*, 3rd edn. Baillière Tindall, London, pp. 490–519.

Mench, J.A. (1988) The development of agonistic behavior in male broiler chicks: a comparison with laying-type males and the effects of feed restriction. *Applied Animal Behaviour Science* 21, 233–242.

Mench, J.A. (1996) Social preferences in laying hens. In: *Proceedings of the International Congress of the International Society for Applied Ethology*. Centre for the Study of Animal Welfare, Guelph, Ontario, p. 38.

Mench, J.A. and Mayeaux, J.D.. (1997) Previous familiarity has little effect on behavior and stress responses in re-grouped laying hens. In: *Proceedings of the 31st International Congress of the International Congress of Applied Ethology*. Prague, Czechoslovakia, p. 95.

Mench, J.A. and Ottinger, M.A. (1991) Behavioral and hormonal correlates of social dominance in stable and disrupted groups of male domestic fowl. *Hormones and Behavior* 25, 112–122.

Mench, J.A., van Tienhoven, A., Marsh, J.A., McCormick, C.C., Cunningham, D.L. and Baker, R.C. (1986) Effects of cage and floor pen management on behavior, production, and physiological stress responses of laying hens. *Poultry Science* 65, 1058–1069.

Meunier-Salaün, M.C. and Faure, J.M. (1984) On the feeding behaviour of the laying hen. *Applied Animal Behaviour Science* 13, 129–141.

Miller, D.B. (1978) Species-typical and individually distinctive acoustic features of the crow calls of red jungle fowl. *Zeitschrift für Tierpsychologie* 47, 182–193.

Millman, S.T. and Duncan, I.J.H. (2000) Strain differences in aggressiveness of male domestic fowl in response to a male model. *Applied Animal Behaviour Science* 66, 217–233.

Mills, A.D. and Faure, J.M. (1991) Divergent selection for duration of tonic immobility and social reinstatement in Japanese Quail (*Coturnix coturnix japonica*) chicks. *Journal of Comparative Psychology* 105, 25–38.

Mills, A.D., Crawford, L.L., Domjan, M. and Faure, J.M. (1997) The behavior of the Japanese or domestic quail *Coturnix japonica*. *Neuroscience and Biobehavioral Reviews* 21, 261–281.

Newberry, R.C. and Hall, J.W. (1990) Use of space by broiler chickens: effects of age and pen size. *Applied Animal Behaviour Science* 25, 125–136.

Nicol, C.J. and Pope, S.J. (1996) The maternal feeding display of domestic hens is sensitive to perceived chick error. *Animal Behaviour* 52, 767–774.

Nicol, C.J. and Pope, S.J. (1999) The effects of demonstrator social status and prior foraging success on social learning in laying hens. *Animal Behaviour* 57, 163–171.

North, M.O. and Bell, D.B. (1990) *Commercial Chicken Production*, 4th edn. Van Nostrand Rheinhold, New York.

Oden, K., Vestergaard, K.S. and Algers, B. (2000) Space use and agonistic behavior in relation to sex composition in large flocks of laying hens. *Applied Animal Behaviour Science* 67, 307–320.

Ottinger, M.A. and Mench, J.A. (1989) Reproductive behaviour in poultry: implications for artificial insemination technology. *British Poultry Science* 50, 431–442.

Pagel, M. and Dawkins, M.S. (1997) Peck orders and group size in laying hens: 'futures contracts' for non-aggression. *Behavioural Processes* 40, 13–25.

Pamment, P., Foenander, F. and McBride, G. (1983) Social and spatial organization of male behavior in mated domestic fowl. *Applied Animal Ethology* 9, 341–349.

Potash, L.M. (1970) Vocalizations elicited by electrical brain stimulation in *Coturnix coturnix japonica*. *Behaviour* 36, 149–167.

Preston, A.P. and Murphy, L.B. (1989) Movement of broiler chickens reared in commercial conditions. *British Poultry Science* 30, 519–532.

Rappaport, S. and Soller, M. (1966) Mating behavior, fertility and rate-of-gain in Cornish males. *Poultry Science* 45, 997–1003.

Reiter, K. (1997) The behaviour of domestic ducks (*Anas platyrhynchos* f. *domestica*). *Archiv für Gefluegelkunde* 61, 149–161.

Roden, C. and Wechsler, B. (1998) A comparison of the behaviour of domestic chicks reared with or without a hen in enriched pens. *Applied Animal Behaviour Science* 55, 317–326.

Rogers, L. (1995) *The Development of Brain and Behaviour in the Chicken*. CAB International, Wallingford, UK.

Rushen, J. (1982) The peck orders of chickens: how do they develop and why are they linear? *Animal Behaviour* 30, 1129–1137.

Salomon, A.L., Lazorcheck, M.J. and Schein, M.W. (1966) Effect of social dominance on individual crowing rates of cockerels. *Journal of Comparative and Physiological Psychology* 61, 144–146.

Salzen, E.A. and Cornell, J.M. (1969) Self perception and species recognition in birds. *Behaviour* 30, 44–65.

Savory, C.J. (1995) Feather pecking and cannibalism. *World's Poultry Science Journal* 51, 215–219.

Savory, C.J. and Mann, J.S. (1999) Feather pecking in groups of growing bantams in relation to floor substrate and plumage colour. *British Poultry Science* 40, 565–572.

Schjelderupp-Ebbe, T. (1922) Beitrage aur Socialpsycholgie des Hauschuhns. *Zeitschrift für Tierpsychologie* 88, 225–252.

Schorger, A.W. (1966) *The Wild Turkey. Its History and Domestication*. University of Oklahoma Press, Norman, Oklahoma.

Shanaway, M.M. (1994) *Quail Production Systems. A Review*. Food and Agriculture Organization of the United Nations, Rome.

Shea-Moore, M.M., Mench, J.A. and Thomas, O.P. (1990) The effect of dietary tryptophan on aggressive behavior in developing and mature broiler breeder males. *Poultry Science* 69, 1664–1669.

Sherwin, C.M. (ed.) (1994) *Proceedings of the Symposium on Modified Cages for Laying Hens.* UFAW, Potters Bar, UK, 102 pp.

Sherwin, C.M. and Kelland, A. (1998) Time-budgets, comfort behaviours and injurious pecking of turkeys housed in pairs. *British Poultry Science* 39, 325–332.

Sherwin, C. and Nicol, C.J. (1993) Factors affecting floor laying by hens in modified cages. *Applied Animal Behaviour Science* 36, 211–222.

Sherwin, C.M., Lewis, P.D. and Perry, G.C. (1999) The effects of environmental enrichment intermittent lighting on the behaviour and welfare of male domestic turkeys. *Applied Animal Behaviour Science* 62, 319–333.

Siegel, P.B. and Hurst, D.C. (1962) Social interactions among females in dubbed and undubbed flocks. *Poultry Science* 41, 141–145.

Siegel, P.B. and Siegel, H.S. (1964) Rearing methods and subsequent sexual behavior of male chickens. *Animal Behaviour* 12, 270–271.

Siegel, P.B., Phillips, R.E. and Folsom, E.F. (1965) Genetic variation in the crow of adult chickens. *Behaviour* 24, 229–235.

Simonsen, H.B., Vestergaard, K. and Willeberg, P. (1980) Effect of floor type and density on the integument of egg-layers. *Poultry Science* 59, 2202–2206.

Skewes, P.A. and Wilson, H.R. (1990) *Bobwhite Quail Production.* Pamphlet.

Sullivan, M.S. (1991) Flock structure in red junglefowl. *Applied Animal Behaviour Science* 30, 381–386.

Tauson, R. (1998) Health and production in improved cage designs. *Poultry Science* 77, 1820–1827.

Tuculescu, R.A. and Griswold, J.G. (1983) Prehatching interactions in domestic chicks. *Animal Behaviour* 31, 1–10.

Vallortigara, G. (1992) Affiliation and aggression as related to gender in domestic chicks. *Journal of Comparative Psychology* 106, 53–57.

Vallortigara, G. and Andrew, R.J. (1991) Lateralization of response by chicks to change in a model partner. *Animal Behaviour* 41, 187–194.

Vallortigara, G. and Andrew, R.J. (1994) Olfactory lateralization in the chick. *Neuropsychologia* 32, 417–423.

Vestergaard, K.S. (1994) Dustbathing and its relation to feather pecking in the fowl: motivational and developmental aspects. PhD thesis, University of Copenhagen, Denmark.

Vestergaard, K. and Lisborg, L. (1993) A model of feather pecking development which relates to dustbathing in the fowl. *Behaviour* 126, 291–308.

Vince, M.A. (1964) Synchronization of hatching in the bobwhite quail (*Colinus virginianus*). *Nature* 203, 1192–1193.

Vince, M.A. (1970) Some aspects of hatching behaviour. In: Freeman, B.M. and Gordon, R.F. (eds) *Aspects of Poultry Behaviour.* British Poultry Science, Edinburgh, pp. 33–62.

Wahlström, A., Tauson, R. and Elwinger, K. (1998) Effects on plumage condition, health and mortality of dietary oats/wheat ratios to three hybrids of laying hens in different housing systems. *Acta Agriculturae Scandinavica Section A, Animal Science* 48, 250–259.

Wakasugi, N. (1984) Japanese quail. In: Mason, I.L. (ed.) *Evolution of Domestic Animals.* Longman, London.

Waldvogel, J.A. (1990) A bird's eye view. *American Scientist* 78, 342–353.

Watts, C.R. and Stokes, A.W. (1971) The social order of turkeys. *Scientific American* 224, 112–118.

Webster, A.B. and Hurnik, J.F. (1994) Synchronization of behaviour among laying hens in battery cages. *Applied Animal Behaviour Science* 40, 153–165.

Wechsler, B. and Schmid, I. (1998) Aggressive pecking by males in breeding groups of Japanese quail (*Coturnix japonica*). *British Poultry Science* 39, 333–339.

Weeks, C.A., Danbury, T.D., Davies, H.C., Hunt, P. and Kestin, S.C. (2000) The behaviour of broiler chickens and its modification by lameness. *Applied Animal Behaviour Science* 67, 111–125.

West, B. and Zhou, B.Z. (1989) Did chickens go north? New evidence for domestication. *World's Poultry Science Journal* 45, 205–218.

Widowski, T.M., Wong, D.M.A.L.F. and Duncan, I.J.H. (1998) Rearing with males accelerates onset of sexual maturity in female domestic fowl. *Poultry Science* 77, 150–155.

Wood-Gush, D.G.M. (1971) *The Behaviour of the Domestic Fowl.* Heinemann, London.

Wood-Gush, D.G.M. and Duncan, I.J.H. (1976) Some behavioral observations on domestic fowl in the wild. *Applied Animal Ethology* 2, 255–260.

Wood-Gush, D.G.M., Duncan, I.J.H. and Savory, C.J. (1978) Observations on the social behaviour of domestic fowl in the wild. *Biology of Behaviour* 3, 193–205.

Ylander, D.M. and Craig, J.V. (1980) Inhibition of agonistic acts between domestic hens by a dominant third party. *Applied Animal Ethology* 6, 63–69.

Yngvesson, J. (1997) *Cannibalism in Laying Hens. A Literature Review.* Department of Animal Environment and Health, Swedish University of Agricultural Universities report no.1, 54 pp.

Zayan, R., Doyen, J. and Duncan, I.J.H. (1983) Social and space requirements for hens in battery cages. In: Baxter, S.H., Baxter, M.R. and MacCormack, J.A.D. (eds) *Farm Animal Housing and Welfare.* CEC Seminar, Martinus Nijhoff, The Hague, the Netherlands, pp. 67–90.

Zuk, M., Thornhill, R. and Ligon, D.J. (1990a) Parasites and mate choice in red jungle fowl. *American Zoologist* 30, 235–244.

Zuk, M., Thornhill, R., Ligon, J.D., Johnson, K., Austad, S., Ligon, S.H., Thornhill, N. and Costin, C. (1990b) The role of male ornaments and courtship behavior in female choice of red junglefowl. *American Naturalist* 136, 459–473.

Zuk, M., Ligon, J.D. and Thornhill, R. (1992) Effects of experimental manipulation of male secondary sex characteristics on female mate preference in red junglefowl. *Animal Behaviour* 44, 999–1006.

Zuk, M., Popma, S.L. and Johnsen, T.S. (1995) Male courtship displays, ornaments, and female mate choice in captive red junglefowl. *Behaviour* 132, 11–12.

Zuk, M., Kim, T., Robinson, S.I. and Johnsen, T. (1998) Parasites influence social rank and morphology, but not mate choice, in female red junglefowl, *Gallus gallus. Animal Behaviour* 56, 493–499.

The Social Behaviour of Sheep 　8

Andrew Fisher and Lindsay Matthews

AgResearch, Ruakura Agricultural Research Centre, PB 3123, Hamilton, New Zealand

---

(Editors' comments: In both domestic and wild sheep, ewes and their juvenile offspring form stable social groups which are based more on gregarious and following behaviour than a dominance order. Ewes only separate from the flock in order to give birth, but, as the lambs are followers from birth, the maternal–offspring groups soon return to the main flock. Ewe flocks maintain the same home range for long periods. In contrast, rams exist in small groups that are organized into a distinct dominance hierarchy. Home ranges for ram groups change frequently and the group itself disbands prior to the breeding season.

The authors identify several aspects of social behaviour in commercial sheep that are unique to sheep. Large flocks of wethers (castrated males) are sometimes assembled for wool production, and these flocks more closely resemble ewe flocks in organization and stability than they do rams. Use of space is dependent not only on the availability and distribution of resources, but also on the breed of the sheep. It appears that, in this gregarious species, isolation in groups of less than four individuals is both stressful and disruptive of management.)

8.1 Basic Social Structure

Sheep (genus *Ovis*) were one of the first animals to be domesticated by humans, over 10,000 years ago. General behavioural characteristics of the sheep are: vigilance, flocking, promiscuous mating and a strong mother–offspring bond in which the young display a following relationship with their dam. Existing wild species of sheep include the

©CAB *International* 2001. *Social Behaviour in Farm Animals*
(eds L.J. Keeling and H.W. Gonyou)

Mouflon (*Ovis musimon*) of Europe and western Asia, the Argali (*Ovis ammon*) and Urial (*Ovis vignei*) in Asia, the North American Bighorn (*Ovis canadensis*) and the Dall sheep (*Ovis dalli*) of Alaska and western Canada. The domestic sheep breeds form the species *Ovis aries* and are widely farmed for their milk, meat, pelt and wool, particularly in temperate zones of the world.

8.1.1 Composition and structure of social groups

Much of the social organization of temperate breeds of sheep under natural conditions is influenced by the seasonality of its reproduction, particularly in temperate zones (see Section 1.6). Ewes are seasonally polyoestrous, with mating occurring in association with decreasing daylength during autumn. Tropical breeds and species may breed during a greater proportion of the year.

Detailed studies have been conducted of the natural flock composition and structure for sheep under wild or feral conditions for a number of species and breeds. These include the Bighorn (Geist, 1971; Festa-Bianchet, 1991; Hass, 1991), feral Soay sheep (Grubb and Jewell, 1966; Grubb, 1974a) and Dall (Murie, 1944). Typically, males and females of reproductive age are segregated during the non-breeding seasons, with separate home ranges. Young rams are an exception, and may remain with the ewe flock for some time after puberty.

In Bighorn, Dall and feral Soay sheep, ewes and older rams are associated during the breeding season in November and December (northern hemisphere) (Fig. 8.1). Younger rams leave the female group when between 2 and 4 years of age in Bighorn sheep, although some associations may occur up to 6 years of age. Although some rams may then spend a few months wandering alone, they generally join with other rams in small groups (Geist, 1971). In Soay and Dall sheep, yearling rams graze with ewe groups until the following breeding season. Soay rams then join ram groups, typically consisting of four to 13 animals of varying ages. Few solitary rams are seen (Grubb and Jewell, 1966). The mean size of Bighorn ram bands is six animals (Woolf *et al.*, 1970), although this varies with the age of the lead ram. A minority of ram groups are led by rams less than 7 years of age, and such groups tend to have fewer animals than groups with older leaders. The mean life expectancy of rams in stable populations of North American wild sheep is 10–12 years. The figure for Asian and European species such as the Mouflon and Arial appears to be slightly lower (Geist, 1971).

Rams tend to form into larger groupings in the month before the breeding season in Bighorn sheep, and an increase in ram association is also seen in the Soay. As the breeding season commences, the rams disperse and animals move in search of oestrous females among the ewe

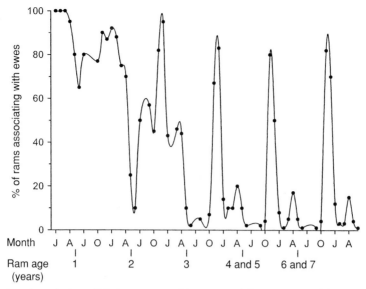

Fig. 8.1. Association of Bighorn rams with ewes at different times of the year and at varying ram ages. The peak in association during the rut in November and December is clearly evident. (Adapted from Geist, 1971.)

groups. Following the breeding season, ram groups are re-established. The membership of ram bands tends to vary from year to year as groups break up and re-form after the mating period. One exception is the Soay, in which the ram groups are highly conserved from year to year, although mortality may have a large impact on group membership (Grubb and Jewell, 1966).

Ram groups have a social structure based on a dominance hierarchy (see Section 1.5). Dominance is correlated with the size of the horns and body, which is usually proportional to age. This holds true both within ram groups and for rams interspersed among the ewe flock during the breeding season, where dominance corresponds with mating success. Bighorn sheep reach full body size and horn development around 8 years of age, and such sheep will almost always be dominant to younger males. Below the dominant male, the hierarchy may not be linear, but generally corresponds to the development of the horns and physical size. In mature ram groups, a younger, more vigorous animal may dominate an older, larger-horned individual. Leadership in established ram groups is a function of dominance, with the other rams following the dominant individual.

In general, ewe groups are much more stable in membership than ram bands, with groups remaining together throughout the year. Group membership is comprised of mothers and daughters over several generations (Hunter and Milner, 1963). However, within a ewe home range group, smaller subgroups may be seen grazing separately at various times. For example, adult Bighorn ewes studied by Festa-Bianchet (1988) migrated to summer grazing areas ahead of the yearling ewes of the same group. Groups of lambs and yearling rams may also form temporary associations within the home range group.

Ewes withdraw temporarily from the group to lamb, with Bighorn ewes, for example, spending up to 5–7 days alone with their lamb, before rejoining the group (Geist, 1971). Mouflon ewes show a similar pattern, spending between 2 and 3 days sequestered with their lamb at the birth site. The size of natural ewe groups varies with species and habitat (see Section 2.5). The mean size of ewe groups of Bighorn sheep in Yellowstone National Park was recorded at eight animals by Woolf *et al.* (1970), but one group comprised 61 individuals. The Soay ewe groups studied by Grubb and Jewell (1966) consisted of between seven and 49 sheep.

The age structure of ewe groups can vary considerably from year to year, depending on lambing rates and mortality, which particularly affects lambs and yearlings. The severity of the winter can be the major determining factor in the size of the lambing crop and the survival of lambs through to the following year. North American species of wild sheep rarely have multiple births, although twins are not unknown among Asian and European species, such as the Mouflon (Briedermann, 1992). The sex ratio of lambs at birth is about 50 : 50.

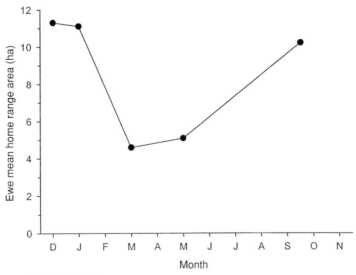

Fig. 8.2. The size of home ranges of wild-living Soay ewes on the Scottish island of Hirta varies widely with season. Home ranges are smaller at the end of winter and during spring. (Data from Grubb and Jewell, 1974.)

Ewe groups have a less well-defined social hierarchy than ram bands (see Section 2.4). Although ewes do not grow in size after reaching full maturity at approximately 2–3 years of age, any dominance orders which are displayed are generally correlated with age. There is no evidence in wild ewes that dominance is related to reproductive success or lamb sex ratio. Studies of Soay sheep showed that there was no dominance hierarchy within ewe flocks (Grubb, 1974a). Bighorn ewe groups may exhibit relatively stable, non-linear dominance hierarchies. The few highest ranking ewes in a group are usually at least 6–8 years of age, and do not establish a position in the upper half of the hierarchy of ewes until about 4 years of age (Hass, 1991).

Leadership is common in ewe groups, but it is not necessarily correlated with dominance. As young rams mature within the ewe flock they will often display dominance over ewes, although leadership of the flock is still provided by the females. Analysis of 23 groups of Bighorn ewes by Geist (1971) indicated that a group would typically follow a mature ewe.

8.1.2 Use of space

Groups of sheep restrict their movements to a particular area (home range) which is not defended (Hunter and Milner, 1963). Home ranges vary in size and habitat between different groups, and can vary with season for a particular group of sheep. The home ranges of separate groups of sheep may overlap, although it is uncommon for two groups to be in the same location concurrently.

The sizes of home ranges of free-ranging sheep exhibit considerable variation. Bighorn home ranges in Canada may range from 50 to 2800 ha (calculated from Geist, 1971), whereas the mean area of Soay home ranges varied from 5 to 16 ha (Grubb and Jewell, 1974). Often, there is a pattern of daily movement within the home range, such that sheep occupy a similar part of the home range at the same time on consecutive days. Wild sheep in their natural hilly or mountainous environment will typically overnight in camping areas in the middle reaches of the home range, moving down in the morning to graze the lower slopes and flats. Within the home range, the social group may separate into smaller 'grazing parties' of varying size and composition, particularly in summer.

In temperate and subarctic regions, the size and/or the location of home range varies with season. With the Soay ewe groups studied by Grubb and Jewell (1974), the home range area was restricted during May, when rapid spring pasture growth was under way and young lambs were present in the flock (Fig. 8.2). As summer extended into autumn, the home range area extended, with sheep grazing high grassy areas on the hills. At the end of winter, when climatic conditions were most harsh, the home range was at its smallest, and sheep foraged close to their shelter sites. Utilization of areas within the home range was closely linked to the availability of preferred pasture species during the year. Similar changes in home range have been observed in Scottish hill sheep, in which the range is at its most extensive in summer and is smallest in winter (Lawrence and Wood-Gush, 1987, 1988).

The size of the home ranges of mountain sheep such as the Dall and Bighorn also varies considerably with season, with summer ranges being much larger. In winter, when snow limits movement, the sheep occupy limited areas, but in summer the sheep may move over areas up to 50 times the size of the winter home range. Although winter and summer home ranges can be close together, sheep often migrate large distances between seasonal home ranges. Bighorn ewes may travel 10 km between winter and summer ranges (Blood, 1963). Rams tend to travel greater distances than ewe groups.

Wild sheep populations may have separate seasonal home ranges. Geist (1971) described five periods during the year when Bighorn sheep move to a separate home range: (i) late September to early October when ewes and rams move to wintering areas; (ii) late October to early November when rams move to their breeding grounds; (iii) late December to early January when rams move from their breeding grounds; (iv) late March to April when ewes and rams move to late winter/spring home ranges; and (v) late May to early July when ewes move to lambing areas and then ewes and rams move to summer home ranges.

Ewe home ranges are highly stable over years, with home range knowledge being acquired by female lambs and yearlings. In contrast, the home ranges of ram bands are less consistent over time, due to the changing size and composition of ram groups. Individual rams, however, often show a high level of fidelity to their home ranges once they are established following departure from the ewe group of their birth.

Although the home ranges of ewe and ram groups can overlap during the non-breeding season, it is rare for the two classes to be in the same area. Segregation is often enhanced by differing habitat utilization within the overlapping ranges. In Bighorn sheep during winter and autumn, rams occupy open slope locations and ewes prefer cliff areas. This may serve to minimize competition between rams and the ewes with which they have mated (Geist and Petocz, 1977).

8.1.3 Communication (see Section 2.3)

Communication between sheep primarily involves olfactory, visual or auditory signals, with communication by tactile means being of lesser importance (except during the breeding season). Tactile signals used by sheep include the nuzzling of the female anogenital region by the ram during courtship (which induces the female to urinate) and the pre-mating nudging and striking with the foot by the ram against the flank of the ewe. Sheep have a well-developed sense of olfaction, which is of considerable importance in social interactions, as well as in predator avoidance and feedstuff recognition. Olfactory social signals originate primarily from the pre-orbital scent glands, urine, amniotic fluid, the tail and anal areas and apocrine secretions on the wool.

The prime purpose of olfactory signals is social cohesion and recognition. Sheep from one social group will sniff and nose an animal from another group (Grubb and Jewell, 1974). Subordinate Dall or Bighorn rams will rub the face of the dominant animal. This action transfers the dominant's pre-orbital scent to their own faces, possibly leading to a 'group scent'. Most importantly, the recognition of its own lamb by a ewe is dependent on olfaction, especially for confirmation of the lamb's identity at close range. Ewes will sniff the tail area of a lamb and reject it if not their own. Ewes deprived of the sense of smell will accept alien lambs (Morgan *et al.*, 1975). When the lamb is born, the smell of the amniotic fluid will attract the ewe and initiate the bonding and recognition process.

Olfactory signals have an important role in facilitating reproduction. Rams use olfactory cues to detect oestrous ewes. The ram will sniff the vulval region of the ewe and any urine that is voided. Often the ram will then exhibit a flehmen response. Volatile compounds, such as oestrogens in the urine of the ewe, are thought to be detected by the ram's vomeronasal organ. Olfactory signals from rams also act as a reproductive signal for ewes. The movement of fully mature rams into the ewe groups at the start of the breeding season helps to stimulate and synchronize the onset of oestrus. In domestic sheep, Knight and Lynch (1980) showed that this 'ram effect' occurred when non-cycling ewes were exposed to the wool and skin wax from rams.

Sheep possess vision of fair to good acuity and have very good perception of movement and depth. The properties of the vision of sheep are reviewed by Piggins (1992). Communication among sheep using visual signals is achieved by the adoption of particular postures or movements. These include alert and display postures and movements associated with sexual and agonistic behaviour.

When sheep are grazing, they maintain other individuals of the group within their visual field. The adoption of an alert posture (head raised and oriented towards the potential threat) by a dam signals her lamb to approach and will cause other sheep also to stop grazing and orient their heads toward the stimulus. The fleeing movement of one individual will also cause the whole flock to flee. Courtship and aggressive intent are usually communicated via visual signals. Rams of both wild and domestic species will approach a prospective opponent with the head and neck in a low stretch posture.

Sheep also use visual cues to recognize their relationship with other individuals. Studies by Kendrick and Baldwin (1987) and Kendrick (1991) measured the response of nerve cells of the temporal cortex of the brain of conscious sheep to photographs of other sheep and potential predators (human and dog). One group of nerve cells responded to the presence and size of horns which are indicators of social dominance. Other cells responded differently to familiar compared with unfamiliar sheep. Frontal views of sheep were much more effective at eliciting nerve cell responses than other profiles. Kendrick (1991) concluded that rapid visual recognition of an individual and its likely dominance status enabled a sheep to quickly make an appropriate response, especially if the other animal was facing it head on.

Sheep are able to perceive a wide range of sound frequencies, extending from 125 Hz to 42 kHz (Heffner and Heffner, 1992). Auditory signals used by sheep vary from low-pitched 'rumbles' emitted by ewes soon after parturition to high-pitched bleats. During dominance fights, wild mountain ram species may growl as they initiate a clash. Mountain rams also grunt or growl during courtship. High-pitched bleats were recorded by Murphy *et al.* (1994) in domestic ewes agitated by separation from conspecifics. Similarly, feral Soay sheep that become separated from their flockmates are reported to bleat in distress (Grubb, 1974a).

The most common use of vocal communication is between ewes and their lambs. Both ewes and lambs bleat when they are separated and searching for the other, and ewes 'rumble' when they are reunited. A study by Shillito-Walser *et al.* (1982) showed that both domestic (Border Leicester, Dalesbred) and feral (Soay) ewes and lambs bleated and answered bleats when separated. Both ewes and lambs bleated more to their own than to alien lambs or ewes. Although ewes have been shown to recognize the bleat of their own lamb (Poindron and Carrick, 1976), hearing is used more to locate the lamb, and final confirmation of its identity is dependent on smell.

8.1.4   Cohesion and dispersion

Sheep as a species are highly gregarious, with relatively small social distances (see Section 2.2). Although group sizes and levels of dispersal within groups vary widely across breeds, habitats and individuals, there are consistent patterns in the factors influencing group cohesion. These factors include season, weather, terrain, feed availability and group composition. Sheep also typically demonstrate a flocking instinct in response to a potential threat. This response is utilized for the handling and movement of domesticated sheep (Fig. 8.3).

Crofton (1958) used aerial photographs to examine the spacing and orientation of domestic sheep in flocks grazing under moderately extensive conditions. Although the field size varied considerably from 5 to 22 ha, group cohesion exhibited less variation, with observations indicating a mean individual distance consistently between 14 and 27 m, regardless of flock or paddock size. More interestingly, regardless of how tightly flocked or otherwise a group of sheep were, most sheep were oriented during grazing such that two other sheep subtended an angle of approximately 110° from its head. Where sheep are grazing on the edge of a flock, they appear to use a stationary object or landmark for the outer reference for the angle (Fig. 8.4). Crofton (1958) hypothesized that positioning within the group was influenced by the width of the visual field, and that an individual's spacing from and orientation towards conspecifics would be determined by the need to have two other sheep at the borders of its vision.

The effect of seasonal factors on the level of cohesion displayed in sheep groups was seen in Soay sheep studied by Grubb and Jewell (1966), which were more widely dispersed during January when grazing was poor than during May when feed was plentiful. The weather on a particular day could override the seasonal trend, however, with the sheep remaining close together in more sheltered areas during

Fig. 8.3. The flocking response of a group of sheep to the approach of a potential predator – a dog. Within the flock, some sheep are facing the dog, while others appear about to institute a fleeing movement by the group.

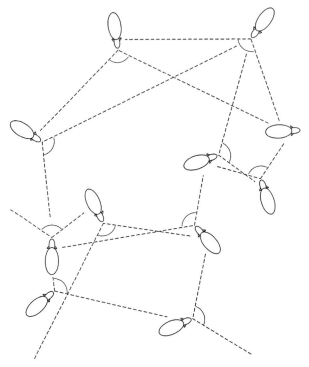

Fig. 8.4.   The distribution and orientation of grazing sheep in a flock. Sheep orient themselves such that two other sheep subtend an angle of approximately 110° – the angle between the optic axes. (Adapted from Crofton, 1958.)

periods of inclement weather. Ewe groups show increased dispersal during lambing time, with ewes moving to seek sheltered areas in which to give birth.

8.1.5   Inter-group interactions

Sheep have the ability to recognize their own group, and therefore identify individuals as coming from a foreign group. Few aggressive interactions between 'foreign' Bighorn ewes were observed by Festa-Bianchet (1986) and, similarly, separate Soay ewe groups studied by Grubb and Jewell (1974) often grazed alongside each other with little overt reaction to one another. Interestingly, when Soay ewes were aggressive towards a foreign ewe, the stranger would be from a group that was not often encountered. Such findings suggest that sheep can recognize individuals from both their own group and/or identify sheep from neighbouring social groups.

8.1.6 Intra-group interactions

Male–male

Social interactions between rams reflect dominance relationships, which are in turn influenced by body size and appearance, especially horn size where present (Geist, 1968, 1971). In general, the patterns of agonistic behaviour exhibited are very similar across sheep species and breeds (Table 8.1). Most intensive agonistic interactions occur between rams of similar size and dominance rank. Mounting behaviour among rams appears to have a function in maintaining social dominance by high-ranking males over subordinates, and is usually initiated by the dominant animal. In the absence of females, other agonistic interactions between rams are usually initiated by the lower-ranking animal challenging the dominant. In a losing encounter, the subordinate ram may flee or display submissive behaviour to the winner. Amicable behaviour may also occur between rams, usually involving rubbing or grooming and the transfer of scent from the pre-orbital glands.

The following agonistic encounters have been described for rams:

- *Low stretch* – a threat display in which the sheep extends its neck forward and horizontal to the ground. This behaviour is commonly displayed by rams towards smaller-horned subordinates (Geist, 1968), and is often accompanied by the *twist*, in which the ram rotates his head about the median axis of his body such that his muzzle faces the other animal (Fig. 8.5).
- *Horn threat* – the ram makes butting movements in the direction of its opponent.

Table 8.1. Agonistic behaviours in different sheep types. (Adapted from Schaller and Mirza, 1974.)

Behaviour	Urial	Bighorn	Soay	Mouflon
Clash	+	+	+	+
Jump	+	+	(+)	+
Shoulder push	(+)	+	+	+
Butt	+	+	+	
Mount	+	+	+	+
Low stretch	+	+	+	+
Twist	+	+	+	+
Kick	+	+	+	+
Horn	+	+	+	
Head shake		+		+

+ = behaviour present; (+) = behaviour present but seldom exhibited.

Fig. 8.5.   Agonistic interactions between rams. A ram aggressively approaches an unfamiliar ram (a), and displays a low neck stretch and twist, together with a front leg kick to its opponent (b). The losing ram then flees, chased by the dominant animal (c).   *Continued*

- *Threat jump* – an intention movement for a clash in which the ram jumps on to its back legs and moves its head as if ready to clash heads with its opponent.
- *Blocking* – described by Grubb (1974b), this term covers a range of behaviours in which two rams stand alongside each other, either head-to-head or tail-to-tail, and push, nudge or butt each other. Blocking may continue for some minutes, with overtly agonistic

Fig. 8.5. *Continued.*

behaviours interspersed with short bouts of grazing, the rams remaining in their parallel positions.

- *Front leg kick* – a common behaviour performed by rams on all subordinate sheep. The ram kicks upward with a stiffened foreleg towards the chest or flank of the other animal, with or without making contact (Fig. 8.5).
- *Clash* – the ram charges at its opponent from a short distance, lowers its head and neck and clashes head and horns with its opponent, which has oriented itself to meet the blow head-on.
- *Mount* – male–male mounting appears to be more common among North American mountain sheep than European and Asiatic species (McClelland, 1991).
- *Chase* – if the loser in a conflict runs away, the dominant ram will often be stimulated to chase it, butting the back legs and hindquarters of the fleeing ram (Fig. 8.5).

The following submissive and amicable behaviours have been described for rams:

- *Rubbing* – the subordinate ram rubs its head against the face, muzzle, horns, chest or shoulders of the dominant animal. This behaviour has also been termed horning (Geist, 1971), but should not be confused with agonistic behaviours. Rubbing may serve to transfer the scent of the dominant to the subordinate animal. Occasionally the subordinate ram will lick the higher-ranking animal.

- *Low neck* – the subordinate lowers his neck and withdraws from an encounter with a dominant ram. This posture has not been observed for American Bighorn or Dall sheep, but occurs in species in Europe and Asia, such as the Mouflon and Urial.

Rams show a high level of agonistic interactions with each other, although in established ram groups more intensive conflicts are less common than low level threats and amicable/submissive behaviours. In Bighorn sheep, dominant rams will commonly approach subordinates in the low stretch posture.

More intensive agonistic interactions arise between rams evenly matched in size and horn development, especially when they are from separate groups. Conflicts among rams are most commonly initiated by a head butt, which may often be followed by a reciprocal head butt or chasing behaviour (McClelland, 1991). Serious dominance fights may last only for a few minutes, or may continue intermittently for up to a day, with repeated clashes, butts and growling from both rams interspersed with blocking behaviours. During the pre-rut period, groups of rams will often form a *huddle*, in which a number of rams will face into a small circle displaying their horns to each other and performing other threat behaviours.

Rams may form associations with other rams, and tend to prefer interacting with other individuals of a similar horn and body size to their own. Cooperative behaviour in defence against predators has been observed among ram groups, with rams uniting to fend off coyotes in North America (Shank, 1977).

Hogg (1984), from studies of Bighorn sheep, suggested that the rams used three separate mating strategies – tending, coursing and blocking. The tending strategy was usually pursued by rams near the top of the dominance hierarchy. A tending ram would consort with a single oestrous ewe and attempt to prevent other rams from gaining access to the ewe. The ram is generally the largest, most dominant male around, and will spend part of his time threatening, blocking and exerting his dominance over the subordinate rams which cluster around, seeking a chance to steal the ewe. Coursing was adopted by the lesser ranking rams, which would congregate near the dominant ram and oestrous ewe and occasionally attack the tending ram or move towards the ewe. If the coursing ram succeeded in getting past the tending ram and gaining access to the ewe, it would immediately try to mount the ewe, which would often move away. The blocking strategy was attempted by rams of a range of ages, including 2-year-old animals. The blocking ram would attempt to sequester a ewe and prevent her from rejoining others in the ewe group by physically positioning itself to block the ewe's path. During one breeding season studied by Hogg (1984), 18% of oestrous ewes were blocked by rams whereas 82% were tended.

Female–female

The level of interaction among female sheep is less than that among rams, and they have a smaller behavioural repertoire, especially agonistic behaviours. Females tend not to show mount, clash and threat-jump types of agonistic behaviours. Although dominance relationships have been described for ewes (Eccles and Shackleton, 1986), it has been proposed that social organization within groups of ewes is less dependent on a dominance structure than on gregariousness and following characteristics (Lynch *et al.*, 1989). Unlike rams, which compete for mating opportunities with oestrous ewes during the rut, female sheep are not often directly competing for the same resource.

Agonistic interactions and social status were measured in female Bighorn sheep by Eccles and Shackleton (1986). Ewes displayed horn threats and threat jumps, but most agonistic interactions were initiated by butting an opponent. The majority of such encounters were effectively ended by the initial butt, but in some cases a period of butting or horning between the two protagonists would result. Most such fights were completed in less than 30 s, with the submissive animal either fleeing, squatting and urinating, or displaying a head shake. Squatting and urinating in response to another sheep is a submissive posture, although it is more common in ewes in interaction with rams. The head shake is performed by ewes in response to an encounter with a dominating ewe or to pestering by an adult male (Geist, 1968). Occasionally, ewes may use the front leg kick against an opponent or rub or nuzzle another ewe in submission. Blocking behaviours are used to prevent other oestrous females from accessing a ram during the breeding season.

Male–female

Adult rams generally treat females in a similar fashion to other subordinates, and will commonly approach them in a low stretch position, often with a twist of the head. If the ewe is near oestrus the ram will follow, tend or try to mount her. A ewe will not challenge a ram, but may show head-shaking behaviour or flee in response to persistent unwelcome attention.

The development of ram dominance over ewes usually occurs during the late yearling stage for young rams present in ewe groups. It is at this time in Bighorn sheep groups that the young rams are closely matched to adult females in size and appearance (Geist, 1971).

Rams higher in the dominance order tended a much greater number of ewes during the rut than lower-ranking rams (Grubb, 1974b). It is uncommon for rams of wild sheep types to form and protect harems of ewes, although rams have been observed to shield pairs of ewes.

During the process of courtship and mating, the low stretch and twist approach of the ram is followed by nosing and sniffing of the ewe's vulva and perineal regions. The mild threat of the low stretch approach by the ram may be a behavioural compromise which allows it to get close while still exerting dominance over the ewe. This low stretch approach to the ewe is common in Bighorn, Mouflon, Soay and domestic sheep. The initial courtship by the ram may be accompanied by a low-pitched vocalization, tongue protrusion and nibbling of the wool of the ewe. If the ewe is recumbent, the ram will often perform a foreleg kick to the flank of the ewe to induce her to stand. Such a foreleg kick, which may also be delivered during courtship to a standing ewe, is generally more ritualized and delivered with less force than during agonistic interactions between rams.

Parent–offspring (see Chapter 3)

Although ewes may show interest in newborn lambs at other times, true maternal responses and acceptance of neonates is only expressed around the time of parturition. Maternal behaviour at this time is strongly correlated with the fall in progesterone and rise in oestrogen which occur just before parturition (Shipka and Ford, 1991). Peri-parturient ewes show a strong attraction to birth fluids, especially amniotic fluid. This attractiveness of birth fluid and subsequent lamb-directed behaviour, such as licking of the neonate and acceptance of suckling, are enhanced by oxytocin release stimulated by the stretching of the ewe's vagina during the birth process (Keverne *et al.*, 1983). The experience of the ewe plays a large part in maternal behaviour and acceptance of the lamb. Multiparous ewes tested by Levy and Poindron (1987) maintained normal levels of maternal behaviour when their newborn lambs were washed to remove amniotic fluids. In contrast, washing the lambs of primiparous ewes substantially reduced normal maternal behaviour and increased aggression towards the lamb.

The presence of birth fluids on the neonate stimulates the ewe, when she rises following the birth, to lick the lamb and remove mucus and membranes from its body. The ewe commences by licking the head area of the lamb and spends most time on the head and anterior body, subsequently concentrating her attention on the hindquarters and tail of the lamb. This licking may serve to stimulate the lamb and also helps the bonding process and recognition of the lamb by the ewe. Ewes will often vocalize with a low rumbling sound while licking the newborn lamb.

Lambs of wild sheep breeds will typically stand within 20 min of birth, with Soay lambs observed to stand within 10 min (Shillito and Hoyland, 1971). Following standing, a healthy lamb will quickly attempt to suck. Interactions between the lamb and the ewe facilitate the finding of the udder and the teat by the lamb. A study by Vince (1987) showed that newborn, unsuckled Soay and Clun Forest lambs responded to tactile contact over the face and eyes by vigorously tilting up their muzzle and lengthening the neck in the direction of the stimulus. The lambs also opened their mouths and displayed reaching and grasping movements with their jaws, lips and tongue. In the same study, contact to the abdomen of the ewe caused her to arch her back upwards, while touch closer to the udder was associated with an outward movement of the hind leg, exposing the teat. Bighorn ewes observed by Geist (1971) also arched their backs in response to tactile stimulation from the lamb, but moving the hind leg was observed only once.

The sensitive period for maternal bonding and subsequent recognition and acceptance of the lamb lasts for only a few hours, and is dependent on hormonal changes occurring at the time of parturition. Alexander *et al.* (1986) suggested that the first hour post-partum was critical for the establishment of the ewe–lamb bond, following a study of Merino ewes and lambs. The establishment of recognition of her own lamb by a ewe is primarily dependent on olfactory cues. During the first few days, it is the ewe which is responsible for maintaining the selectivity of the mother–young relationship. Upon rejoining the group, the lamb may be vulnerable to being poached or butted by other ewes. In wild species, the duration of the seclusion period following parturition is generally sufficient to allow the development of recognition of the ewe by the lamb. The development of this recognition commonly takes between 3 and 6 days.

The location and recognition of the lamb by the ewe are aided by visual, auditory and olfactory senses. Although confirmation of the identity of the lamb is dependent on smell, sight and hearing are used by the ewe to locate the lamb. After a couple of weeks, the lamb no longer remains continuously in close association with the ewe, and ewes and their lambs will search to find each other, bleating loudly.

Lambs display a follower relationship with their dams, and do not hide singly in cover. Bighorn ewes and lambs studied by Shackleton and Haywood (1985) remained close together upon rejoining the flock after parturition, but then, after a few days, spent less time together as the lamb became more active. The mean distance between the ewe and the lamb subsequently decreased again, coinciding with the commencement of grazing by the lamb (Fig. 8.6). This close spatial relationship during grazing is thought to assist learning by the lamb. Lambs have been shown to learn to eat new foods by observation (Lynch *et al.*, 1983), and it is likely that the proximity of the lamb to its mother as its starts to graze helps the lamb in learning appropriate diet selection.

When the ewe and lamb are separated, it is usually the lamb which returns to the ewe, particularly as the lamb ages, and almost always to suck. Suckling bouts fulfil two functions – nutritional and social. The duration of suckling bouts decreases during the course of lactation,

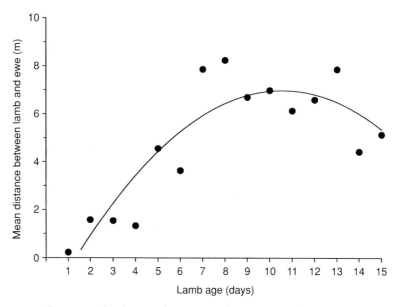

Fig. 8.6. Changes in the distance between Bighorn ewes and their lambs as a function of lamb age. (Adapted from Shackleton and Haywood, 1985.)

with a steeper decline during the first month after birth, and a gradual decline thereafter until weaning. Suckling bouts of less than 10 s duration are often terminated by the lamb, and are thought to be primarily social in function, strengthening the ewe–lamb bond. Longer, more nutritive suckling bouts are almost always terminated by the ewe, usually by walking away.

Weaning is also prompted by the ewe, which permits fewer and fewer suckling bouts. Geist (1971) commonly observed 3- to 4-month-old Bighorn lambs sucking in September, but the same lambs were suckled only occasionally in October and rarely in November. The last observed suckling bout occurred in December. The weaning process coincides with the rut, but more significantly also coincides with a reduction in feed resources for the ewe. Nutritional studies in domestic (Merino) sheep indicate that the milk yield of the ewe is a major determinant of the duration of lactation, and that below a certain threshold the lamb is weaned (Arnold *et al.*, 1979). This concurs with the results of studies of wild sheep (Berger, 1979a), in which ewes weaned their lambs in conjunction with energetically stressful periods.

There is conflicting evidence as to whether any association between ewes and their offspring persists during the yearling stage and into adulthood. Rams disperse from the ewe group of their birth, but, in Soay sheep, ewes have been reported to maintain associations with their dams (Grubb, 1974a). In studies with domestic breeds, Merino ewes have been observed to maintain long-term spatial associations with their offspring (Hinch *et al.*, 1990), whereas no such associations were observed in a study of Scottish hill sheep (Lawrence, 1990).

Juvenile interactions

Within the ewe home range group, the lambs soon form juvenile bands when they are not with their dams. Geist (1971) observed that Bighorn lambs could be seen grouped in juvenile bands from around 2 weeks of age. The juvenile band would exist when the ewes were stationary or feeding, but the lambs would return to their mothers to suck or when the ewes were moving or resting. Babysitting of Bighorn lambs did not occur, as juvenile bands were not observed to associate with any particular ewe. Occasionally ewes would move some distance away to feed, leaving their lambs in the juvenile band for several hours. Soay lambs studied by Grubb (1974a) were most commonly grouped into juvenile bands late in the afternoon, when they would play. Soay lambs also grouped together to graze, rest or to explore.

Although lambs which encounter each other from a few days after birth may display mutual investigative behaviour, bouts of play develop from several weeks of age. Play in lambs includes components of sexual, agonistic and allelomimetic behaviours, such as gambolling (running and jumping together). Movements such as neck twists, heel kicks and gambolling indicate an intention to play (Berger, 1980).

Most play behaviours in lambs closely mimic adult behaviour patterns. A comprehensive study of play in Bighorn lambs by Hass and Jenni (1993) recorded play interactions categorized as sexual (mount, twist and front leg kick), agonistic (butt, clash, threat jump, horn threat, present, face rub, shoulder push and low stretch) and play-specific (head touch, neck wrestle and neck twist). Play bouts were short, averaging 1.5–2.3 min, and the incidence of play peaked in mid-summer when the lambs were between 9 and 11 weeks of age. Butting, shoulder pushing and mounting were the most commonly observed social interactions. Male lambs mounted more and generally played more than female lambs. This result is similar to studies of play in domestic sheep. In Dorset sheep, male lambs were more likely to mount, whereas females were more commonly observed to gambol (Sachs and Harris, 1978). Play behaviour in sheep mainly occurs within age and size groupings, and is uncommon in yearling or older sheep. The environment appears to influence play behaviour, as the frequency of play differs between mountain and desert-living Bighorn sheep (Berger, 1979b).

Interactions among juvenile siblings are uncommon among wild sheep types, as the incidence of twinning is rarer than in domestic breeds (Geist, 1971; Schaller and Mirza, 1974). Studies in domestic sheep indicate that the strength of sibling bonds may vary between breeds. Arnold and Pahl (1974) recorded that only 27% of twin Merino lambs formed a strong association. In contrast, Shillito-Walser *et al.* (1981) observed that sibling pairs of Dalesbred and Jacob sheep formed associations which existed up to 14 weeks of age.

8.2 Social Behaviour Under Commercial Conditions

8.2.1 Social groupings

The size of farmed groups of sheep varies widely, from small flocks of just a few animals held in small areas, to farms with hundreds or even thousands of animals grazing extensively. The ewe group forms the basis of many commercial sheep enterprises. Breeding ewe groups are used for milk and wool production and breeding replacement. Ewe flocks are often of mixed age, with younger animals being introduced into the breeding group and older ewes removed due to age, teeth wear or some other disability. Depending on the environment and production system, young ewes are commonly mated when they are 1.5–2.5 years of age, and are then typically kept for four to five breeding seasons.

The degree of cohesion between animals varies with breed. Mediterranean types (e.g. Merino) are tight flocking and maintain close contact in large groups whereas English lowland types are less gregarious and disperse more widely. Scottish hill types of sheep are some of the least gregarious, forming small subgroups with relatively large inter-animal distances (Dwyer and Lawrence, 1999).

Although dominance hierarchies have been described within commercial ewe groups, these are often not stable, especially if the ewes are similar in age. Lynch *et al.* (1989) investigated social organization in groups of yearling Merino ewes. Although dominance rankings were established during each observation period, these were not consistent between observation periods. Similarly, flock leadership was not consistent within the same-age groups. It was proposed by Lynch *et al.* (1989) that social organization among Merino ewe groups of similar age and size was more dependent on the gregarious and following characteristics of sheep than on social dominance or leadership. In contrast, studies with Scottish Blackface ewes revealed a stable hierarchy over a 3-month period (Lynch *et al.*, 1985). It is possible that domestic sheep breeds that are less gregarious may exhibit stronger dominance hierarchies.

When farmed ewes are lambing, the withdrawal period from the flock is shorter or absent compared with wild species. Ewes are often lambed in smaller, sheltered paddocks where they can be supervised, or lambed indoors. Because domestic sheep breeds have often been selected for fecundity, multiple births are much more common among domestic sheep than wild types. Ewes may display a preference towards one lamb of the litter, or even reject an offspring. Most commercial farming operations wean lambs by removing them from the ewe group. This commonly occurs between 12 and 16 weeks of age, but can be as early as 4 weeks in some specialized systems.

Rams are kept mainly for breeding purposes at a ratio of about 1 : 50 ewes. The males are kept separate from the females in relatively small groups apart from during the breeding season. The social organization of domestic rams in groups is very similar to that of wild sheep in their bachelor groupings. With many commercial rams not having horns, dominance is often based on size and weight, which are generally proportional to age. Commercial rams are usually culled before they become old. Fights will occur more often between rams of similar size, especially if they are the largest rams in the group. Mounting of other rams in the group is commonplace. Ram groups are often managed so as to avoid regroupings and the introduction of new animals, which result in an increase in aggressive interactions.

The social interactions between rams when mixed with ewes will be influenced by the number of males in the group. When the rams are mixed with the ewe groups for breeding, the number of rams used will depend on the size of the ewe flock and the terrain and dispersal of the sheep. Although one ram can mate with well over 100 ewes (Allison, 1975), a ram inclusion level of 2% or more is more common. Where young or less experienced rams are among those introduced into the ewe flock, the inclusion rate is increased.

Although a single ram will be the dominant individual in a mixed-sex group of sheep (Stolba *et al.*, 1990), most of the activity of the ram directed towards the ewes will be sexual, with the ram seeking oestrous females. Under more intensive conditions, or where the ram inclusion rate is relatively high, rams may spend more time fighting. Under such conditions a dominant ram will prevent a subordinate from mating. Subordinate rams may perform a large amount of ewe searching and may get to mate with ewes that are in early or late oestrus, but the dominant ram will mate with the ewe in peak oestrus. Dominant, experienced rams will often concentrate their activity around important areas such as watering points.

Social interactions between ewes can influence the mating process. Occasionally, ewes in mid-oestrus will interfere with the tending of another ewe by the ram. Younger ewes may be displaced by this type of activity and may also be less overt than older ewes in their behavioural signalling of receptivity.

The existence of large flocks of wethers (castrated males) in commercial farming has no parallel in wild sheep populations. Castration is usually performed before the sheep reach puberty. Although dominance hierarchies have been described for groups of wethers, their social behaviour, like ewes, appears not to be strongly influenced by dominance relationships. Squires and Dawes (1975) found a near-linear dominance hierarchy among Merino wethers at a feed trough, but a Border Leicester wether flock had a less clear hierarchy. In both groups there was a strong correlation between position in the flock when it was moving and social dominance, with more dominant animals being closer to the front of the group. In contrast to these results, Dove *et al.* (1974) determined a linear dominance hierarchy among housed Corriedale wethers that did not correlate with position during flock movement. It is likely that dominance and leadership in wethers (as with ewes) may be context-specific, and that dominance only becomes a significant factor when access to a valuable resource is restricted.

8.2.2   Effects of group size and space allowance on social behaviour

In general, four or five animals may be considered the lower limit for the number of sheep to constitute a socially stable group. At group sizes lower than this, behaviour may not be species typical and feed intakes may be reduced. This lower limit varies with breed, with sheep types with a strong flocking instinct such as the Merino being less comfortable in very small groups than less gregarious breeds. Scottish Blackface sheep, for example, under extensive conditions, may split into groups consisting of as few as four animals (Hewson and Wilson, 1979). Increases in group size do not have large impacts on social behaviour in sheep, unless space is restricted and resources (particularly feed) are limiting. Dove *et al.* (1974) recorded more rigid dominance hierarchies in housed sheep with reduced space allowance. These effects can lead to increased competition at feeding troughs as group size is increased relative to the amount of space available.

At pasture, increases in group size (as determined by the farmer) may lead to the establishment of a single larger foraging group of sheep with no discernible change in social behaviour, or the animals may split into subgroups. It is uncertain how many other sheep an individual sheep can recognize, but as social organization in large commercial flocks of sheep appears to be more dependent on following behaviour than on social hierarchy, lack of individual recognition may not alter social behaviour within large groups. Squires (1974) studied a flock of 1000 Merino ewes, which remained as a single group. The sheep foraged as a dispersed flock, and was led by a few sheep on journeys to watering points, with the rest of the flock following in a broad triangular formation. When sheep from different flocks are mixed together, they may remain as separate subgroups for some time afterwards. Two groups of Merino ewes of the same age which were put together by McBride *et al.* (1967) did not fully integrate for 20 days following mixing (Fig. 8.7). Sheep from different breeds may never fully integrate when kept in the same fields (Arnold and Pahl, 1974).

The formation of subgroups of sheep at pasture is partly determined by feed availability. Whereas sheep are likely to remain as a single group or as a few large groups when feed is not limiting, flocks may split into a number of smaller, widely distributed subgroups during periods of feed scarcity. The formation of subgroups is also more likely in undulating terrain, or where there are wooded areas. The introduction of rams into ewe flocks may influence the formation of subgroupings. Fletcher and Nicolson (1976) observed that a flock of Merino ewes during mating was usually split into numbers of subgroups during the day. Each subgroup generally consisted of less than 30 ewes and at least one ram.

The size of sheep subgroups is influenced by breed and age. Studies by Arnold and Pahl (1967) indicate that Merino sheep form larger subgroups than British breeds, and that subgroup size increases with age (Tables 8.2 and 8.3).

As outlined earlier, decreasing space allowance for housed sheep increases the influence of dominance effects as expressed by increased

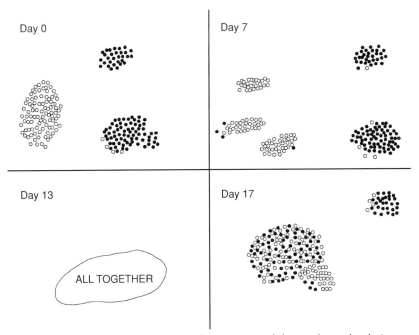

Fig. 8.7.   The progressive integration of two groups of sheep, shown by their positions as they rested. Each group consisted of 100 Merino ewes of the same age. (Adapted from McBride *et al.*, 1967.)

Table 8.2. Mean subgroup sizes for different breeds of sheep under the same conditions. (From Arnold and Pahl, 1967.)

Sheep breed type	Mean number in subgroups
Merino (adult)	44.9
Merino (9–12 months)	32.8
Border Leicester × Merino (9–12 months)	29.4
Dorset Horn × Merino (9–12 months)	13.9

Table 8.3. Mean subgroup sizes within the same Merino sheep flock at different ages. (From Arnold and Pahl, 1967.)

Sheep age (months)	Mean number in subgroups
4	8.4
7	18.0
8	25.3
11	34.0

competition for resources. At pasture, altered space allowance can influence flock dispersal, but this appears to be dependent on breed. Arnold and Maller (1985) found that Merino wethers actually decreased the space that they occupied from 60 m² per individual sheep to 30 m² per sheep as the available paddock area was increased from 345 m² to 1250 m² per sheep. In contrast, Corriedale wethers did not alter their dispersal with changing paddock size.

So, although dispersal patterns vary with topography and feed availability, certain gregarious breeds (e.g. Merino) are more suited to grazing pastures where the resources are relatively evenly distributed. Other breeds which more readily split into subgroups are better adapted for grazing in environments where food is distributed patchily or at higher altitudes (Dwyer and Lawrence, 1999).

8.2.3 Social effects on production

In general, competition within groups of sheep at pasture is minimal, and exerts little influence on feeding and productivity. Where feed quality or availability is reduced, sheep tend to increase their level of dispersion during grazing. In addition, sheep adjust their social organization and dispersal patterns to suit changes in topography or other geographical features. Competition effects can occur in sheep where a valuable resource, usually food, is presented within a confined space. This may happen during housing, or when sheep on poor pasture are supplemented with concentrated feeds such as grain. Arnold and Maller (1974) found that competitive displacement occurred at supplementary grain troughs among both ewe and wether flocks. As the available trough space per animal was reduced, the level of competition increased, until some individual sheep were effectively 'non-feeders'. Not all inhibition of feeding occurs through displacement; many sheep passively avoid feeding in a competitive situation. The critical trough length per animal for both ewes and wethers of varying breeds was approximately 16 cm per animal. Greater trough lengths of approximately 40–50 cm per animal are recommended where sheep are continuously confined (Kilgour and Dalton, 1984; O'Toole, 1984).

Social competition may not always be the reason for non-feeding individuals within groups of confined sheep. Studies investigating the problem of inappetence among Australian sheep in feedlots during preparation for shipment have shown that reducing the potential for competition by drafting out non-feeders reduces inappetence in some, but not all, animals (Norris *et al.*, 1990).

Pure dominance effects on productivity in sheep are largely restricted to rams (see Section 1.5). Because of the relationship between dominance rank and access to mid-oestrous ewes among groups of rams, a sub-fertile but dominant ram can be responsible for a lowered pregnancy rate among a ewe flock. The use of defined and limited mating periods in most management systems can enhance this adverse effect. Fowler and Jenkins (1976) investigated the pregnancy rates of ewe flocks joined with ram groups in which either the dominant or subordinate rams were infertile. The pregnancy rate among ewes joined with rams in which the dominant individuals were infertile was 72%. In contrast, ewe flocks joined with ram groups of either normal fertility, or with infertile subordinates, had pregnancy rates of 90%. Therefore, it is recommended that rams are changed part way through the breeding season to reduce the risk of an infertile dominant ram.

Because of the strong social nature of sheep, almost all farming systems endeavour to keep a number of individuals together wherever possible. Sheep can be strongly stressed by social isolation, unless they have been reared as 'pets'. Such stress effects have been shown to be partially ameliorated by the use of a mirror to reflect the image of an isolated animal (Parrott *et al.*, 1988). Studies on groups of sheep ranging from one to 15 animals have shown that feeding is reduced among groups of less than four individuals (Penning *et al.*, 1993).

One aspect of social behaviour that has a large impact on the productivity of sheep farming is the strength and quality of maternal behaviour (see Chapter 3). Lamb death between birth and weaning is one of the major components of production loss in many sheep farming systems. Ewes which are less likely to be disturbed from the birth site during the immediate post-parturient period, or to desert their lambs when disturbed at other times, are more likely to rear their offspring successfully (O'Connor *et al.*, 1985).

In common with other species, multiparous ewes more commonly exhibit appropriate maternal behaviour than primiparous animals. In addition, domestic sheep breeds are known to vary in their maternal behaviour. Observations indicate that Merino ewes exhibit a lower incidence of good mothering behaviours than other breeds such as the Perendale, Cheviot and Border Leicester-Romney (Whateley *et al.*, 1974). Evidence for a genetic component to sheep maternal ability is supported by the capacity for breeders to select and breed lines of sheep differing in this characteristic. Selection for mothering ability may also be possible through the use of indirect tests, such as behaviour of a ewe in an arena in the presence of a human (Kilgour and Szantar-Coddington, 1995). Different breeds of lambs also vary in their ability to recognize their mother. Border Leicester × Merino lambs have been shown to be more attracted to ewes and more effectively to discriminate their own dam than pure-bred Merino lambs (Nowak and Lindsay, 1990).

The fostering of an orphan lamb on to a newly lambed ewe is a manipulation of sheep social behaviour that has been undertaken by shepherds throughout the ages (see Section 3.5). Approaches to fostering are generally based on (either singly or in combination): (i) giving the orphan lamb the odour of the ewe's natural lamb; (ii) giving the orphan lamb the odour of birth fluids; (iii) manipulating the birth tract of the ewe to induce maternal behaviour; (iv) masking the odour of the orphan lamb with a strong-smelling substance; (v) rendering the ewe anosmic or sedated; and (vi) restraining or otherwise inhibiting the ewe from rejecting the orphan. Experiments examining fostering strategies in sheep indicate that an approach incorporating the transfer of odour from the ewe's own lamb are the most successful (Alexander and Bradley, 1985; Alexander and Stevens, 1985a, b; Alexander *et al.*, 1985, 1987).

Many sheep management procedures require handling of the animals through yards, and enhanced farm efficiency and animal welfare may be achieved by improvements in the movement and behaviour of sheep during handling. Studies examining the influence of sheep social characteristics on movement through yards have shown that sheep will move much more readily when they are following or moving towards other sheep (Hutson, 1981; Franklin and Hutson, 1982). Sheep movement into unfamiliar or empty areas can be facilitated by the use of trained leader sheep (Bremner *et al.*, 1980).

8.3 Social Behaviour, Management and Welfare

8.3.1 Grouping and separation problems

Because dominance hierarchies do not exert a strong influence on the behaviour of ewe or wether groups of sheep under most commercial conditions, regrouping of these sheep classes does not usually lead to significant behavioural problems. It may take considerable time before different groups fully integrate after mixing. However, ram groups do exhibit strong dominance hierarchies, and ram regroupings can intensify the level of aggression which occurs among rams.

As lambs are usually weaned earlier under commercial practices than would be the case under natural conditions, the process of separation at weaning can induce some stress for lamb and ewe. Obvious signs of weaning-induced stress generally wane after a few days. Studies by Orgeur *et al.* (1998) recorded high levels of bleating and other indications of disturbance in ewes and lambs after separation at weaning, although these effects ceased after 2 days. Similarly, Rhind *et al.* (1998) measured increased plasma cortisol concentrations in lambs during the 72-h period following weaning. Guidelines for weaning lambs suggest ensuring that ewes and lambs are not within hearing of each other. Alternative strategies studied include progressive separation. Orgeur *et al.* (1998) found that a programme of daily separation for increasing periods of time during the 2 months leading up to weaning decreased behavioural disturbance in lambs and ewes following the final separation at 3 months of age. The presence of familiar juvenile conspecifics may give some form of social support to recently weaned sheep. There is evidence that stress in lambs caused by separation from their ewe is reduced when they are placed with their familiar conspecifics compared with being placed with unfamiliar lambs (Porter *et al.*, 1995).

8.3.2 Social isolation and facilitation

One potential problem with early weaning of lambs is that they may not have had sufficient time to learn appropriate grazing strategies from their dams. Studies with conditioned feed aversions have indicated that lambs learn which feeds to avoid by grazing in association with their mothers (Thorhallsdottir *et al.*, 1990). Sheep grazing extensively in hill country acquire information about the location and seasonality of resources through social learning (from dam to offspring). Thus, extra care is required when sheep are transferred to unfamiliar home ranges under extensive farming conditions.

Because of their strong social nature, sheep should not be penned alone. One exception is that housed ewes are often moved into an individual pen at the point of lambing, and remain alone with their offspring while the bonding period occurs. This procedure follows naturally occurring behaviour. Occasionally, sheep are penned individually for scientific studies. Although, many sheep undergo apparent adaptation to this situation, some individual animals do not adjust, and should be removed to be with a group of animals (Kilgour and Dalton, 1984). In operant conditioning studies, sheep do not perform consistently and predictably unless a social companion is present.

There is strong social facilitation of activities such as feeding, drinking and resting among groups of sheep (Arnold and Dudzinski, 1978). Where there is limited access to food, water or fresh air, as can sometimes occur in close confinement, then animals may smother each other when they all attempt to gain access to these resources at the same time. Such problems can be avoided by providing continuous availability of and/or greater space for access to the essential resources.

8.3.3 Dominance-related problems

Dominance-related behaviour may become a problem for sheep where feeding or lying space is restricted during housing. This may be exacerbated when animals of different sizes, classes or horn development are mixed together. However, most problems associated with dominance in sheep groups occur with rams. Rams can spend periods of time fighting, especially immediately preceding and during the breeding season. The fighting may be exacerbated where there are two or more rams which are evenly matched in horn, body size and dominance status. Rams may be injured during fights, and wounds may become infected by clostridia and other bacteria. Dominance behaviour in ram groups is also associated with mounting. Usually, it is the subordinate rams which are mounted. In intensive housing, such as during shipping, mounting behaviour may be a serious problem in ram pens, leading to heat stress and smothering. The inclusion of wethers in ship pens at a ratio of 1 ram to 1.8 wethers has been shown to reduce mounting behaviour to negligible levels (Black, 1997).

8.3.4 Abnormal behaviour

One of the most serious types of abnormal social behaviour in sheep is that of poor maternal behaviour by ewes, which was observed to be directly responsible for 16% of lamb deaths under farming conditions in Australia (Arnold and Morgan, 1975). Disturbance from the birth site is a critical influence in the failure of maternal behaviour in sheep at pasture, and may be caused by human interference, or by other sheep in lambing paddocks which are too densely populated. Studies with Merino ewes have shown that time spent at the birth site is inversely related to lamb separation and mortality (Putu *et al.*, 1988a). The incidence of wandering from the birth site and lamb desertion by ewes varies with breed. Lamb desertion and rejection are also more common where there are multiple offspring, and the ewe mothers only one lamb. Alexander *et al.* (1983) studied the separation of ewes from twin lambs and observed that pure-bred Merino ewes were much more likely to lose contact with at least one lamb during the neonatal period than Dorset Horn, Romney and Border Leicester × Merino ewes. Strategies to minimize ewe desertion of lambs include minimizing human disturbance of lambing ewes, providing adequate space in lambing paddocks to minimize interference and lamb poaching by other ewes, and ensuring an adequate level of pre-lamb feeding. In a study by Putu *et al.* (1988b), a reduced level of feeding during late pregnancy in twin-bearing ewes resulted in a desertion rate post-lambing of 19%, compared with a rate of 4% for well-fed ewes.

The failure of rams to copulate with oestrous ewes may be classed as abnormal behaviour, particularly when the ram is of full maturity. Young, sexually inexperienced rams often show reduced mating capacity when first introduced to ewes, especially if they have had no contact with females between weaning and puberty (Casteilla *et al.*, 1987). These effects are often transient, but may persist into later matings. In addition, some rams fail to exhibit any sexual interest in ewes, even after repeated or prolonged exposure. A study by Price *et al.* (1988) recorded that 18.5% of young rams failed to show any sexual activity during mounting tests. A further 7.4% of rams preferred to mount males rather than females, and were classified as being homosexual in orientation. This homosexual orientation was independent of male–male mounting exhibited in all-ram groups, which was equally performed by heterosexually oriented rams. Although the proportion of 'non-worker' rams may be reduced by increasing the level of contact with females during post-weaning development, practical management considerations may make this approach unsuitable.

Sheep are among the most strongly social of farm animals, requiring close association with numbers of conspecifics to reduce stress, to permit normal behaviour and to facilitate management and productivity. However, this requirement for close contact is not accompanied by a complex social organization within sheep flocks. Apart from rams, where dominance hierarchies strongly influence social interactions, gregariousness and follower characteristics are typical of sheep groups. These characteristics have undoubtedly contributed to the early domestication of sheep and cognizance of them is essential even today for successful sheep husbandry.

References

Alexander, G. and Bradley, L.R. (1985) Fostering in sheep. IV. Use of restraint. *Applied Animal Behaviour Science* 14, 355–364.

Alexander, G. and Stevens, D. (1985a) Fostering in sheep. II. Use of hessian coats to foster an additional lamb on to ewes with single lambs. *Applied Animal Behaviour Science* 14, 335–344.

Alexander, G. and Stevens, D. (1985b) Fostering in sheep. III. Facilitation by the use of odorants. *Applied Animal Behaviour Science* 14, 345–354.

Alexander, G., Stevens, D., Kilgour, R., de Langen, H., Mottershead, B.E. and Lynch, J.J. (1983) Separation of ewes from twin lambs: incidence in several sheep breeds. *Applied Animal Ethology* 10, 301–317.

Alexander, G., Stevens, D. and Bradley, L.R. (1985) Fostering in sheep. I. Facilitation by use of textile lamb coats. *Applied Animal Behaviour Science* 14, 315–334.

Alexander, G., Poindron, P., Le Neindre, P., Stevens, D., Levy, F. and Bradley, L. (1986) Importance of the first hour post-partum for exclusive maternal bond in sheep. *Applied Animal Behaviour Science* 16, 295–300.

Alexander, G., Stevens, D. and Bradley, L.R. (1987) Fostering in sheep. V. Use of unguents to foster an additional lamb onto a ewe with a single lamb. *Applied Animal Behaviour Science* 17, 95–108.

Allison, A.J. (1975) Flock mating in sheep. I. Effect of number of ewes joined per ram on mating behaviour and fertility. *New Zealand Journal of Agricultural Research* 18, 1–8.

Arnold, G.W. and Dudzinski, M.L. (1978) *Ethology of Free-ranging Domestic Animals.* Elsevier Scientific Publishing Company, Amsterdam, 193 pp.

Arnold, G.W. and Maller, R.A. (1974) Some aspects of competition between sheep for supplementary feed. *Animal Production* 19, 309–319.

Arnold, G.W. and Maller, R.A. (1985) An analysis of factors influencing spatial distribution in flocks of grazing sheep. *Applied Animal Behaviour Science* 14, 173–189.

Arnold, G.W. and Morgan, P.D. (1975) Behaviour of the ewe and lamb at lambing and its relationship to lamb mortality. *Applied Animal Ethology* 2, 25–46.

Arnold, G.W. and Pahl, P.J. (1967) Sub-grouping in sheep flocks. *Proceedings of the Ecological Society of Australia* 2, 183–189.

Arnold, G.W. and Pahl, P.J. (1974) Some aspects of social behavior in domestic sheep. *Animal Behaviour* 22, 592–600.

Arnold, G.W., Wallace, S.R. and Maller, R.A. (1979) Some factors involved in natural weaning processes in sheep. *Applied Animal Ethology* 5, 43–50.

Berger, J. (1979a) Weaning conflict in desert and mountain bighorn sheep (*Ovis canadensis*): an ecological interpretation. *Zeitschrift für Tierpsychologie* 50, 188–200.

Berger, J. (1979b) Social ontogeny and behavioural diversity: consequences for Bighorn sheep *Ovis canadensis* inhabiting desert and mountain environments. *Journal of Zoology* 188, 251–266.

Berger, J. (1980) The ecology, structure and functions of social play in bighorn sheep (*Ovis canadensis*). *Journal of Zoology* 192, 531–542.

Black, H. (1997) Sexual and agonistic behaviour modification associated with improved welfare of rams transported on long sea voyages. *New Zealand Veterinary Journal* 45, 123–124.

Blood, D.A. (1963) Some aspects of behaviour in a bighorn herd. *Canadian Field Naturalist* 77, 79–94.

Bremner, K.J., Braggins, J.B. and Kilgour, R. (1980) Training sheep as leaders in abattoirs and farm sheep yards. *Proceedings of the New Zealand Society of Animal Production* 40, 111–116.

Briedermann, L. (1992) Ergebnisse von Untersuchungen zur Reproduktion des Mufflons (*Ovis ammon musimon*). *Zeitschrift für Jagdwissenschaft* 38, 16–25.

Casteilla, L., Orgeur, P. and Signoret, J.P. (1987) Effects of rearing conditions on sexual performance in the ram: practical use. *Applied Animal Behaviour Science* 19, 111–118.

Crofton, H.D. (1958) Nematode parasite populations in sheep on lowland farms. VI. Sheep behaviour and nematode infections. *Parasitology* 48, 251–260.

Dove, H., Beilharz, R.G. and Black, J.L. (1974) Dominance patterns and positional behaviour of sheep in yards. *Animal Production* 19, 157–168.

Dwyer, C.M. and Lawrence, A.B. (1999) Ewe–ewe and ewe–lamb behaviour in a hill and a lowland breed of sheep: a study using embryo transfer. *Applied Animal Behaviour Science* 61, 319–334.

Eccles, T.R. and Shackleton, D.M. (1986) Correlates and consequences of social status in female bighorn sheep. *Animal Behaviour* 34, 1392–1401.

Festa-Bianchet, M. (1986) Seasonal dispersion of overlapping mountain sheep ewe groups. *Journal of Wildlife Management* 50, 325–330.

Festa-Bianchet, M. (1988) Seasonal range selection in bighorn sheep: conflicts between forage quality, forage quantity and predator avoidance. *Oecologia* 75, 580–586.

Festa-Bianchet, M. (1991) The social system of bighorn sheep: grouping patterns, kinship and female dominance rank. *Animal Behaviour* 42, 71–82.

Fletcher, I.C. and Nicolson, A. (1976) Group structures within a flock of Merino ewes and rams on a large paddock in semi-arid pastoral country. *Proceedings of the Australian Society of Animal Production* 11, 145–148.

Fowler, D.G. and Jenkins, L.D. (1976) The effects of dominance and infertility of rams on reproductive performance. *Applied Animal Ethology* 2, 327–337.

Franklin, J.R. and Hutson, G.D. (1982) Experiments on attracting sheep to move along a laneway. III. Visual stimuli. *Applied Animal Ethology* 8, 457–478.

Geist, V. (1968) On the interrelation of external appearance, social behaviour and social structure of mountain sheep. *Zeitschrift für Tierpsychologie* 25, 199–215.

Geist, V. (1971) *Mountain Sheep. A Study in Behavior and Evolution.* University of Chicago Press, Chicago, 383 pp.

Geist, V. and Petocz, R.G. (1977) Bighorn sheep in winter: do rams maximize reproductive fitness by spatial and habitat segregation from ewes? *Canadian Journal of Zoology* 55, 1802–1810.

Grubb, P. (1974a) Social organization of Soay sheep and the behaviour of ewes and lambs. In: Jewell, P.A., Milner, C. and Morton-Boyd, J. (eds) *Island Survivors – the Ecology of the Soay Sheep of St Kilda*. Athlone Press, London, pp. 131–159.

Grubb, P. (1974b) The rut and behaviour of Soay rams. In: Jewell, P.A., Milner, C. and Morton-Boyd, J. (eds) *Island Survivors – the Ecology of the Soay Sheep of St Kilda*. Athlone Press, London, pp. 197–223.

Grubb, P. and Jewell, P.A. (1966) Social grouping and home range in feral Soay sheep. *Symposium of the Zoological Society of London* 18, 179–210.

Grubb, P. and Jewell, P.A. (1974) Movement, daily activity and home range of Soay sheep. In: Jewell, P.A., Milner, C. and Morton-Boyd, J. (eds) *Island Survivors – the Ecology of the Soay Sheep of St Kilda*. Athlone Press, London, pp. 160–194.

Hass, C.C. (1991) Social status in female bighorn sheep (*Ovis canadensis*): expression, development and reproductive correlates. *Journal of Zoology* 225, 509–523.

Hass, C.C. and Jenni D.A. (1993) Social play among juvenile Bighorn sheep: structure, development and relationship to adult behavior. *Ethology* 93, 105–116.

Heffner, H.E. and Heffner, R.S. (1992) Auditory perception. In: Phillips, C. and Piggins, D. (eds) *Farm Animals and the Environment*. CAB International, Wallingford, UK, pp. 159–184.

Hewson, R. and Wilson, C.J. (1979) Home range and movements of Scottish Blackface sheep in Lochaber, north-west Scotland. *Journal of Applied Ecology* 16, 743–751.

Hinch, G.N., Lynch, J.J., Elwin, R.L. and Green, G.C. (1990) Long-term associations between Merino ewes and their offspring. *Applied Animal Behaviour Science* 27, 93–103.

Hogg, J.T. (1984) Mating in Bighorn sheep: multiple creative male strategies. *Science* 225, 526–529.

Hunter, R.F. and Milner, C. (1963) The behaviour of individual, related and groups of south country Cheviot hill sheep. *Animal Behaviour* 11, 507–513.

Hutson, G.D. (1981) An evaluation of some traditional and contemporary views on sheep behaviour. *Wool Technology and Sheep Breeding* 29, 3–6.

Kendrick, K.M. (1991) How the sheep's brain controls the visual recognition of animals and humans. *Journal of Animal Science* 69, 5008–5016.

Kendrick, K.M. and Baldwin, B.A. (1987) Cells in temporal cortex of conscious sheep can respond preferentialy to the sight of faces. *Science* 236, 448–450.

Keverne, E.B., Levy, F., Poindron, P. and Lindsay, D.R. (1983) Vaginal stimulation: an important determinant of maternal bonding in sheep. *Science* 219, 81–83.

Kilgour, R. and Dalton, C. (1984) *Livestock Behaviour. A Practical Guide*. Methuen, New Zealand, 320 pp.

Kilgour, R.J. and Szantar-Coddington, M.R. (1995) Arena behaviour of ewes selected for superior mothering ability differs from that of unselected ewes. *Animal Reproduction Science* 37, 133–141.

Knight, T.W. and Lynch, P.R. (1980) Source of ram pheromones that stimulate ovulation in the ewe. *Animal Reproduction Science* 3, 133–136.

Lawrence, A.B. (1990) Mother–daughter and peer relationships of Scottish hill sheep. *Animal Behaviour* 39, 481–486.

Lawrence, A.B. and Wood-Gush, D.G.M. (1987) Social behaviour of hill sheep; more to it than meets the eye. *Applied Animal Behaviour Science* 17, 382.

Lawrence, A.B. and Wood-Gush, D.G.M. (1988) Home range behaviour and social organization of Scottish Blackface sheep. *Journal of Applied Ecology* 25, 25–40.

Levy, F. and Poindron, P. (1987) The importance of amniotic fluids for the establishment of maternal behaviour in experienced and inexperienced ewes. *Animal Behaviour* 35, 1188–1192.

Lynch, J.J., Keogh, R.G., Elwin, R.L., Green, G.C. and Mottershead, B.E. (1983) Effects of early experience on the post-weaning acceptance of whole grain wheat by fine wool Merino lambs. *Animal Production* 36, 175–183.

Lynch, J.J., Wood-Gush, D.G.M. and Davies, H.I. (1985) Aggression and nearest neighbours in a flock of Scottish Blackface ewes. *Biology of Behaviour* 10, 215–225.

Lynch, J.J., Hinch, G.N., Bouissou, M.F., Elwin, R.L., Green, G.C. and Davies, H.I. (1989) Social organization in young Merino and Merino × Border Leicester ewes. *Applied Animal Behaviour Science* 22, 49–63.

McBride, G., Arnold, G.W., Alexander, G. and Lynch, J.J. (1967) Ecological aspects of behaviour of domestic animals. *Proceedings of the Ecological Society of Australia* 2, 133–165.

McClelland, B.E. (1991) Courtship and agonistic behaviour in mouflon sheep. *Applied Animal Behaviour Science* 29, 67–85.

Morgan, P.D., Boundy, C.A.P., Arnold, G.W. and Lindsay, D.R. (1975) The roles played by the senses of the ewe in the location and recognition of lambs. *Applied Animal Ethology* 1, 139–150.

Murie, A. (1944) *The Wolves of Mount McKinley.* United States Government Printing Office, Washington, DC, 238 pp.

Murphy, P.M., Purvis, I.W., Lindsay, D.R., LeNeindre, P., Orgeur, P. and Poindron, P. (1994) Measures of temperament are highly repeatable in Merino sheep and some are related to maternal behaviour. *Proceedings of the Australian Society of Animal Production* 20, 247–250.

Norris, R.T., McDonald, C.L., Richards, R.B., Hyder, M.W., Gittins, S.P. and Norman, G.J. (1990) Management of inappetent sheep during export by sea. *Australian Veterinary Journal* 67, 244–247.

Nowak, R. and Lindsay, D.R. (1990) Effect of genotype and litter size on discrimination of mothers by their twelve-hour-old lambs. *Behaviour* 115, 1–13.

O'Connor, C.E., Jay, N.P., Nicol, A.M. and Beatson, P.R. (1985) Ewe maternal behaviour score relationships with lamb survival. *Proceedings of the New Zealand Society of Animal Production* 53, 201–202.

Orgeur, P., Mavric, N., Yvore, P., Bernard, S., Nowak, R., Schaal, B. and Levy, F. (1998) Artificial weaning in sheep: consequences on behavioural, hormonal and immuno-pathological indicators of welfare. *Applied Animal Behaviour Science* 58, 87–103.

O'Toole, M. (1984) Sheep wintering systems and handling facilities. In: Geoghegan, P.V. (ed.) *Sheep Production. An Foras Taluntais, Dublin,* pp. 91–95.

Parrott, R.F., Houpt, K.A. and Misson, B.H. (1988) Modification of the responses of sheep to isolation stress by the use of mirror panels. *Applied Animal Behaviour Science* 19, 331–338.

Penning, P.D., Parsons, A.J., Newman, J.A., Orr, R.J. and Harvey, A. (1993) The effects of group size on grazing time in sheep. *Applied Animal Behaviour Science* 37, 101–109.

Piggins, D. (1992) Visual perception. In: Phillips, C. and Piggins, D. (eds) *Farm Animals and the Environment.* CAB International, Wallingford, UK, pp. 131–158.

Poindron, P. and Carrick, M.J. (1976) Hearing recognition of the lamb by its mother. *Animal Behaviour* 24, 600–602.

Porter, R.H., Nowak, R. and Orgeur, P. (1995) Influence of a conspecific agemate on distress bleating by lambs. *Applied Animal Behaviour Science* 45, 239–244.

Price, E.O., Katz, L.S., Wallach, S.J.R. and Zenchak, J.J. (1988) The relationship of male–male mounting to the sexual preferences of young rams. *Applied Animal Behaviour Science* 21, 347–355.

Putu, I.G., Poindron, P. and Lindsay, D.R. (1988a) Early disturbance of Merino ewes from the birth site increases lamb separations and mortality. *Proceedings of the Australian Society of Animal Production* 17, 298–301.

Putu, I.G., Poindron, P. and Lindsay, D.R. (1988b) A high level of nutrition during late pregnancy improves subsequent maternal behaviour of Merino ewes. *Proceedings of the Australian Society of Animal Production* 17, 294–297.

Rhind, S.M., Reid, H.W., McMillen, S.R. and Palmarini, G. (1998) The role of cortisol and beta-endorphin in the response of the immune system to weaning in lambs. *Animal Science* 66, 397–402.

Sachs, B.D. and Harris, V.S. (1978) Sex differences and developmental changes in selected juvenile activities (play) of domestic lambs. *Animal Behaviour* 26, 678–684.

Schaller, G.B. and Mirza, Z.B. (1974) On the behaviour of Punjab Urial (*Ovis orientalis punjabiensis*). In: Geist, V. and Walther, F. (eds) *The Behaviour of Ungulates in Relation to Management.* IUCN, Morges, Switzerland, pp. 306–323.

Shackleton, D.M. and Haywood, J. (1985) Early mother–young interactions in California bighorn sheep, *Ovis canadensis californiana. Canadian Journal of Zoology* 63, 868–875.

Shank, C.C. (1977) Cooperative defense by Bighorn sheep. *Journal of Mammalogy* 58, 243–244.

Shillito, E.E. and Hoyland, V.J. (1971) Observations on parturition and maternal care in Soay sheep. *Journal of Zoology* 165, 509–512.

Shillito-Walser, E., Willadsen, S. and Hague, P. (1981) Pair association between lambs of different breeds born to Jacob and Dalesbred ewes after embryo transplantation. *Applied Animal Ethology* 7, 351–358.

Shillito-Walser, E., Walters, E. and Hague, P. (1982) Vocal communication between ewes and their own and alien lambs. *Behaviour* 81, 140–151.

Shipka, M.P. and Ford, S.P. (1991) Relationship of circulating estrogen and progesterone concentrations during late pregnancy and the onset phase of maternal behavior in the ewe. *Applied Animal Behavior Science* 31, 91–99.

Squires, V.R. (1974) Grazing distribution and activity patterns of Merino sheep on a saltbush community in South-East Australia. *Applied Animal Ethology* 1, 17–30.

Squires, V.R. and Dawes, G.T. (1975) Leadership and dominance relationships in Merino and Border Leicester sheep. *Applied Animal Ethology* 1, 263–274.

Stolba, A., Hinch, G.N., Lynch, J.J., Adams, D.B., Munro, R.K. and Davies, H.I. (1990) Social organization of Merino sheep of different ages, sex and family structure. *Applied Animal Behaviour Science* 27, 337–349.

Thorhallsdottir, A.G., Provenza, F.D. and Balph, D.F. (1990) The role of the mother in the intake of harmful foods by lambs. *Applied Animal Behaviour Science* 25, 35–44.

Vince, M.A. (1987) Tactile communication between ewe and lamb and the onset of suckling. *Behaviour* 101, 156–176.

Whateley, J., Kilgour, R. and Dalton, D.C. (1974) Behaviour of hill country sheep breeds during farming routines. *Proceedings of the New Zealand Society of Animal Production* 34, 28–36.

Woolf, A., O'Shea, T. and Gilbert, D.L. (1970) Movements and behavior of Bighorn sheep on summer ranges in Yellowstone National Park. *Journal of Wildlife Management* 34, 446–450.

The Social Behaviour of Horses 9

Natalie K. Waran

Institute of Ecology and Resource Management, The University of Edinburgh, Edinburgh EH9 3JG, UK

(Editors' comments: Our knowledge of the social behaviour of wild horses is based on observations of feral and free-ranging animals, as no true wild horses currently exist. The basic social groups are bands of mares with a single stallion, and bachelor male groups. Bands move about within overlapping home ranges. However, the author points out that there are exceptions to these generalizations, depending upon the resources available and social conditions.

Under commercial conditions, horses often change social groupings, sometimes within a month of birth when mare and foal leave the breeding farm to return to their home. However, feral horses may also change social groups several times during their lifetime. Perhaps more than any of our other domestic species, horses are kept in partial or complete social isolation for much of their lives. Often when this is the case, humans become a significant part of the social environment of the animal. The management practices employed for this primarily recreational species affect the social lives of these animals and may result in unique problems due to isolation and human interactions.)

Horses kept for commercial reasons are mainly used for sporting or recreational purposes, although in some countries they are used as a source of meat, for production of milk and for draught or traction. Methods of keeping and managing horses under commercial conditions, therefore, vary considerably between countries and tend to be based mainly on traditional practices. However, although horses have proved very successful in their ability to cope with intensive management practices, there is an increasing awareness of the high percentage of such horses exhibiting abnormal behaviour, often at a cost to their health and performance.

9.1   Basic Social Characteristics

9.1.1   Composition and structure of social groups

Interestingly there are no truly wild horses living today, and so our understanding of the horse's natural behaviour under free-living conditions has been gained through studies of the social behaviour of those horses living in feral conditions. More recently attempts have been made to reintroduce Przewalski horses to the wild, with some success, and studies of their behaviour are currently being made. Most studies agree in their descriptions of the typical equine social structure, the most common social unit being the single male harem band or herd (Keiper, 1986) where one mature male lives in close proximity with several mature females and their offspring under 3 years of age (see Section 1.6). The multi-male band (Franke Stevens, 1990) has also been reported, but in general only one stallion (the dominant stallion) has been seen to mate with the females. Also common is the bachelor band, generally consisting of immature males of age 2 years or more, which have left their natal (or birth) herd and have not joined with another harem herd. Harems tend to be relatively stable units, especially where the mares are concerned. In the Pryor mountains of Montana, only 7.6% of adult females changed bands in one year and, on the island Shackleford Banks, 10.8% of mares changed herds (Keiper, 1986). However, in the Rachel Carson Estuarine Sanctuary in North Carolina, 1.5 km from the nearby island Shackleford Banks, up to 30% of the adult females appeared to change harems during the months preceding the breeding season in 1985 and 1986 (Franke Stevens, 1990).

9.1.2   Use of space

Herds and bachelor bands occupy specific undefended non-exclusive geographic areas of their environment, which are termed home ranges. A home range contains the resources important for survival such as watering holes, suitable grazing areas and protected areas, for avoidance of biting insects and thermal extremes. Home ranges vary in size between and within areas where feral horses have been studied. This seems to depend on the availability of the important resources and the available area; ranges have been reported from 0.9 to 48 km^2 in different study sites. Seasonal movements within the home range of the horses of the Red Desert (Miller, 1983) result in a rotational grazing system with some areas only being utilized at certain times of the year. Although home ranges may overlap, where different herds are forced to share a common resource such as a watering hole, group cohesion is maintained by the stallions. Group dominance tends to be determined by group size, and it is this that determines who has first claim on the resource (Berger, 1977).

Various studies show that horse groups vary in size and behaviour according to the type of land they occupy and the climate in which they live (see Section 1.4 and Table 9.1).

In the dry, open mountains of Montana, small bands of horses have large overlapping ranges and a few mares drift from group to group; in this harsh country, infant and old-age mortality is high. In the desert of the Grand Canyon, the ranges do not overlap so much and the bands remain separate, each probably in its own side canyon. Horses in this area tend to graze widely through the late winter and early spring and then concentrate their movements around their water source during the summer. But on the wetter islands the ranges are much smaller, probably due to the greater amount of food and water. On the relatively lush Shackleford Bank, the population density is so high that on one end of the island the stallions with large harems defend their territories. Such defence of a territory seems to be a result of the particular geographical features of the island. Where the island is narrow, the horses have an uninterrupted view of their environment, and defence of grazing areas is possible. This is most unusual among feral horses,

Table 9.1. Composition of four groups of feral horses.

	Pryor Mts, Montana[a] (open, mountain, fairly dry)	Grand Canyon, Arizona[b] (steep, desert)	Sable Island, east coast, Canada[c] (sandy, flat)	Shackleford Bank, east coast, US[d] (sandy and marshy)
Total no.	225	78	240	104
Density (km^{-2})	2	0.2	6.3	11
Age structure:				
Adults	58%	–	64%	61%
Youngsters	28%	–	21%	21%
Foals	13%	–	15%	19%
Home range	25 km^2	20 km^2	under 7 km^2	6 km^2
Defended territories	no	no	no	yes: 3 km^2
Herd stability:				
Mare changes per year	7.6%	none	–	10.8%

Source: [a]Feist and McCullough, 1976; [b]Berger, 1986; [c]Welsh, 1975; [d]Rubenstein, 1981.

since in most cases defence of the typically large open area that makes up the home range is neither efficient nor effective.

On the east coast Sable Island, as in the marshy Camargue (southern France), bands may group together in large herds when loafing or resting (up to 80 in the Camargue), especially in summer (Duncan and Vigne, 1979). Camargue horses have been shown to get bitten by horseflies less often when they are in groups. The fact that horses group in marshy places in the summer but not in the winter and do not group in drier areas suggests that they do so as a protection against bloodsucking flies. Although most adult stallions will not let other adult stallions near their mares, especially during the breeding season, they will tolerate such closeness if it means suffering less from flies.

9.1.3   Communication (see Section 2.3)

Sight and body language

As a prey species, the eyes of horses are large and sited at either side of its head. This means that the horse can achieve a wide visual field, almost 357°, when its head is held high. It also means that the blind spot directly behind the horse coincides with the place normally occupied by a rider, and the blind spot directly in front of its nose ensures that it cannot see exactly what it is eating when it is actually eating (Fraser, 1992). The horse's panoramic view also depends on monocular vision, that is, it has the ability to see separate views with each eye at the same time.

A wide field of vision has obvious survival advantages in terms of detecting predators and also for ensuring good visual contact with the rest of the herd. However, this wide field of vision may be at the expense of visual acuity at close ranges. The muscles around the optic lens of the horse are relatively weak, and so, although some accommodation can be achieved, much is likely to occur through movements of the head and neck. This means that, if the horse lowers its head, it can see close objects in sharper view, but if it raises its head it can see well over long distances. The movements of the head appear to ensure that the image of an object falls on the most sensitive area of the retina, the visual streak (Hebel, 1976). This is a sort of elongated fovea, which is the area directly along the main axis, on the same horizontal plane as the eye. It is suggested that this phenomenon explains why horses will often shy away from an object that appears to have been in their visual field for some time. It is likely that a change in head position may have resulted in the object suddenly being clearly visible, almost as if it appeared from nowhere (see Budiansky, 1998). In addition, a horse's ability to see a particular stimulus will depend on factors such as breed, gait, training and state of arousal (Saslow, 1999), since all of these are likely to influence head and neck carriage.

However, when an object is in focus it appears that the horse is extremely good at detecting very small detailed movements; much of its communication system relies on its ability to perceive very small changes in body posture, such as slight changes in the position of the ears. However, the work that has been carried out on the visual acuity of horses is limited, and dependent upon being able to train horses to distinguish between unnatural stimuli for a food reward (Timney and Keil, 1992). Conclusions drawn from the testing of three horses using this sort of methodology suggest that the visual acuity of the horse is limited by ganglion cell density in the temporal portion of the narrow visual streak (Timney and Keil, 1992).

The horse has good night vision due to the tapetum, which is a layer of cells that acts like a mirror, reflecting light back into the eye to allow the optical cells to use all the available light. In addition the horse's retina is rich in rods, the cells that are most sensitive to dim light. Although horses are not nocturnal by nature, they are active for some of the night. The importance of the ability to see in dim light is probably due to the importance of being able to detect ambushes by nocturnal predators, and for group cohesion both at night and in dimly lit habitats such as those inhabited by some of the ancestors of the domestic horse.

The horse's retina also consists of cells called cones, which are associated with colour vision. Earlier research suggested that horses were dichromates, with limited colour vision in the red–blue colour spectrum (Pick, 1994). However, there appears to be some disagreement between researchers in the colour-seeing abilities of horses. Recent research (Smith and Goldman, 1999) has indicated that horses can discriminate the colours red, yellow, green and blue from various shades of grey. This disagreement between researchers may be because of the ways in which colour vision has been tested.

However, there seems no doubt that horses can distinguish between some colours, and one possible function may be that of predator detection, which may be made more effective if the animal is able to see through camouflage, such as is the case if an animal can detect colour (Budiansky, 1998). Little is known about the role colour vision may play in individual recognition of members of a group.

Most social communication is through a wide array of highly sophisticated visual signals. Some body postures or outlines are more obvious than others to the human observer, such as the flattened ears indicating aggressive behaviour. Others are more subtle, such as facial movements: relaxing or tensing of the muscles around the nostril, mouth and chin. As with other species of animals it is the warnings of

LIVERPOOL
JOHN MOORES UNIVERSITY
AVRIL ROBARTS LRC
TEL. 0151 231 4022

aggressive behaviour that have received most attention in the literature. The horse, like other social animals, shows escalating warnings of aggression, which are often ritualized. Mild bite threats are displayed by laying back both ears and moving the mouth suddenly towards the stimulus (Waring *et al.*, 1975). In some cases, if the desired response does not occur, the horse lunges with an extended head and neck and will bite. Bite threats have been shown to account for 74% of the 488 agonistic interactions observed by Montgomery (1957) in a group of horses. Another social signal is the tail swish, which seems to indicate irritation, and if ignored can lead to escalation such as a lift of the hind leg and even a kick. In addition to the ears and tail, the horse appears to signal using its whole body. Horses in a high state of arousal exhibit a high postural tonus with head and tail held high, and often elevated paces. This posture makes the horse appear much bigger than he is, and is often seen in stallions. In contrast a drooping head and tail in a horse showing a low body posture is indicative of pain, distress and depression. The horse also expresses itself using facial expression; the wrinkling of the nose usually precedes a bite, and the drooping lower lip is observed in resting or relaxed horses. In addition, the shape of the mouth also changes during certain encounters, and the shape of the nostrils and eyelids all characterize certain visual expressions (Waring *et al.*, 1975). Although horses appear much more reactive than many of the larger domesticated species, the purpose of the displays is of course to ensure that individual space is maintained and damage is limited. By far the most common behaviour that occurs within a social group of horses is affiliative in nature. However, very little research has been carried out to identify the postures and signals that are used for this type of communication. This may be because of the importance to human handlers of predicting aggressive behaviour in their horses, and also because of the preoccupation of early animal behaviour scientists with characterizing dominance hierarchies.

Hearing and vocalizations

It is thought that horses hear the bulk of the frequency range audible to humans; however, the extent of their hearing capabilities is not known. It is thought that horses hear the low frequencies, or 'p' waves, that are apparent before an earthquake, but this is purely anecdotal. The audiogram for horses (see Heffner and Heffner, 1983) does suggest that they are able to hear sounds in the ultrasonic range that are inaudible to humans. Sound does not appear to be located very effectively despite the horse's large and mobile ears. Both are independently moveable, and are therefore able to rotate up to 180° to pick up the maximum amount of sound. The horse should be able to determine the localization of the sound, which could be of great importance to the horse in locating possible danger or the whereabouts of the rest of the herd (Fraser, 1992). Heffner and Heffner (1983) suggest that horses localize sounds no more accurately than a rat, despite their much larger inter-aural (between ears) distance. However, these results must be viewed with caution since the sample of horses was small and the testing procedure was limited to frontally placed sounds.

The language of the horse is a subtle one. As with most animals they rely much more on body language than do humans. Although vocalizations do appear to confer some information they are perhaps less specific and more limited. This is probably because horses living in herds are rarely out of sight of each other, and so can make use of visual signals. In addition, being adapted for life on the open plains, it may be that vocal signals were a less effective means of communicating and would have been more likely to attract the attention of predators.

There appear to be four basic types of horse vocalization: the nicker, the whinny (or neigh), the squeal and the groan. All of these have been recorded from horses in a variety of situations (Kiley-Worthington, 1987). The nicker is a low-pitched pulsating sound, about 100 Hz, and is formed with the mouth shut (Waring *et al.*, 1975). It is usually used as a greeting, for maintaining contact, especially between dam and young, and in anticipation of a pleasurable event such as feeding. The whinny (or neigh) is usually fairly loud, up to 2000 Hz (Budiansky, 1998). It appears to be used in a wide variety of circumstances ranging from social isolation to aggressive threats, and so may be thought of as a means of establishing contact since it appears to travel well over long distances.

The squeal is generally associated with aggressive or threatening encounters with other horses; it appears to act as a warning signal to ward off unwelcome attention. The groan is used by horses in anguish or discomfort, such as by a mare struggling to deliver a foal. Other sounds made by the horse are non-laryngeal, that is, they are not produced using the larynx. Of these, the snort appears in conflict situations, but also when the horse is clearing its airways and when it encounters an interesting odour. The snort of the herd stallion often appears to act like an alarm call, attracting the attention of the members of the herd. The blow seems to indicate high arousal when the horse is in a state of anxiety, perhaps due to a novel object in its path. This sound can carry up to 200 m (Budiansky, 1998) and appears to have similar characteristics to the bark of the anxious dog.

It seems likely that the horse's vocalizations serve more to attract attention and to relay information about the general state of arousal of the horse than as a means of transmitting specific information (Veekman and Odberg, 1978; Rubenstein and Hack, 1981; McCall, 1991). The context of the sounds and the visual signals given by the producer are probably used by the receiver for interpreting the more specific meaning of the communication.

Smell and taste

The sense of smell is important to the horse for exploring its environment, identifying feeding material and in group and individual recognition (Marinier and Alexander, 1988). The horse's long nasal passages allow it, through sniffing actions, to intensify the odour, thus enabling it to detect odour molecules. In addition, the horse's vomeronasal organ, situated on the floor of the nasal cavity, detects pheromones (Whitten, 1985). Through the behaviour called flehmen, in which the horse apparently curls its top lip to allow the air to 'drop' on to the organ, additional screening for the information transmitting pheromones can take place (Crowell-Davis and Houpt, 1985). This is often seen performed by the stallion during mating (Stahlbaum and Houpt, 1989).

Smell is also important in communication. Horses scent mark, deposit a group smell over their bodies and use scent to recognize their young. Feral stallions will defecate on top of the dung of members of their herd or other stallions. They also tend to defecate in certain selected areas called dung piles. Although these do not mark territory boundaries as with some other equid species, they do appear to be used to signal the presence of a particular stallion in an area. Detection of pheromones through the smelling of dung and urine imparts information regarding identity, sexual receptivity, perhaps even status, since depositing the last layer of dung on a pile is something that stallions have been seen to compete for (Kiley-Worthington, 1987).

Communication through odours is important for group recognition; horses deposit the group smell on themselves by rolling in communal rolling areas. Information about the sexual receptivity of mares is passed to stallions via pheromones in the mare's urine, which is produced frequency as the mare solicits the attention of the stallion (Fraser, 1992). Horses who have recently met or been parted for a time will usually spend a great deal of time sniffing one another. The odours of the breath and body must confer a great deal of information about the food that has been eaten, the state of health and other horses it has been in contact with (McGreevy, 1996).

9.1.4   Cohesion and dispersion

A herd is defined as a structured social unit, made up of groups (harems or bands) of horses that follow similar movement patterns within a common home range (Miller, 1981). Changes in the herd composition are due mainly to death and birth. In addition, within a herd, changes in group composition may occur through natal dispersal, that is, the movement, particularly of young horses, from one maternal group to another, as occurs in the Misaki horses of Kyushu Island in Japan (see Khalil and Kaseda, 1997). Emigration or the permanent movement of horses from birth site to first or potentially first breeding site (Greenwood, 1980) occurs regularly among young mares and colts. In the mares it is generally through abduction by other stallions or being driven out by older mares (Klingel, 1969, 1982). Young males will either leave voluntarily due probably to a reduction in food availability (Welsh, 1975), or they are forced out due to a new offspring or by the herd stallions (Franklin, 1974; Berger, 1986). Khalil and Kaseda (1997) found that on Kyushu Island the Misaki horses have arranged themselves into two areas defined by the two hills in the centre of the Misaki range. Dispersal between these areas has been shown to occur in both males and females, with 80% of young males apparently leaving their natal herd voluntarily and 20% being forced out when their younger siblings are born. In general, it has been shown that dispersal tends to be density dependent, with young mares leaving the herd when the density of young females in the natal herd is high, whereas males tend to leave if the density of young males at their destination is not too high.

The dispersal of young females is an important means of regulating interbreeding. In the harem system, there is usually one stallion with exclusive breeding rights, and as such reproductive success is improved through dispersal. However, where dispersal occurs less frequently, perhaps due to physical constraints such as the suitability of available land, other mechanisms must play a role in controlling interbreeding. On the Assateague Island off the US Atlantic coast (Keiper, 1985), the foaling percentage of mares bred to their father was found to be 23% as compared with the 62% found in mares bred to an unrelated stallion. This suggests that intrinsic mechanisms for preventing interbreeding exist.

9.1.5   Inter-group interactions

As stated previously, the harem stallion defends his females vigorously since his lifetime reproductive success depends on his ability to maintain his harem. In general the resident stallion will defend his mares from an approaching stallion, by herding his mares away. Home ranges vary in size depending on conditions (see Table 9.1). Where ranges do not overlap, inter-group interaction will be low, but, where overlapping occurs, harem stallions will be forced to defend their females. Generally fighting between stallions is confined to mock battles (Fraser, 1992), involving threats (rearing on hind legs and laying back of ears). However, they may escalate to real contests, especially in restricted conditions in captivity where horses are unable to escape. In this situation, injury is a likely outcome.

9.1.6   Intra-group interactions

Dominance relationships

The ability to form close social bonds has been identified in both feral and domestic groups of horses. In feral populations, dominance relationships or hierarchies result in a reduction in aggressive behaviour and an increase in group cohesion. In a stable social group, members learn early in life their place in the social hierarchy, position in that hierarchy being closely related to age, with younger members of the group occupying a lower position in the order (Houpt *et al.*, 1978; Houpt and Keiper, 1982). Other determinants of dominance are less clear: in some feral populations, males are dominant over females (Feist and McCullough, 1976); in others mares are dominant (Berger, 1977; Houpt and Keiper, 1982); while, in domestic horses, geldings have sometimes been shown to be dominant (see Section 2.4). Expressions of dominance tend to be subtle and, in general, aggressive behaviour is avoided since it expends energy and increases the risk of injury.

Individual dominance order (see Section 1.5) is unidirectional, but may not be linear throughout the group (Houpt *et al.*, 1978). The group order may be complex, but is usually fairly stable. In smaller groups, the linear hierarchy is usual, whereas, in large free-ranging herds, more complicated relationships are apparent. Rank appears to be influenced by a number of factors including prior fighting experience, skill, strength and stamina. In addition, particularly where the mares are concerned, rank may be inherited. This may be through the young learning the successful strategies of the dam. At certain times such as when the mares have a foal at foot, or if a mare is in season, rank appears to change temporarily. In general a high rank confers advantages only in times of conflict, such as competition for a favoured feeding site, for an oestrous mare or for proximity to the stallion.

Although dominance and subordinance relationships play an important part in the social organization of herds of horses, it is now recognized that social order is also maintained through tolerance and attachment relationships (Kolter, 1984). This theory of equine rank suggests that the degree of tolerance applied to a herd member by the highest ranking individual generates a subsystem of preferred status. Preferred horses are more tolerated by higher status individuals, and this modifies the whole of the social system. Preferred associates, indicated by reciprocal following and standing together, are also evident. It is likely that a horse's preferred grooming partner will be such an attached individual. This type of social system, where dominance (based on fighting ability) and tolerance (based on preferences) systems operate, better explains the manner in which horses appear to organize themselves.

Male–male interactions

Most harem mares are covered by the resident stallion, but some younger mares may be covered by the 2-year-old colts who are tolerated reproductively active members of the herd (McDonnell, 1986). The harem stallion ensures his reproductive success, by trying to keep his mares together by directing their movements away from other harems, and actively defending his mares against invading stallions. He maintains his herd by patrolling the edges while they are grazing or resting, and by following behind, or herding, while they are moving.

Male–female interactions

During the reproductive season, the stallion shows increasing interest in the mares in the days preceding oestrus. Approach behaviour by the stallion is characterized by a high posture and exaggerated gait, usually accompanied by whinnies and nickers. Nevertheless the early courtship advances tend to be aggressively responded to by the mare. She often clamps her tail over her perineum, and may display threats, even escalating to kicks. The stallion tends to move away in these circumstances. However, once the mare is in full oestrus, she actively seeks the stallion, maintains proximity to him and solicits matings. Pre-copulatory behaviour may begin with some aggressive interactions and accompanying squeals, but as the stallion sniffs the mare from the muzzle to the perineal area she becomes more responsive, and will show standing behaviour with the tail raised (Fraser, 1992). The stallion will also show a flehmen response after sniffing the vaginal fluids. Once mounted, the stallion may grasp and nip at the mares mane and neck while attempting intromission. Ejaculation is achieved after several deep thrusts and the stallion normally then dismounts.

Female–female interactions

Apart from the mother–young bond, the other associations that form within the herd include those between related and unrelated mares within the herd, the close relationships that occur between young of similar ages and the temporary close relationship that occurs between the oestrous mare and usually the herd stallion. Pair bonds are a common social strategy in horses, often identified through observation of grooming partnerships and nearest grazing neighbours. These seem to be particularly important for maintaining group cohesion, since it is due to the strong relationships between mares that herds stay together.

Mother–offspring interactions (see Chapter 3)

Development of social behaviour begins almost immediately after birth – the instinctive 'following' behaviour aids the foal's survival. As soon as the foal joins the rest of the herd with its mother, it begins to learn how to be a social animal and how to live in a group. Observational or social learning plays a large part in the young animal's behavioural development. If unmanaged, mares will wean their young shortly before the birth of their next foal, some 1 to 2 years later.

The foal, once born, attempts to stand almost immediately (Crowell-Davis, 1986). During this time the mare will lick the foal and fetal membranes extensively, after which she seems to be able to distinguish it from other foals (see Sections 3.4 and 3.5). The first nursing bout usually takes place soon after standing, the foal showing repeated attempts to locate the udder, until it grasps a teat and sucks usually 30–120 min after birth. Unless the mare is particularly aggressive towards her foal, she will aid this searching and sucking process by standing still with her back legs extended or one hind leg flexed. During the early days of life the foal will suckle frequently, up to seven bouts per hour during the first week, falling to one per hour at 6 months of age (Kiley-Worthington, 1987). During the first week of life the mare and foal relationship is almost exclusive: 90% of their time is spent within 5 m of each other. Proximity is maintained through the actions of both mare and foal. The mare repeatedly leaves the foal perhaps during a nursing bout and moves a short distance away, while the foal follows. This following behaviour, important in such a ranging animal, is an instinctive tendency in a foal that is refined through learning during the first week of life. Mare–foal recognition seems to be through smell when close to one another, but also through recognition of coat colour, as well as vocal cues such as neighs and nickers. Much of a foal's early experience while with its dam equips it for life with the herd. It will begin sampling grass even from a few hours old, and by the age of 1 year it may be grazing for 44 min h^{-1} with the rest of the herd.

Social behaviour is learned from an early age, as the foal begins to play with other similar aged foals and has to learn to avoid other mares with young and the herd stallion. By 8 weeks of age it almost never plays with its mother, all play activity being directed towards other foals or itself (Waring *et al.*, 1975).

Juvenile interactions

Finally the main behaviour observed between foals and juveniles is that of play. It seems to serve a number of purposes, although the exact benefits are not known. Play either alone or between young appears to be important in neuromuscular development, learning ability and social relationships (Crowell-Davis, 1986). Horses will run, leap and mimic adult combat and sexual behaviour, and manipulate objects such as sticks and stones. Although both colts and fillies engage in similar amounts of play, they tend to play quite differently. Colts tend to engage in more interactive play, often with the same sex, whereas fillies tend towards more general motor play, which is characterized by bursts of speed and exaggerated postures and gaits.

The social behaviour of feral horses is relatively well documented, and studies such as those described in this section, in which the horse's behaviour is subject to evolutionary and more natural environmental pressures, can be used to gain some insight into the behavioural needs of domestic horses kept under commercial conditions. Caution must be used when making comparisons, however, since selective breeding of our domestic breeds, which may lead to differences in their behavioural responses, does not necessarily mean that there are welfare problems. A more practical approach is to use the knowledge gained through studies of the feral horse to identify the sorts of behaviour that might be important to the domestic horse, and to test how important these are through further behavioural research.

9.2   Social Behaviour Under Commercial Conditions

Studies of domestic horse behaviour in managed conditions indicate that their behaviour has changed little through domestication (see Chapter 4). Despite selective breeding that has resulted in the present range of horse types, each with its own behavioural characteristics, much of the behaviour observed in wild or feral populations of horses can also be seen in domestic populations. For example, even when domestic horses are confined in restricted paddocks, they will attempt to use the space in the same way as can be observed in feral populations, using some areas for elimination, some for grazing and others for body-care such as rolling. Fraser (1992) states that 'there has been no convincing proof of the elimination of any equine behaviour essential to survival'.

9.2.1   Social groupings

There are numerous different ways in which horses are housed and managed under commercial conditions (Fig. 9.1). These range from tethering in stalls (straight stalls) to the relative freedom of open ranges. The way in which a horse is kept depends upon a variety of factors including its purpose, cultural traditions and availability of space. Typically most horses are used for one of four different purposes: (i) for sporting purposes such as dressage, racing, show jumping and polo; (ii) for recreation; (iii) for farm work; and (iv) for meat production.

There are a number of different methods for managing sports horses; however, the demands of competition and training are often associated with controlled exercise, restricted feeding regimes and

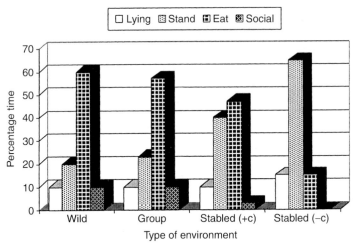

Fig. 9.1. Average time budgets for horses in different environments.
'Wild' = Camargue horses; Group = group-housed horses on a yard with *ad libitum*
hay and straw; Stabled (+c) = individually stabled horses fed *ad libitum* hay and
straw and able to see and touch each other; Stabled (–c) = individually stabled
horses fed restricted fibre and unable to see or touch each other. (Adapted from
Kiley-Worthington, 1987.)

housing for at least part of the day. Housing needs and performance
demands are quite different for sports horses as compared with the
food-producing animals, although many of the recommended housing
practices have been extrapolated from intensively housed livestock
(Clarke, 1994).

There are three basic types of housing for horses: the stall (known
as the straight stall in the US), in which the horse is usually tethered, is
the most restrictive. The stall is usually the length and width of a horse.
The horse usually stands facing a wall or partition, but sometimes faces
other stalled horses. Generally there is very limited access to other
horses, and no space to turn or move around voluntarily. The horse is
limited to some forward and backward movement, and lying down on
its brisket and sometimes its side if the tether allows. Although stalls
are less popular now than during the era of the working draught horse,
they are still used where space is limited or where traditional methods
of housing are valued.

The most common type of housing in Europe is the stable or
loose-box (also called the box stall in the US), where the horse has
limited freedom of movement, some external stimulation, and differing
degrees of access to conspecifics depending on the internal partitions
(Fig. 9.2). These may be solid, in which case there is no tactile contact
between neighbouring horses and visual contact with other horses is

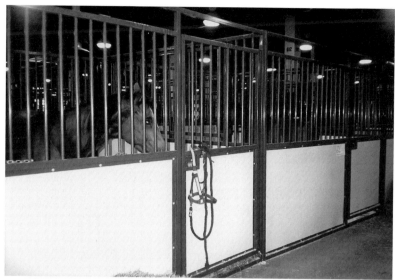

Fig. 9.2. The vertical bars in some loose-boxes (box stalls) allow limited social contact with neighbouring horses.

limited to a sideways view of the horse next door over a half stable door. In the US, horses are likely to be limited to the view they have between the vertical bars of their sliding door. In some cases the internal partitions may be solid to halfway, in which case metal bars are often used to prevent horses having too much contact. Loose-boxes are usually built side by side around a courtyard design, where horses may face a row of stables, or back to back in long lines. However, they may also be built into a large barn, so that horses share the same airspace, can see horses on all sides of them, and staff have an indoor working environment.

The third type of housing used for the performance horse is group or loose housing in barns or yards. Group housing is mainly used for keeping breeding mares or young stock together, since these are rarely handled, and so access to each individual every day is not necessary, and they do not require careful monitoring of exercise and feeding. This system works best where group composition can be kept stable, and where there is space available for horses to escape any unwanted attention from others in the group. Its advantages are that horses can interact freely, move fairly freely and live a more natural life. Group sizes vary from two or three to 60 or more and space allowances of the same dimensions as a loose-box per individual are recommended.

There are other variations on the above themes used for housing horses, and probably the best from the horse's point of view is the field shelter (run-out shed) with continual access to pasture or paddock (K. Houpt, personal communication).

9.2.2   Effects of group size and space allowance on social behaviour

Under commercial conditions an individual horse is unlikely to spend its life in the same commercial establishment. Horses are frequently sold and moved to new locations and often with this move in premises comes a different system of management. Even within one establishment a horse is unlikely to be maintained in the same social group throughout its life. Horses that are stabled for part of the day and turned out to grass with different horses for a few hours a day can be considered to be part-time group members. It may, where groups are constantly changing, be necessary for these horses to re-establish their social order each time they are turned out together. This as well as the restrictive nature of the domestic environment, e.g. limited space and access to feed, may account for the higher levels of aggressive interactions often reported in domestic horses. Groups of domestic horses given restricted supplementary feeding averaged 47 aggressive acts per hour compared with the 1.3 acts recorded among free-living horses (Houpt and Keiper, 1982), although this is not a usual feeding practice. However, where members of a part-time group remain the same for a period of more than a year, it has been shown that the hierarchy remains relatively stable (Houpt and Wolski, 1980). Using paired feeding tests it was shown that dominance hierarchies changed little over a period of 2 years.

Additionally, horses are often moved between groups, resulting in a constantly changing social order. Aggressive interactions and injury tend to be greatest where horses are offered supplementary feed under these conditions (Houpt and Wolski, 1980; Wood-Gush and Galbraith, 1987). Aggressive interactions away from the feed area were found to be three times less frequent than the number recorded at the feeder (Wood-Gush and Galbraith, 1987).

In captivity, space tends to be a limiting factor. Horses kept either in part-time groups or in changing groups are relatively confined. Interactions between horses are therefore more likely and aggressive interactions less easy to avoid. The greater frequency and intensity of aggressive behaviour recorded among horses kept in small paddocks as compared with open ranges (see Keiper, 1988) suggests that as with other animals aggressive behaviour increases as space decreases and invasion of individual space becomes unavoidable.

Group composition also tends to be unnatural in captive conditions. Stallions are rarely kept with mares. Breeding mares are usually kept together and their foals are removed at 6 months of age, but they tend to be kept in groups of similar age and are not generally kept with the young stock. After weaning, if young animals are kept together it is usually without the influence of older animals. Young horses are likely to be kept in single-sex groups. For colts this is perhaps not that unnatural, bearing some resemblance to the bachelor bands seen under feral conditions. In most cases such groups contain castrated males (geldings), a type of horse that rarely exists under feral conditions. However, gelding groups seem to form stable hierarchies and social behaviour between them appears to be similar to that recorded for other groups of horses (Wood-Gush and Galbraith, 1987). Mares kept with geldings will often solicit their attention when in oestrus (Kiley-Worthington, 1987), and many commercial studs use geldings as 'teasers' to stimulate and test the mare for oestrus. Entire males (stallions) tend to be segregated from an early age and are usually kept in for most of the time in loose-boxes (box stalls), out in grassless paddocks or sand arenas, although some may have limited access to grazing, usually alone.

9.3　Social Effects on Behaviour, Welfare and Production

The term 'behavioural problem' can be used to describe a whole variety of behavioural responses. These may be behaviours that are problematic particularly from the owner's point of view, for example they may be aesthetically disturbing or financially expensive, or they may also be a problem for the horse, perhaps affecting its welfare in some way. The problem with the term is that in essence it is only describing behaviour that differs from the norm, which assumes that there is a normal way for a domestic horse to behave. There are certainly some behavioural responses that are likely to affect the horse's biological fitness. These include such behaviours as self-mutilation by stallions, behavioural anorexia and hyperactivity. Behavioural problems occur so commonly among commercially housed horses (up to 35% of racehorses) that it is clear that these environments represent a distinct challenge to the adaptability of the animal. Comparing the behaviour of animals in such systems with the 'normal' behaviour of free-ranging horses may not be the best means of judging what is 'abnormal'.

In this section, some of the behavioural problems that are considered to be associated with commercial practices are described and possible causal factors are discussed.

9.3.1　Separation problems

Breeding practices

Although breeding seasons vary between latitudes and often in the individual horse, free-living horses tend to breed in the spring and summer. Where photoperiod and temperature regimes can be controlled under commercial conditions, mares may cycle all year around.

In these circumstances mares can be mated early in the year to enable the subsequent foal to have the whole of the spring and summer to grow. This is common practice on commercial studs where horses are reared for racing or showing. The birth of all racing thoroughbreds is registered as 1 January in their year of birth, whether born in February or July. Thus there are severe disadvantages for those born later when racing against other '2-year-olds'.

Mares usually foal away from home at the stud farm, where the stallion she is next due to be mated to is kept. Stallions very rarely travel to the mare (although more recently there has been an increase in the use of semen, collected at the stud farm and transported to the mare for artificial insemination). The traditional arrangement enables her to be mated to the stallion as soon as possible after foaling. During the first few days after birth it is usual for mare and foal to be left alone to bond, after which she may be mixed with other mares and their foals out at grass. Once successfully mated and tested in foal some 3 weeks after mating, the mare and foal are transported back to the home stud. Thus during the first few weeks of life most foals born under commercial conditions experience some changes in their physical and social conditions, which may be advantageous in preparing them for coping with the changes they will experience during their lives.

When a mare dies during or following foaling, it is common practice to attempt to foster the foal on to either a mare who has lost her foal and therefore has milk, or a tolerant older non-lactating mare or 'nanny'. The lactating mare is encouraged to accept the orphan by smothering it in her amniotic fluid or covering it in the dead foal's skin. In general foals reared by surrogate mothers appear to suffer few emotional disturbances later in life. However, in many cases a substitute mother cannot be found, and orphans are hand reared. Bottle-fed foals often exhibit behavioural problems later in life. They appear to become over dependent on humans and fail to learn normal social behaviour. This often results in them relating sexually to humans rather than to other horses, colts often show increased aggressive behaviour and this may result in them becoming problematic to handle and train (Fraser, 1992).

Early management

When there are no problems with foaling, it is usual for mares and foals, after a few days together, to spend the majority of the day out at grass with other mares and their similar aged foals. During this time foals of both sexes can engage in play behaviour, so necessary for development of motor skills and normal social behaviour. Pair bonding may also develop, with two foals establishing a closer relationship with each other and choosing to spend time grooming each other and playing together. Weaning usually takes place at approximately

6 months of age, allowing the mare to continue through pregnancy without the energy demands of the rapidly growing foal. The normal practice of abrupt weaning and separation from the mares, often into the isolation of a stable or pen, is considered to be a major trauma for the young horse (Kiley-Worthington, 1987). The enforced separation often results in a large amount of general locomotion and vocalizations (Crowell-Davis, 1986). When foals are weaned in pairs, vocalizations are reduced (Houpt *et al.*, 1984), although later separation from the companion can result in a second trauma. In addition, problems associated with abnormal sucking and dominance are sometimes apparent. Initial partial weaning before final separation of mare and foal, so that the foal has experienced separation for a limited time before being returned to the mare, has been shown to aid the weaning process, with foals showing less post-weaning walking, trotting and whinnying (McCall *et al.*, 1985). Other methods include interval weaning, where mares are successively removed from the group in order of their foals' age, eventually leaving all the foals together.

As with other farm animals, weaning involves a change in both social and physical conditions. The changes in diet, environment and social conditions associated with weaning are thought to be related to the development of various behavioural problems. Under more natural conditions the foal remains with the mare until the next foal is born, during which time it is likely that it learns, through observing its mother's behaviour, what to feed upon, what to avoid and where to find food. Many of the feeding problems following weaning are likely to be due to poor management practices. If mares and foals are not exposed together to any novel feeds and feeding regimes prior to weaning, the foal may fail to recognize the novel food as food, or may experience difficulties in learning to operate any automatic feeding or drinking devices. In this situation the foal may become undernourished, may experience behavioural anorexia and even death (Crowell-Davis, 1986). Competitive feeding practices also lead to problems in group-reared foals, where foals of low status have difficulties gaining access to the communal feeder, and particularly aggressive foals spend more time in defence than feeding. Abnormal feeding behaviour may also result from mineral deficiencies, the diet change or separation anxiety following weaning. Foals may habitually ingest unnaturally large quantities of undesirable substances such as dirt (pica) or fresh faeces (coprophagy), which, although a natural behaviour in younger foals prior to weaning, may cause digestive problems later in older foals. The housing and feeding of young foals are considered to contribute to the development of various oral and locomotory stereotypies, such as wind-sucking, crib-biting and weaving (McGreevy *et al.*, 1995).

During the rearing period following the weaning process, management of horses varies widely depending on the future use of the horse.

Apart from those bred for racing at an early age, most young horses will be turned out to grass until sufficiently physically developed for riding or driving. Horses (apart from racing horses) usually begin their training under saddle from 3 years onwards, and so they may spend the first period of their life in the company of other horses under various systems of management. These systems include: in similar age groups of single or mixed sex, housed in barns or out at grass depending on the climate; or in mixed age groups at grass, spending some time housed individually.

In contrast, the management of most foals in racing yards tends to be highly restrictive (McGreevy et al., 1995). Of 33 trainers surveyed in the UK who reared young stock, 73% stated that they kept all their young horses individually stabled, and only 45% of them were fed forage on an *ad libitum* basis. These management factors are thought to play a considerable part in the development of abnormal behaviour in racehorses. Horses in racing yards where less forage per day was fed had a higher incidence of abnormal oral-based activities such as crib-biting and wind-sucking than those with less restrictive feeding. Further, a subsequent study of trotting horses and racehorses in Sweden (Redbo et al., 1998) reported that the thoroughbreds, who spent more time stabled and less time in social contact with other horses, as well as being fed less roughage, performed more abnormal behaviour than the trotters. These results are applicable to other types of performance horses such as dressage horses and show jumpers.

It appears from recent evidence that redirected oral activity in restricted housed and fed horses is likely to develop following weaning at a time when the motivation to perform oral behaviour is likely to be high and frustrated.

Problems with reproduction

Social stress resulting from the relationship an animal has with its peers can markedly influence reproductive success, particularly in confined conditions where social interaction between peers becomes frequent and intense. Although this may be because dominant animals have greater access to the male, or have greater access to the most desirable environmental conditions, there is some evidence, at least in other species, that social rank can affect fertility by directly altering the animal's neuroendocrine regulation of reproduction (Moberg, 1985). Although little is known about whether and, if so, how so-called stress hormones may influence reproduction in mares, there is some evidence that this may occur, at least in domestic horses (van Niekerk and Morgenthal, 1982).

Isolation of stallions from an early age can lead to behavioural problems later. In addition to the potential problems associated with social stress, reproductive behaviour in domestic horses may also be influenced by management factors; for example, stallions tend to be individually housed from an early age with little or no opportunity to learn about normal social interactions and courtship techniques.

Oestrus detection in mares in domestic populations can be problematic, requiring the presence of a stallion, or at least some tactile or auditory stimulation (Veeckman and Odberg, 1978), and, in the absence of a stallion, there may be few behavioural signs. Under controlled conditions, there is little possibility of normal courtship behaviour and pre-copulatory behaviour is often inhibited by the need to protect the socially naive stallion from the mare. Mares are commonly restrained during mating and the stallion may be muzzled to prevent him from biting the mare. The abnormal nature of the sexual interaction may lead to certain behavioural problems, such as low sexual interest among young stallions (due possibly to their lack of social experience with their own species) and excessive aggression towards mares and handlers. It has been estimated that almost 25% of stallions may experience some sexual behavioural problem that limits their fertility (McDonnell, 1986).

9.3.2 Problems due to confinement and isolation

The confined conditions in which most performance and recreation horses are kept at least for part of a 24-h period restrict their normal behaviour. The stable appears to conflict with many of the horse's survival instincts, by making it vulnerable through isolation, restricting its sensory abilities, preventing detection of potentially threatening events and preventing escape behaviour. The associated effects of restricted feeding and exercise are thought to lead to a number of problems for the horse. The inability to perform 'normal social behaviour' often leads to performance of undesirable behavioural problems, e.g. self-mutilation, such as flank biting in stallions, who are traditionally kept confined and isolated for breeding purposes (Dodman *et al.*, 1994). Other somatic or body-based forms of abnormal behaviour (Fraser, 1992) are stereotyped pacing or box walking, weaving and pawing. All of these activities are associated with the stable environment, and some are for obvious reasons associated with a reduction in health.

Horses who are housed for long periods of time and isolated from conspecifics have also been shown to exhibit what may be termed isolation-induced aggression (Houpt, 1981). Such antisocial behaviour is often directed at both horse and handler alike, and is probably a result of mismanagement, particularly during early development. It is thought to be due to them having little chance to learn the rules of social relationships due to their restricted upbringing (Kiley-Worthington, 1987). In addition, the prevention of normal social contact and resulting frustration may lead to a chronic arousal state

(Luescher *et al.*, 1991). The horse may then vent its frustration in inappropriate behaviour, such as increased aggression.

9.3.3 Dominance-related problems

Under commercial conditions the main dominance-related problems are seen between horses kept in unstable groups in confined conditions. The resulting escalation in aggressive behaviour can be the cause of stress and injury to some group members.

Breed, temperament and rearing environment all influence an animal's response to management procedures (Fraser, 1992). Horses raised under group conditions where space is not limited tend to have larger flight distances and be more fearful of humans than those reared in more intensive environments. However, close confinement may also lead to increased aggressive behaviour leading to problems with handling. Often dominance-related problems in such horses are due to unintentional reinforcement of the behaviour by the handler. One area where dominance-related problems are often reported is during feeding (McGreevy, 1996). In a more natural environment defence of a patch of food is rare, since grass is widely distributed, but horses fed infrequently in a stable will often exhibit aggressive behaviour, show-ing threatening behaviour as the handler approaches with the feed. Rapid withdrawal from the horse at this time serves to reinforce the behaviour, so inadvertently shaping the horse's behaviour, so that it behaves in a particular way at feed times.

However, the main area where dominance problems are most likely to occur is in the relationship between humans and horse during the training process. In no other domestic farm species do humans attempt to control and manipulate behaviour as they do with horses. Horses experience a particularly intense relationship with their human carers due to their training schedules and management regimes. This intense relationship is made possible because of the social nature of the horse.

Emotionality (see Chapter 13) in many animals tested under labora-tory conditions has been shown to be influenced by the frequency and manner in which those animals are handled during early development. Despite the obvious importance of such knowledge in the training of horses, it has received little scientific interest. It is known that orphan foals reared by humans are less emotional when placed in a novel environment than normally reared foals (Houpt and Hintz, 1983), which suggests that early handling can influence reactivity later in life (see Chapter 14). Miller (1989) suggests the use of 'imprint training' for newly born foals. By intensively handling foals, he claims early bonding between human and horse takes place, which enhances the ability of the horse to cope with training and management later in life.

The amount of handling required to improve trainability has been further investigated by Mall and McCall (1996). This method of training does appear to fit with the natural behavioural traits of young horses in free-living situations, which are constantly absorbing information during their time with the natal herd before reaching maturity at 3 years. Despite this information, horses under commercial conditions are rarely intensively handled until they are considered sufficiently physically developed to carry a rider at about 3 years. The racing industry train and back their horses much earlier (a year old in some cases), and, despite objections by many concerning the ethics of their being ridden and raced at such a young age, they may benefit from the earlier handling and intense training that will result.

9.3.4 Abnormal behaviour

Stereotypic behaviour in horses is generally classified as either loco-motory or oral. Locomotory stereotypies are likely to be associated with a lack of social contact leading to separation anxiety, and frustration due to inadequate housing and prevention of certain behavioural requirements. Oral-based stereotypies tend to be associated with feeding practices. Examples of locomotory stereotypies in the horse are: weaving (swinging the head and neck from side to side, sometimes including the whole body, and shifting the body weight from foreleg to foreleg) and box or stall walking (pacing or circling around the stable). Oral-based stereotypies include; wind-sucking (where the horse opens its mouth, contracts the pharyngeal musculature and appears to swallow or gulp air, although it is likely that the horse is actually only moving air into the mouth and pharynx (McGreevy, 1996), and crib-biting (which is the same as wind-sucking but the horse also grasps a solid object with its front incisors).

Confinement, isolation from social contact, restricted access to grazing and exercise areas and lack of environmental stimulation may all be factors that lead to the development of locomotory stereotypies in horses. Horses kept for commercial purposes have little control over their environments. The inability to perform species-specific behaviour is thought to lead to motivational conflict in the horse. It might be highly motivated to escape from its confined conditions, it might be motivated to seek social contact and to move about more in search of food; but it is physically prevented from so doing, either by a closed stable door or by a rider or handler. Its response is to develop a behav-iour that enables it to live within its situation: the behaviour is likely to be channelled by the limited opportunities available to it; it may, for example, begin to shift its weight from leg to leg and to wave its head about as it peers over the stable door; or it may walk repeatedly round

LIVERPOOL JOHN MOORES UNIVERSITY
LEARNING & INFORMATION SERVICES

its box. It may find this satisfying in that it reduces frustration, level of arousal or motivation. The behaviour may then become a habit (i.e. fixed) and emancipated (i.e. no longer associated with the initial cause) as the horse learns that the weaving or box-walking behaviour is associated with rewards, such as perhaps a lowered level of frustration, self-stimulation or even attention from humans.

Some authors believe that weaning at an abnormally early age may encourage the development of stereotypies, especially oral types such as cribbing. It is thought that the abrupt end of the suckling phase, as well as the change in diet, may be traumatic for some foals and that the comfort behaviour of sucking is replaced by other oral fixations when a horse is stressed in later life. It has been shown that 93.7% of yearlings started exhibiting abnormal behaviours immediately following weaning (Borroni and Canali, 1993). Others believe that horses learn about these behaviours through observing other horses performing them, perhaps their mothers, or others within their view. There might appear to be some truth in this in that a number of horses within one yard are generally found performing similar behaviour abnormalities, but this has never been proved. Indeed, experiments on observational learning have shown that horses are not good at learning discrimination tasks from observing the behaviour of others (Baer *et al.*, 1983; Lindberg *et al.*, 1999). The truth might be simply that horses housed in the same yard are exposed to similar stresses and are therefore just as likely to develop a behavioural problem.

While one of the best means of solving a problem is to prevent the behaviour from developing in the first place, sometimes the incidence of abnormal behaviour can be reduced by changing a management system. For example, if a horse only performs the behaviour when it is isolated from social contact, such as is the case with weaving, this suggests that either the horse needs to be housed in the company of other horses, or that it needs to be gradually conditioned to accept the situation through constant positive reinforcement.

An alternative and preferable approach to the problem is to prevent the behaviour from developing. This may be done through improving the quality of housing conditions through environmental enrichment. Enrichment of the horse's housing environment can be made by, for example, modifying the way in which it is fed or the amount of social contact it has. The small amount of time spent feeding in modern husbandry routines has already been mentioned as a factor contributing to the development particularly of oral stereotypies. One solution could be to turn a horse out in the company of conspecifics to pasture, to graze for many hours.

We tend to regard our care of horses as superior to our care of farm animals, due to their fiscal and emotional values and presume that their welfare is, therefore, better. It is more likely, however, that welfare would be improved by meeting the horse's need to perform certain behaviours and behavioural patterns which can be identified from studies of feral populations. It seems from the studies carried out on both feral and domestic horses that, although the physical characteristics of domestic horses may have altered from those of their wild ancestors, little of their species-specific behaviour has changed.

Studies of feral horses indicate that social behaviour is a prime requirement. Horses are capable of forming close social bonds and a lack of social contact is likely to be one of the most serious stressors for commercially kept horses. Social needs are unlikely to be met in the relatively confined and restricted commercial systems in which they are kept, which is likely to lead to behavioural problems often indicative of reduced welfare. In designing commercial housing and management practices for ensuring the health and the proper behavioural development of horses, it is necessary to strike a balance between the needs of the horse and its keeper. Although the basic needs of the horse are more than adequately covered in commercial systems, there needs to be a move towards the recognition of the importance of other equally important needs, such as behavioural opportunity, space and social behaviour.

References

Baer, K.L., Potter, G.D., Friend, T.H. and Beaver, B.V. (1983) Observation effects on learning. *Applied Animal Ethology* 11, 123–129.

Berger, J. (1977) Organizational systems and dominance in feral horses in the Grand Canyon. *Behavioural Ecology and Sociobiology* 2, 131–146.

Berger, J. (1986) *Wild Horses of the Great Basin: Social Competition and Population Size.* The University of Chicago Press, Chicago, pp. 181–195.

Borroni, A. and Canali, E. (1993) Behavioural problems in thoroughbred horses reared in Italy. In: Nichelmann, M., Wierenga, H.K. and Braun, S. (eds) *Proceedings of International Congress on Applied Ethology.* Berlin, pp. 43–46.

Budiansky, S. (1998) *The Nature of Horses.* Weidenfeld and Nicolson, London, pp. 111–120.

Clarke, A.F. (1994) Stables. In: Wathes, C.M. and Charles, D.R. (eds) *Livestock Housing.* CAB International, Wallingford, UK, pp. 379–403.

Crowell-Davis, S.L. (1986) Developmental behaviour. In: Crowell-Davis, S.L. and Houpt, K.A. (eds) *Veterinary Clinics of North America: Equine Practice* 2(3), 557–571.

Crowell-Davis, S.L. and Houpt, K.A. (1985) The ontogeny of flehmen in horses. *Animal Behaviour* 33, 739–745.

Dodman,N.H., Normille, J.A., Shuster, L. and Rand, W. (1994) Equine self mutilation syndrome (57 cases). *American Journal of the Veterinary Medicine Association* 204, 1219–1223.

Duncan, P. and Vigne, N. (1979) The effect of group size in horses on the rate of attacks of blood-sucking flies. *Animal Behaviour* 27, 623–625.

Feist, J.D. and McCullough, D.R. (1976) Behaviour patterns and communication in feral horses. *Zeitschrift für Tierpsychologie* 41, 337–371.

Franke Stevens, E. (1990) Instability of harems of feral horses in relation to season and presence of subordinate stallions. *Behaviour* 112, 149–161.

Franklin, W.L. (1974) The social behaviour of *Vicuna* Ln. In: Geist, V. and Walther, W. (eds) *The Behaviour of Ungulates and its Relation to Management*. IVCN (new series) Vol. 24, pp. 477–487.

Fraser, A.F. (1992) *The Behaviour of the Horse*. CAB International, Wallingford, UK.

Greenwood, P.J. (1980) Mating systems, philopatry and dispersal in birds and mammals. *Animal Behaviour* 28, 1140–1162.

Hebel, R. (1976) Distribution of retinal ganglion cells in 5 mammalian species (pig, sheep, ox, horse, dog). *Anatomical Embryology* 150, 45–51.

Heffner, H.E. and Heffner, R.S (1983) The hearing ability of horses. *Equine Practice* 5(3), 27–32.

Houpt, K.A. (1981) Equine behaviour problems in relation to humane management. *International Journal on the Study of Animal Problems* 2, 329–336.

Houpt, K.A. and Hintz, H.F. (1983) Some effects of maternal deprivation on maintenance behaviour, spatial relationships and responses to environmental novelty in foals. *Applied Animal Ethology* 9, 221–230.

Houpt, K. and Keiper, R. (1982) The position of the stallion in the equine dominance hierarchy of feral and domestic ponies. *Journal of Animal Science* 54, 945–950.

Houpt, K.A. and Wolski, T.R. (1980) Stability of equine hierarchies and the prevention of dominance related aggression. *Equine Veterinary Journal* 12(1), 15–18.

Houpt, K.A., Law, K. and Martinisi, V. (1978) Dominance hierarchies in domestic horses. *Applied Animal Ethology* 4, 273–283.

Houpt, K.A., Hintz, H.F. and Butler, W.R. (1984) A preliminary study of two methods of weaning foals. *Applied Animal Behaviour Science* 12, 177–181.

Keiper, R.R. (1985) *The Assateague Ponies*. Tidewater Press, Cambridge, Maryland.

Keiper, R.R. (1986) Social structure. In: Crowell-Davis, S.L. and Houpt, K.A. (eds) *Veterinary Clinics of North America: Equine Practice* 2(3), 465–483.

Keiper, R.R. (1988) Social interactions of the Przewalski horse (*Equus przewalskii* Poliakov, 1881) herd at the Munich Zoo. *Applied Animal Behaviour Science* 21, 89–97.

Khalil, A.M. and Kaseda, Y. (1997) Behavioural patterns and proximate reason of young male separation in Misaki feral horses. *Applied Animal Behaviour Science* 45, 281–289.

Kiley-Worthington, M. (1987) *The Behaviour of Horses*. J.A. Allen, London.

Klingel, H. (1969) Reproduction in the plains zebra: *Equus burchelli boehmi*. Behaviour and ecological factors. *Journal on Reproductive Fertility* Suppl. 6, 339–346.

Klingel, H. (1982) Social organisation of feral horses. *Journal on Reproductive Fertility* 32 (suppl.), 89–95.

Kolter, L. (1984) Social relationship between horses and its influence on feeding activity in loose housing. In: Unshelm, J., van Putten, G. and Zeeb,

K. (eds) *Proceedings of the International Congress of Applied Ethology in Farm Animals.* KTBL Darmstadt, Kiel, pp. 151–155.

Lindberg A.C., Kelland, A. and Nicol, C.J. (1999) Effects of observational learning on acquisition of an operant response in horses. *Applied Animal Behaviour Science* 61, 187–201.

Luescher, U.A., McKeowin, D.B. and Halip, J. (1991) Reviewing the causes of obsessive-compulsive disorders in horses. *Veterinary Medicine*, May.

Mal, M.E. and McCall, C.A. (1996) The influence of handling during different ages on a halter training test in foals. *Applied Animal Behaviour Science*, 50, 115–121.

Marinier, S. L. and Alexander, A.J. (1988) Flehmen behaviour in the domestic horse: discrimination of conspecific odours. *Applied Animal Behaviour Science* 19, 227–237.

McCall, C.A. (1991) Utilising taped stallion vocalisations as a practical aid in estrus detection in mares. *Applied Animal Behaviour Science* 28, 305–310.

McCall, C.A., Potter, G.D. and Krender, J.L. (1985) Locomotor, vocal and other behavioural responses to varying methods of weaning foals. *Applied Animal Behaviour Science* 14, 27–35.

McDonnell, S. (1986) Reproductive behaviour of the stallion. *Veterinary Clinics of North America: Equine Practice* 2, 535–555.

McGreevy, P. (1996) *Why does my Horse...?* Souvenir Press, UK.

McGreevy, P.D., Gripps, P.J., French, N.P., Green, L.E. and Nicol, C.J. (1995) Management factors associated with stereotypic and redirected behaviour in the thoroughbred horse. *Equine Veterinary Journal* 27(2), 86–91.

Miller, R. (1981) Male aggression, dominance and breeding behaviour in Red Desert feral horses. *Zeitschrift für Tierpsychologie* 57, 340–351.

Miller, R. (1983) Seasonal movements and home ranges of feral horse bands in Wyoming's Red Desert. *Journal of Range Management* 36, 190–201.

Miller, R.M. (1989) Imprint training the newborn foal. *Large Animal Veterinarian* 44(4), 21.

Moberg, G.P. (1985) Influence of stress on reproduction: measure of wellbeing. In: *Animal Stress.* American Psychology Society, Waverley Press, Baltimore, Maryland, pp. 245–269.

Montgomery, G.G. (1957) Some aspects of the sociality of the domestic horse. *Transactions of the Kansas Academy of Science* 60, 419–424.

Pick, D.F. (1994) Equine colour perception revisited. *Applied Animal Behaviour Science* 42, 61–65.

Redbo, I., Redbo-Torstensson, P., Odberg, F.O., Hedendahl, A. and Holm, J. (1998) Factors affecting behavioural disturbances in racehorses. *Animal Science* 66, 475–481.

Rubenstein, D.I. (1981) Behavioural ecology of island feral horses. *Equine Veterinary Journal* 13, 27–34.

Rubenstein, D.I. and Hack, M.A. (1992) Horse signals: the sounds and scents of fury. *Evolutionary Ecology* 6, 254–260.

Saslow, C.A. (1999) Factors affecting stimulus visibility for horses. *Applied Animal Behaviour Science* 61, 273–284.

Smith, S. and Goldman, L. (1999) Color discrimination in horses. *Applied Animal Behaviour Science* 62, 13–25.

Stahlbaum, C.C. and Hoput, K.A. (1989) The role of the flehmen response in the behavioural repertoire of the stallion. *Physiology of Behaviour* 45, 1207–1214.

Timney, B. and Keil, K. (1992) Visual acuity in the horse. *Vision Research* 32, 2289–2293.

Van Niekerk, C.J. and Morgenthal, J.C. (1982) Foetal loss and effect of stress on plasma progesterone levels in pregnant mares. *Journal of Reproductive Fertility* Suppl. 32, 453–457.

Veeckman, J. and Odberg, F.O. (1978) Preliminary studies on the behavioural detection of oestrus in Belgian 'warm-blood' mares with acoustic and tactile stimuli. *Applied Animal Ethology* 4, 109–118.

Waring, G.H., Wierzbowski, S. and Hafez, E.S.E. (1975) The behaviour of horses. In: Hafez, E.S.E. (ed.) *The Behaviour of Domestic Animals*, 3rd edn. Williams and Wilkins, Baltimore, Maryland, pp. 330–369.

Welsh, D.A. (1975) Population, behavioural and grazing ecology of the horses of Sable Island, Nova Scotia. PhD thesis, Dalhousie University, Halifax, Nova Scotia.

Whitten, W.K. (1985) Vomeronasal organ and the accessory olfactory system. *Applied Animal Behaviour Science* 14, 105–109.

Wood-Gush, D.G.M. and Galbraith, F. (1987) Social relationships in a herd of 11 geldings and 2 female ponies. *Equine Veterinary Journal* 19, 129–132.

The Social Behaviour of Fish   10

Eva Brännäs, Anders Alanärä and Carin Magnhagen

Department of Aquaculture, Swedish University of Agricultural Sciences, 90183 Umea, Sweden

(Editors' comments: The authors point out that 'fish' refers to many species and this chapter mentions more species than any other in this section of the book. Even within species, social behaviour evidences considerable variation, with some individuals living in several social conditions in their lifetime. Unlike terrestrial animals, fish live in a three-dimensional medium, in which volume may be the appropriate measure of space. However, for some species, such as bottom feeders, area remains the appropriate means of describing space. Space restriction under commercial conditions may lead to the adoption of synchronized schools as the best social strategy.

Social hierarchies may exist in groups of fish, but the importance of dominance weakens as group size increases. Indeed, the authors suggest that one way to overcome problems of disparate growth due to social competition is to increase group size. Once group size reaches a point when the fish adopt a schooling social structure, more uniform growth returns.)

Fish covers a wide range of species from the seahorse to the tuna and consequently there is a great diversity in behaviour. There are several new and thorough reviews that focus on fish behaviour in general such as *Behaviour of Teleost Fishes*, edited by Tony J. Pitcher (1993), *Ecology of Teleost Fishes*, by Robert J. Wootton (1998) and *Behavioural Ecology of Teleost Fishes*, edited by Jean-Guy Godin (1997).

This chapter will focus on the aspects of social behaviour that are relevant for commercially interesting species in rearing conditions (Table 10.1). According to international statistics from 1996 on the

©CAB *International* 2001. *Social Behaviour in Farm Animals*
(eds L.J. Keeling and H.W. Gonyou)

Table 10.1. Species mentioned in Chapter 10.

Common name	Scientific name
Salmonids	
Atlantic salmon	*Salmo salar*
Brown trout	*Salmo trutta*
Chinook salmon	*Oncorhynchus tschawytscha*
Chum salmon	*Oncorhynchus keta*
Coho salmon	*Oncorhynchus kisutch*
Masu salmon	*Oncorhynchus masou*
Rainbow trout	*Oncorhynchus mykiss* (freshwater form)
Steelhead trout	*Oncorhynchus mykiss* (migrating form)
Arctic char	*Salvelinus alpinus*
Brook char	*Salvelinus fontinalis*
Lake char	*Salvelinus namaycush*
Other species used in fish farming	
Sea bream	*Sparus aurata*
Sea bass	*Dicentrarchus labrax*
Eel	*Anguilla* spp.
Carp	*Cyprinus carpio*
Grass carp	*Ctenopharyngodon idella*
Channel catfish	*Ictalurus punctatus*
Yellowtail	*Seriola quingueradis*
Others	
Bluntnose minnow	*Pimephales notatus*
Cod	*Gadus morhua*
Haddock	*Melanogrammus aeglefinus*
Perch	*Perca* spp.
Pike	*Esox lucius*
Pumpkinseed sunfish	*Lepomis gibbosus*
Three-spined stickleback	*Gasterosteus aculeatus*

annual production in thousands of tonnes, various species of carps dominate world aquaculture (9410), followed by salmonids (470), sea bream (32), sea bass (19) and catfish (0.7). Other species of interest are yellowtail (0.6), eel (0.7), tilapia (0.3), some species of flatfish (0.3) and sturgeon (0.5) (Food and Agriculture Organization of the UN (FAO); Federation of the European Agricultural Producers (FEAP)). The first part of this chapter will focus on the basic social structure in fish and how it is influenced by internal as well as external factors. Especially the differences in social behaviour related to the biology of the exemplified species will be considered. The second part will focus upon the impact of these topics on social behaviour in farming conditions and the third part will describe the types of problem that are related to social interactions in fish, the causes for the problems and suggestions how these can be tackled. Further, the chapter will, as far as possible, attempt to describe the various aspects of social behaviour in farmed fish according to the guidelines in the other chapters and species. As cows and fish are totally diverse animals both in their behaviour and in the media they live in, some modifications had to be made. For instance parental care does indeed exist in teleosts but not among the species that are used in aquaculture. Further, social interactions that are related to reproductive behaviour are beyond the scope of this chapter, because the development in most species strives towards artificial breeding.

Salmonids are the most investigated species, in particular on social behaviour. Consequently, this family of teleost fish will be focused upon in this chapter. Whenever relevant, examples from other non-commercial species will be cited.

10.1 Basic Social Characteristics

10.1.1 Composition and structure of social groups

The social structures (see Section 1.3) of fish can range from a completely solitary lifestyle to a synchronized schooling behaviour. A solitary lifestyle, often including an aggressive defence of a feeding territory, is typical of pike lurking in the reeds. Herring or mackerel are, on the other hand, typical examples of species that swim in schools, sometimes composed of thousands of extremely synchronized individuals (Pitcher and Parrish, 1993). These two opposite social lifestyles make illustrative examples of how the fish's behaviour is evolved by factors in its environment, such as vulnerability to predation and food availability. The distribution of resources (e.g. food or shelter), in both space and time, and whether or not these are defendable can explain the behavioural difference in different environments (Grant, 1997; Fig. 10.1). Thus, territorial and aggressive behaviour are beneficial in some environments, whereas a schooling and apparently friendlier behaviour has been beneficial in others.

The pike's solitary lifestyle is suitable in shallow lakes with many refuges where the fish can use a sit-and-wait strategy for catching prey that passes close by. As predators can turn into prey, refuges are also important as hiding places. A sit-and-wait feeding strategy is also fundamental for stream-living species like most juvenile salmonids, which defend a feeding station to get access to prey drifting through the territory (Kalleberg, 1958). The food is generally scarce but more predictable in streams and fish can monopolize an area, thus making a territorial behaviour beneficial.

In large volumes of water, like a lake or the ocean, a school of conspecifics may be the only refuge for a single fish as a protection

Environmental factor

Economic defendability

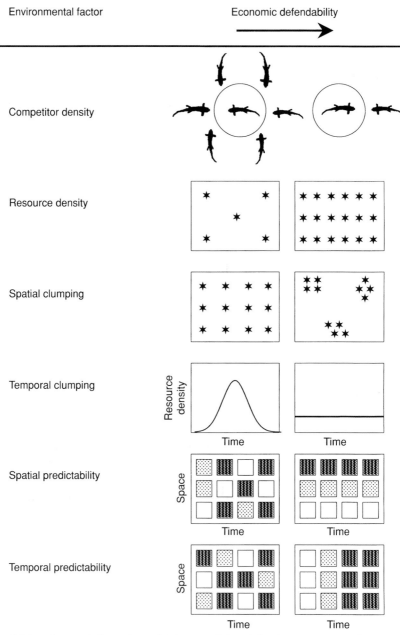

Competitor density

Resource density

Spatial clumping

Temporal clumping

Spatial predictability

Temporal predictability

Fig. 10.1. Factors influencing a fish's potential to defend a resource, such as a feeding territory. An increased defensibility is predicted to be accompanied by an increase in aggression by dominant fish. Economic defensibility will eventually decrease at an extremely low density of competitors or extremely high levels of resources or spatial clumping. (Modified after Grant, 1997.)

from predation. Fish in schools have an increased probability to detect an approaching predator in time, as many eyes make detection of enemies easier. A school also protects the individual by the dilution effect, with each fish reducing its probability of being eaten with an increasing number of fish (Pitcher and Parrish, 1993). In addition, the predators are confused by a large group of prey with similar appearance, which is a well-known phenomenon both for animals and humans (Milinski, 1990). Another benefit of living in a group is the higher probability of finding food with many foraging individuals. In contrast to streams, prey are distributed more randomly in time and sometimes clumped in the pelagic region of lakes and oceans. The disadvantage of schooling, on the other hand, may be a higher competition for food (Pitcher and Parrish, 1993).

There is a scale of different degrees of 'social gathering' in between a territorial and a schooling behaviour. The formation of shoals is defined as social aggregation of fish in general without the tight synchronization that occurs in schools (Pitcher and Parrish, 1993). Within species, the social structure may change over the lifetime, depending on what environment they live in (Elliott, 1994). The life histories of several salmonid species are good examples of this, for example Atlantic salmon and brown trout (Kalleberg, 1958). Atlantic salmon fry emerge in a synchronized manner from the river gravel as a predator avoidance strategy (Godin, 1982). After a few days, they become territorial and apply a sit-and wait strategy feeding on drifting prey. The fry themselves use stones or other structures on the bottom of the river as refuges. After 1–6 years, depending on the temperature and feeding conditions, they become physiologically and behaviourally ready to enter the sea as smolts. The salmon then gather in shoals when migrating to the sea (Hvidsten *et al.*, 1995), again to increase the protection from predators but also to find food more easily. As they grow larger in the sea, they become less vulnerable to predation and they change to a more solitary behaviour. After a few years the sexually mature salmon return to their native rivers and at least the males become territorial again, this time to defend the spawning ground and to compete for females.

Some salmonid species have different life cycles and lack the territorial stream-living phase, e.g. the pink salmon, which migrates in shoals to the sea as newly emerged fry, and Arctic char, which are generally a lake-dwelling species throughout their lives.

It is important to realize that not all salmon in a stream are successfully defending territories and that not all herring within a school have the same opportunities to forage and grow (Magurran, 1993). The number of territories is often a limiting factor for stream-living juvenile salmonids, forcing fish to adopt different feeding strategies. Juvenile salmonids can be divided into territorials, non-territorial shoaling fish and floaters (Puckett and Dill, 1985). The dominant territorial individuals are the most successful ones and defend their feeding position aggressively (Metcalfe, 1986); the floaters may switch between shoaling and territorial behaviour depending on the territory availability. The non-territorial ones are the lowest in the hierarchy, and may move to other areas when the feeding conditions are poor (Elliot, 1994). Also for fish in schools or shoals, some individuals are more successful in catching food and have a higher growth while, consequently, other individuals grow at a lower rate, which leads to increased size differences between individuals (Fausch, 1984).

10.1.2   Use of space

The use of space depends to a high degree on species and foraging strategies, also discussed elsewhere in the text (see Section 1.4). Some species feed mainly on prey from the bottom. Other species forage in a more three-dimensional habitat, feeding on organisms in the water column. Thus, in some cases the area and in others the volume of water is the measure used for estimating habitat availability. Stream-living fish usually occur close to the bottom and can often tolerate shallower water than lake populations, which use the whole volume of water. Even for stream-living fish, however, an increasing water depth serves as a protection from predatory birds, and these fish are found to go deeper as they grow. Finally, several salmonid species, like the Atlantic salmon, are bottom dwellers when small but move out to the ocean (or to a lake in some species) as they grow older.

In many freshwater fishes, individuals stay within a restricted area for long periods of time (Gerking, 1959). This was found, for example, in grass carp that were radio-tracked for a year (Chilton and Poarch, 1997). Also, juvenile sea bass appear to remain near their nursery areas throughout the year (Pawson *et al.*, 1987). Sometimes these limited movements depend on territory defence, but may also be due to the occurrence of home ranges, which can be shared between individuals. The advantage of using a restricted area could be to obtain a good knowledge of the surroundings (Wootton, 1998). The size of a territory or a home range usually increases with the size of the fish and decreases with increasing food density (Keeley and Grant, 1995).

Furthermore, the space is used for different types of activities. Adult sea bream rested close to some structure and made short feeding bouts into the open water (Bégout and Lagardère, 1995). Similarly, stream-living salmonids typically move between a foraging area and a hiding area (Bachman, 1984). The foraging site can be characterized as a position with high water velocities and high food availability (drifting food items) (Fausch, 1984). Since these sites are often found in the central parts of streams, they may lack cover, which increases the predation risk (Bachman, 1984). The metabolic costs of swimming against the water current may also be of importance (Bachman, 1984).

10.1.3　Cohesion and dispersion

How fish are dispersed in space much depends on their foraging behaviour. As mentioned above, some fish are solitary foragers, sometimes utilizing a 'sit-and-wait' strategy, feeding on prey moving through the area. In other species, they forage in groups, actively searching for food. Sit-and-wait predators are thus dispersed in their habitat and avoid crowding by territory defence. Shoals of foraging fish may be more or less cohesive, with some species schooling in a synchronized manner tightly together, while others are more randomly spread out. Cohesiveness increased with shoal size but decreased with hunger in the bluntnose minnow, for example (Morgan, 1988). Also, light may influence cohesiveness, with shoals becoming more dispersed at low light levels (Shaw, 1961).

Predation risk (see Section 1.3) may cause even loosely grouped fish to get closer to each other, a behaviour assumed to decrease the risk of being eaten (Pitcher and Parrish, 1993). Also, several species of fish, for example the three-spined stickleback, 'sort' themselves into shoals that consist of similar sized individuals. This probably reduces even further the conspicuousness to predators of each individual (Peuhkuri *et al.*, 1997).

10.1.4　Environmental factors influencing social behaviour

Like most animals, fish show rhythmic fluctuations in their activity that are governed by abiotic factors like temperature and daylength (Helfman, 1993). These fluctuations are both seasonal and daily. Seasonal fluctuations allow fish to cope with predictable changes in the environment (Daan, 1981), while daily patterns result from the need to concentrate vital activities to the time of day when the balance between food availability and predation risk is the best (Helfman, 1993). Typical visual feeders, like salmonids, have been considered to be diurnal (Hoar, 1942), with peaks in activity around dawn and dusk (Eriksson, 1973).

When not feeding, juvenile salmonids often move to a refuge site; these are characterized by lower water velocities and a larger degree of cover (see above) (Jenkins, 1969). This behaviour is especially important when the temperature drops below 10°C and salmonids becomes more nocturnal. In general, during wintertime, juvenile salmonids hide in cover during daytime and are active at night, whereas they are more flexible during the rest of the year (Fraser *et al.*, 1993). This temperature-based switch may be a common feature in many fish species, and serves to reduce the predation risk, since fish, being ectotherms, are sluggish at low temperatures whereas some of their predators (day-active birds and mammals) are not (Fraser *et al.*, 1993). Sea bass have been shown to switch towards more night activity when the temperature drops in the autumn, indicating a similar behaviour to that in salmonids (Sánchez-Vázquez *et al.*, 1998).

The daily activity patterns in fish are also affected by biotic factors like competition from other individuals in the group. For example, competition for food may lead to segregation between individuals in times of foraging. In groups of rainbow trout and Arctic char, some individuals become nocturnal feeders, which may be a good strategy for subordinate fish to increase their food intake as the dominant fish often feed during daytime (Brännäs and Alanärä, 1997).

Predation risk may also influence the behaviour of fish in different ways, leading to changes in social interactions. In the coho salmon, agonistic behaviour decreased in the presence of predators compared with in their absence and the use of cover, which is an important anti-predator tactic, increased (Reinhardt and Healey, 1997). Juvenile sea bass have been observed to burrow themselves in soft substrate when faced with predation risk (Pickett and Pawson, 1994), a behaviour comparable to the use of cover in salmonids.

10.1.5   Physiological factors influencing social behaviour

It is well known that dominant individuals are often bigger than subordinates in groups of fish with dominance hierarchies. It is likely that the size advantage is more a consequence of their status than its cause (Huntingford *et al.*, 1990). Recent findings suggest that social status may result from different physiological factors, some of which may be inherent. Evidence of a relationship between status and standard metabolic rate (SMR) has been demonstrated in Atlantic salmon. Individuals with a high SMR were more aggressive than individuals with a low SMR (Metcalfe *et al.*, 1995). Also in masu salmon, SMR was positively correlated with social position (Yamamoto *et al.*, 1998).

10.2   Social Behaviour Under Commercial Conditions

Generally, the social behaviour a fish species applies, territorial versus schooling, is determined by the type of habitat it lives in together with a cost–benefit analysis (from the behavioural ecologist's point of view) of applying either behaviour. Similarly, economic and environmental factors determine the fish farmer's choice of species to rear and which rearing system to use.

Rearing systems for fish for human consumption range in complexity from simple earth ponds or extensive lagoon systems with no or limited external input in nutrients to net-pens and indoor or outdoor tanks where the fish are fed with pelleted high-energy feed. The possibility to control abiotic factors is limited in both ponds and net-pens whereas indoor tanks offer a more or less unlimited (technically) and costly advanced control system. Both tank and net-pen systems may include a complex feeding system that adjusts the feeding to match the expected appetite of the fish without wasting any food (Shepherd and Bromage, 1992). Such variance in rearing systems corresponds to pig or poultry farming where, in the most extensive system, a few individuals are kept more or less free around the houses and the most intensive farming is done in highly technical meat factories. Accordingly, irrespective of which type of animal is domesticated for human consumption, the productivity in terms of growth rate and number of individuals in each unit must increase with increasing costs of the rearing system.

Until 30 years ago, fish were mainly reared in freshwater ponds, for example carp (China, former USSR, Eastern Europe and India), channel catfish (USA) and rainbow trout (USA, Denmark and Central Europe). Sea bass and sea bream were reared in salt water lagoon or pond systems, mainly around the Mediterranean and the latter also in Japan. These species are still reared extensively in ponds but have been subjected to an intensification of production. The intensification mainly has two directions: (i) to increase the productivity in the ponds or raceways or (ii) to intensify the production rate by rearing the fish in tanks and/or net-pens. To rear fish in land-based tank systems until slaughter is much more expensive than rearing them in net-pens. A common routine is to rear the young more sensitive stages in tanks and then move the fish into net-pens at sizes between 20 and 500 g, depending on species and achievement with the production (Shepherd and Bromage, 1992).

Together with the increasing biological knowledge and technical possibilities in the aquaculture industry, there has been a development for using new species in aquaculture such as halibut in Norway and the UK, Japanese flounder in Japan, some species of Pacific salmon in Canada (coho, chum), perch and sturgeon in Central Europe. There is also a development to rear common species, e.g. Atlantic salmon, in locations well outside their natural range, such as in Chile and off the west coast of Canada.

10.2.1 Social groupings under restricted conditions

In extensive rearing systems, such as ponds and lagoons, the social grouping of fish is unlikely to be different from that in the wild, at least in highly productive areas where the densities of fish are high. The reports of social behaviour and grouping in extensive systems are scarce and mainly rely on visual reflections. Such observations give a picture of fish forming small schools while some are seen as solitary. Cannibalism is the main type of 'social behaviour' that has been documented in extensive rearing systems (see below). By using acoustic telemetry it is possible to study how fish swim around in a pond depending on the number of conspecifics. The movements of adult sea bream were tracked either kept as single fish in the pond, as five fish all together or with numerous conspecifics (Bégout and Lagerdère, 1995). Isolated fish had a low swimming activity and showed a nocturnal habit, while tagged fish kept with only four other individuals increased their activity but were still active mainly at night. Fish with many conspecifics were schooling and showed a high swimming activity during daytime. A fish farmer would hardly keep fish isolated (unless they were huge!) but the result points out the plasticity in fish behaviour where both the level and the daily peak of activity are affected by schooling. Nocturnal activity was probably an anti-predator response of isolated fish or very few individuals that found themselves without cover of a school of conspecifics.

Most studies on the behaviour of fish when confined to restricted conditions are done in experimental set-ups due to the difficulty of studying fish in full-scale rearing conditions. Some of these laboratory experiments are aimed to scale the situations down from farming conditions due to the difficulty of separating individual behaviour among thousands of other fish in net-pens or tanks. When groups of fish are observed in aquariums under restricted conditions and at low densities, a dominance hierarchy (see Section 1.5) occurs, at least in salmonids. In a group of Arctic char, the feeding activity of 15 individuals exemplifies development of such a hierarchy (Brännäs and Alanärä, 1993) (Fig. 10.2). These fish were new to self-feeders where the fish themselves regulate the feed supply by biting on a trigger in the water. During the first days after they were put into the tank, nearly all individuals 'tested' the system, which gave food as soon as a fish bit on the trigger. After a few days one individual dominated the trigger activity and one or two other individuals exhibited some activity while the rest did not hit the trigger at all. In such situations, successful competitors have higher feed intakes and consequently higher growth rates than less competitive ones (Adams et al., 1995; see also Fig. 10.3).

One way to classify individuals in dominance hierarchies is as dominant, subdominant and subordinate (Symons, 1970), which is equivalent to the classification of stream dwelling salmonids in Section 10.1.1 above. In groups of self-fed Arctic char the dominant individuals had the highest self-feeding activity, the highest growth rates, and

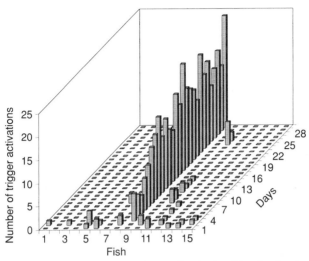

Fig. 10.2. The trigger-biting activity of 15 individuals of Arctic char kept in a fish-rearing tank for 29 days with a demand feeder. The individual trigger activity was monitored by PIT-tagging the fish under the chin and by placing an antenna close to the pendulum. (Modified after Brännäs and Alanärä, 1992.)

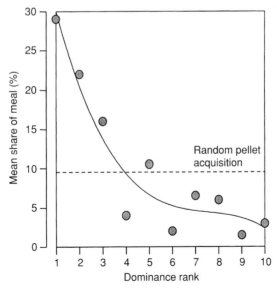

Fig. 10.3. The proportion of pellets eaten during feeding trials plotted against fish ranked by frequency of initiated attacks (1 = highest, 10 = lowest). (Modified after Adams *et al.*, 1995.)

showed low signs of stress as measured by levels of stress hormone (Alanärä *et al.*, 1998). Subdominant fish have an intermediate growth rate and low levels of stress hormones. The subordinate fish show very low swimming and feeding activity. They suffer from weight loss and display elevated levels of stress hormones.

When given no chance to escape, the subordinate fish are likely to 'keep a low profile strategy' as an adaptive response, since it will reduce the risk of initiating attacks from dominant and aggressive individuals (see review by Winberg and Nilsson, 1993). The activity of these individuals is inhibited physiologically or by a behavioural 'cost and benefit analysis' suggesting that the best way for a subordinate fish to survive is to spend as little energy as possible rather than to compete unsuccessfully for food (Metcalfe, 1986).

10.2.2 Social groupings under farm conditions

So, do social hierarchies exist in full-scale condition with thousands of fish held at high densities in net-pens and tanks? Luckily for farmers of salmonids, fish in general exhibit some adaptive flexibility in their behaviour and even species that are typically territorial reduce their aggression if the conditions make such behaviour less profitable (Dill, 1983). It has been shown that the significance of social dominance is reduced or diminishes when group sizes increase (see Section 1.4). The reason is that neither repeated attacks nor defence of a favourable area or food resource can be sustained by dominant individuals under such conditions (Brown *et al.*, 1992). Fish farmers generally have much higher densities of fish than in experimental conditions where the social behaviour has actually been studied. In a study on Arctic char and rainbow trout it became increasingly difficult for a single fish to monopolize the feeding area with increasing density (Fig. 10.4). As densities increased from low to high, about 25% of the total fish group became collectively dominant and grew well (Alanärä and Brännäs, 1996).

Still, no matter how skilful the farmer is, there will always be size discrepancies between individuals. This is found in more or less all fish farms today, where a varying proportion of the fish exhibits higher growth rates than the remaining stock (Shepherd and Bromage, 1992). The reason for this may be genetically determined and/or an effect of competition between individuals governed by their social status. In commercial rearing conditions the differences in social status are probably more related to aggressiveness than to the development of dominance hierarchies where highly ranked fish are more competitive during food release.

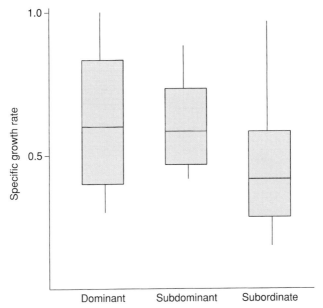

Fig. 10.4. The growth rate of individuals in six groups of Arctic char classified by behavioural observations as dominants, subdominants and subordinates. The classification was done in the aquarium and the groups were then kept in circular tanks with a water current and given a surplus of feed to stimulate schooling behaviour. (E. Brännäs, unpublished results.)

By measuring position in aquarium and aggressive interactions, individual groups of Arctic char were classified into one of the three dominance hierarchies listed above (E. Brännäs, unpublished results). These groups were then raised for 4 months in conditions that stimulated schooling behaviour, i.e. in a constant water current (Jørgensen *et al.*, 1993), and with high feed levels (McCarthy *et al.*, 1992). The variance in growth was less extreme in this situation and subordinate individuals did grow well although significantly more slowly than the dominant and subdominant individuals. There was no difference in growth rate between the latter ranks (Fig. 10.5). This scenario probably reflects the variance in growth rate that is found in full-scale rearing systems where the conditions promote decreased aggressive interactions, even schooling.

The net-pens can physically hold very high densities without 'forcing the fish closer together'. As opposed to farmed animals on land, fish are reared in a three-dimensional space and voluntarily form tight schools. However, the schools of fish need space so they can alter their swimming depth depending on hunger level or strong sunlight (review by Juell, 1995). Studies on social behaviour in fish at full-scale rearing conditions suggest that most fish species that are reared in net

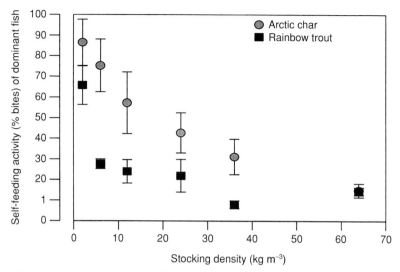

Fig. 10.5. A comparison of the effect of stocking density on the strength of dominance hierarchies in groups with Arctic char and rainbow trout. The dominance was determined by the per cent trigger-biting activity on a self-feeder. (Modified after Alanärä and Brännäs, 1996.)

cages gather in schools. These schools appear as a ring structure with few fish near the centre or close to the cage wall. For instance, Atlantic salmon and rainbow trout form circular polarized schools during daytime and at night the schooling groups disperse (review by Juell, 1995).

10.3 Social Behaviour, Management and Welfare

Several problems related to the behaviour of fish can be identified in culture conditions. The causes of these problems are often linked to aggressive interactions and insufficient rearing conditions, but also to predatory and anti-predator behaviours. Visible signs of behavioural problems are often a reduced or skewed growth rate, stress, fin damage, mortality and abnormal behaviour. Some of these problems are fairly easy to detect early and therefore possible to solve, whereas others are more difficult.

10.3.1 Aggression-related problems

As stated in previous sections, aggressive behaviour is a very common feature in many fish species in the wild, especially if aggression or territorial behaviour have, through evolution, been of advantage for the individual fitness, as for most salmonids. Conflicts and establishment of social status are solved by a repertoire of activities that function as social signals. During escalated conflicts, aggressive displays are replaced by attacks against the fins (Abbott and Dill, 1989). It is unlikely that territorial species will change their behaviour in farming conditions, just modify it. There are a number of factors that will affect the degree of aggression in farmed fish stocks, such as the abundance and distribution of food (Grant, 1997).

10.3.2   Crowding-related problems

Crowding or high stocking density is consequently likely to be more stressful for some species than for others depending on if it is a schooling or territorial species. Schooling species like sea bass and sea bream are examples of species that are likely to show higher tolerance to crowding than territorial salmonids.

High stocking densities are known to depress feeding activity and growth rate in rainbow trout, at least in laboratory conditions (Baker and Ayles, 1990). A peak in the trout's feeding activity was found at a stocking density of 30 kg m^{-3} but the activity decreased with further crowding (Alanärä and Brännäs, 1996). Accordingly, the growth rate in rainbow trout was highest at a stocking density of 10 kg m^{-3}, then declined linearly, with the lowest growth at 85 kg m^{-3} (Baker and Ayles, 1990).

High density conditions have also been reported to reduce feeding activity in coho salmon (Schreck *et al.*, 1985) and brook char (Vijayan and Leatherland, 1990). The studies suggested that stress was the prime reason for this decrease.

In contrast to most other salmonids, growth rates in Arctic char seem to be positively correlated to stocking density when food and oxygen supplies are adequate (Baker and Ayles, 1990). Even at densities well over 100 kg m^{-3}, growth rates do not seem to be suppressed (Jørgensen *et al.*, 1993; see also Fig. 10.6). The reason for the char's positive reactions to crowding is probably that they more easily apply a schooling behaviour because of their origin as a non-territorial lake-living species (Brown *et al.*, 1992). The other examples of salmonids that respond to crowding with stress symptoms are typical stream-living territorial species as juveniles. When they are forced into schooling they are probably stressed by being too close together.

10.3.3   Feeding-related problems

A fish farmer has to maintain a balance between the size of the feed portion and the number of pellets the fish is able to catch before the

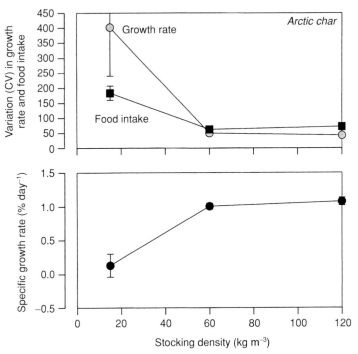

Fig. 10.6. The variation in growth rate and feed intake in Arctic char raised in different stocking densities. X-rays monitored the feed intake after giving labelled feed. (Modified after Jørgensen *et al.*, 1993.)

feed passes out of the rearing system. To avoid feed waste the best strategy would be to deliver very small portions, which ensures that the fish are able to catch all the pellets. Such a feeding strategy may be very stressful, since many fish will compete for a limited number of feed items at each delivery. An illustrative example is a study on rainbow trout held in cages and fed according to two different feeding regimes where one group was frequently fed every 5 min during the whole day and the other group was fed using self-feeders (Alanärä, 1992). Both groups were fed the same daily amount of pellets. The results showed that frequently fed trout grew 20% less and had lower feed efficiency than self-fed fish. Visual observations made during the study indicated that each feed portion resulted in a dramatic increase in swimming activity, where the water surface 'boiled' with fish. This activity decreased only to some extent until the next feed portion was delivered 5 min later. Repeated feeding at short intervals during the whole day seems to induce stress and energy losses. Self-fed trout ate most of their daily ration during 2–4 intense hours in the morning. After this feeding period, the fish swam at deeper depths without activity bursts.

10.3.4   Predatory environment

Fish held in culture are very sensitive to 'apparent' riskiness and if they are frightened too often or subjected to constant threats they may reduce their feeding activity. Juveniles especially attempt to avoid environments that may put them at risk of predation, even if no real predator is present. Low water depths, high light intensities, light-coloured tanks, lack of cover and too frequent handling routines are examples of environmental factors that are risky and may put the fish under stress. Larger fish in cages are also subjected to factors they may experience as predation risk. Many fish species avoid swimming close to the surface to avoid being attacked by birds. In cages, the feeding activity may decrease after encounters with seals or larger fish in the surrounding water. Farmed fish may also increase their oxygen consumption (a stress response) when people are walking past the cages, as demonstrated in a Norwegian fish farm (Gjedrem, 1993).

10.3.5   Cannibalism

In rearing conditions, cannibalism has been observed in several species (Smith and Reay, 1991). However, cannibalism is rare in salmonid culture unless the size difference between fish is very large. Among those species where cannibalism is a problem in rearing conditions, it is mainly found in the early stages. Conditions that increase the likelihood of cannibalism include low food availability, high fish densities, large size variation and the absence of hiding places. Since conspecifics constitute an optimal combination of nutrients needed, growth rates are often high in cannibals. At the same time population densities may be regulated in this way, and numbers of competitors decreased. Thus, the individuals that become cannibals probably have a great advantage over others due to a higher growth rate and a higher chance of surviving (Smith and Reay, 1991).

10.3.6   Signs of behavioural problems

Reduced or skewed growth rate
Two causes for reduced or skewed growth can be identified in farming conditions: genetic variation in the efficiency of converting feed into body weight and differences in competitive ability, which is a behaviourally mediated problem. For fish farmers, it is possible to reduce the behaviour problems. Jobling (1995) suggested some tools for assessing the social environment by observing weight gain, size variation and feed conversion ratios:

1. A high growth rate and little variation in body size indicate good rearing conditions with little or no depression of growth due to aggressive behaviour.

2. Suboptimal and disparate growth coupled to poor food utilization reflect a poor social environment, which may be the result of competition due to underfeeding or because of bad temporal or spatial distribution of feed.

3. Poor growth with little variation indicates a general growth depression resulting from a poor environment, and a poor feed conversion ratio would reflect feed waste.

Thus, by keeping precise track of fish growth and feed consumption, a fish farmer can evaluate the social situation in the rearing environment and thereby avoid behavioural problems that may lead to severe growth variation.

Stress

Fish in all environments sometimes suffer from stressful conditions. Factors such as quick changes in temperature, handling (e.g. size grading and transport) and bad water quality adversely affect fish health (see review by Anderson, 1990). Stress has been defined in many different ways but mainly in physiological terms (Adams, 1990). Examples of physiological signs are elevated corticosteroid production, blood glucose, lactate levels and brain hormones like serotonin (Winberg and Nilsson, 1993).

Social stress is more complex than stress in general, and most probably the stress experienced by subordinate individuals initially results from losing fights, whereas later on it probably relates to constant threats by dominant individuals (Zayan, 1991). The stress leads to a suppressed aggressive behaviour (Abbott *et al.* 1985), reduced food intake (Metcalfe *et al.*, 1990) and lowered swimming activity (Winberg and Nilsson, 1993) relative to that shown by dominants. Long-term exposure to stress will decrease the condition of fish, which will probably have a negative effect on health.

Health and welfare

One of the first signs of bad health in farmed fish is the occurrence of fin damage through aggressive behaviour. The dorsal and caudal fins are frequently damaged in rainbow trout (Abbott *et al.*, 1985) but the degree of fin damage might vary between species, probably as an effect of differences in aggression (Bosakowski and Wagner, 1995). Bites on the fins are not the only cause of damage. Another form of damage, fin erosion, can occur because of abrasion on the tank bottom or sides or other substrates in the rearing environment. This damage is possible to distinguish from fin bites because it usually occurs on the pectoral and pelvic fins.

Stress is likely to have negative effects on the immunocompetence of fish, even if very few studies have been done in this area. Fish may seem healthy before, during and immediately after a period of stress, but there may be disease problems later on. They may be asymptomatic carriers of pathogens that under normal conditions are held back by the immune system. When that system is impaired or suppressed by stress, the disease-causing agent may start to grow, gain control and kill the fish (Anderson, 1990). In addition, stress may suppress the immune system and increase the vulnerability of the fish to invading pathogens (Wedemeyer, 1976). Evidence for this was found in stressed rainbow trout of subordinate rank, which showed lowered immunocompetence compared with higher rank conspecifics (Thompson, 1993).

Surface activity is frequently found in fish that are held in culture, and situations with hyperactivity may indicate that something is wrong in the rearing environment. Leaping (fish jumping with their whole body breaking the water surface) and rolling (the dorsal part of the body breaking the surface) are well-known behaviours in salmonids (Furevik *et al.*, 1993). In nature, leaps are used to pass falls and obstructions during migration in running water, but what does the behaviour mean in culture conditions? In a study on Atlantic salmon held in cages, leaping occurred all year round, but most frequently during summer and high temperature conditions. The cause for leaping seems to be related to infections of ectoparasites (e.g. lice), exposure to acute stress and the presence of predators (Furevik *et al.*, 1993). The rolling activity may be explained by buoyancy compensation (gas bladder filling) (Furevik *et al.*, 1993). Fish often lose air from the swimbladder during stress exposure and it is important that the neutral buoyancy is restored within a relative short period of time.

Measuring stress on fish is difficult but there are signs of stress and bad environment that can be detected in culture conditions:

- Abnormal swimming behaviour (e.g. leaping, rolling)
- Low swimming activity
- Low and variable growth
- Fin damage, wounds and scale loss
- High number of external parasites.

Non-desirable genetic selection

In most farmed animals, domestication has a considerable effect on the behaviour (Hemmer, 1991) and an increase or decrease in aggression level between individuals may be an undesired or desired goal in breeding programmes (Hemmer, 1991; Ruzzante, 1994). In fish, there is a lack of knowledge on how aggression and correlated behaviours are affected by domestication (Ruzzante, 1994). Culture conditions might lead to increased aggressiveness in salmonids, probably as an indirect result of selection for rapid growth. Juvenile coho salmon cultured for five consecutive generations were more aggressive than conspecifics from wild parents (Swain and Riddell, 1989). If aggressiveness is a heritable trait in fish, selection for rapid growth may indirectly select for aggressive behaviour rather than physiological growth rate (Doyle and Talbot, 1986). Little is so far known about the implications of this but both negative and positive effects can be expected.

The negative effects are increased social interactions, stress and fin damage. A genetic improvement in competitive ability (aggressiveness) results in an increase in the mean competitive ability of all contemporaries, and such selection may lead to a general increase in aggression (Kinghorn, 1983).

The positive effect, on the other hand, is a decreased number of individuals that display a subordinate behaviour. The effect could then be a more homogeneous growth.

Whatever consequences genetic selection has on the behaviour of fish, it is important that the genetic correlation between aggressiveness and growth rate is evaluated. Doyle and Talbot (1986) suggest that behavioural criteria should be incorporated into selection programmes and that such criteria may be important for their success. A selection strategy in breeding programmes may be to screen fish for social status and to select subdominant fish. As described in the previous section these individuals are less aggressive but grow just as well as the dominants if the conditions promote a non-aggressive behaviour. If there is a link between aggressiveness and growth, high growth performance is still included in the selection and the most 'aggressive genes' should be sorted out.

10.3.7   Reducing behavioural problems

Stocking density

As described in the previous section, stocking density may influence the behaviour of fish in various ways. Increasing the density may reduce problems with social hierarchies, but on the other hand may lead to increased problems with stress and health. Different species seems to react differently to stocking density, and it is therefore important to find out the optimal stocking density for the particular species that is being cultured.

The effect of stocking density has been studied in most of the commercially important salmonid species, including rainbow trout, Atlantic salmon, brook char and Arctic char. In Swedish fish farms where they rear fish in net-pens, char is generally kept at densities of between 50–60 kg m^{-3} and rainbow trout at 30–40 kg m^{-3}. On the other hand typical schooling species like sea bream and sea bass are generally not reared in densities that exceed 10 kg m^{-3} (Gimfer-Artigas, 1998). It is important to point out that other factors like water quality and regulation of the oxygen level also determine the optimal stocking density.

Water current

Several studies have shown that sustained exercise (i.e. swimming against a low current) improve growth rates in several species of salmonid fish (brown trout: Davidson and Goldspink, 1977; brook charr: East and Magnan, 1987; rainbow trout: Houlihan and Laurent, 1987; Atlantic salmon: Totland *et al.*, 1987; Arctic char: Christiansen and Jobling, 1990). A growth enhancement effect has also been reported for juvenile striped bass (Young and Cech, 1994).

The reason of this growth improvement is debated. One hypothesis is that active swimming increases the protein synthesis at the expense of the fat deposition, which results in increased weight gain (Houlihan and Laurent, 1987).

A second possibility is that the increased growth is the result of decreased aggressive behaviour. Several studies on Arctic char have shown that fish are less aggressive when forced to swim for prolonged periods at moderate speeds of about one to two body lengths (BL) per second (Christiansen and Jobling, 1990). One very obvious change in behaviour at increasing water velocities is the change from irregular swimming activity (passivity to bursts) to pronounced schooling behaviour at exercise conditions (Christiansen and Jobling, 1990). The dominant individuals may be less aggressive when they are forced to 'exercise' due to increased energetic costs of aggression and fighting in a current (Grant and Noakes, 1992). Consequently, the subordinates increase their growth rate either by an increase in their feed intake (Christiansen and Jobling, 1990; see Fig. 10.7) and/or by a decrease in their stress-related metabolic costs (Adams *et al.*, 1995).

Whatever mechanisms are involved, it is very clear that exercise at moderate speeds (e.g. 1–2 BL s^{-1}) reduces the level of social interactions and improves growth in salmonid fish. The effect on other important species in culture is unknown, but it is likely that the effects are similar in those species showing aggressive behaviour.

In net-pen rearing, it is not possible to create a circular current, but fish generally form schools anyway. The shape of the cages is important for the willingness of the fish to form tight schools. It is easy for fish to swim around in circular cages whereas square cages tend to disrupt the group structure when the fish school hits the corners (Kils, 1989).

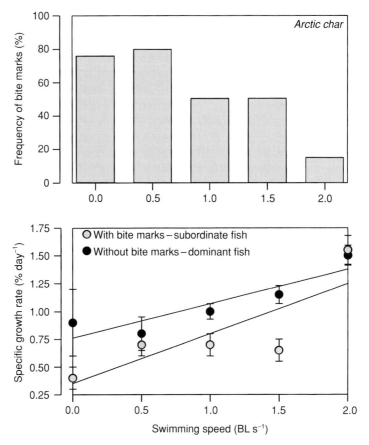

Fig. 10.7. Aggression (frequency of bite marks) and growth rate depending on the swimming speed in Arctic char. (Modified after Christiansen and Jobling, 1990.)

Feeding regime and feed distribution

Behavioural problems that are related to the feed ration only are of minor importance for fish farmers that feed the fish to near satiation. However, the spatial and temporal distribution of the feed is important for decreasing growth variations within a fish stock.

If feed is delivered to a single location in the rearing unit, dominant fish are likely to defend and hold a position close to that area, giving them greater opportunities than lower ranked fish (Ryer and Olla, 1995). Aggression was less frequent in groups of juvenile chum salmon if the feed was dispersed compared with those groups which were fed in a restricted area (Fig. 10.8). When Atlantic salmon in a fish rearing cage received feed distributed locally, 48% was consumed by only 18% of the fish, whereas spreading of the feed gave even feed consumption among all individuals (Thorpe *et al.*, 1990).

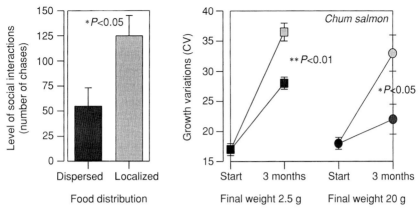

Fig. 10.8.   The level of aggressive interactions (number of chases) and variation in growth in chum salmon that were fed the same amount of feed dispersed or clumped. (Modified after Ryer and Olla, 1995.)

There is also a temporal aspect of feed defensibility, meaning that, if feed presentation were made less predictable in time, it would be more difficult for competitive individuals to monopolize the feeding area (Kadri *et al.*, 1996). Feeding in fish farms often occurs at the same time of day over long periods, and highly competitive individuals may learn this and position themselves at the best spots at the right time of the day.

The feeding of fish in culture always results in some stress and increased metabolic costs. To feed the fish intensively during short periods of the day could be one way of overcoming this problem. The feeding periods should be not more than two or three per day, at least for larger fish (above 100 g). Two intensive feeding periods of 1–2 hours' duration, one in the morning and one in the evening, with a longer period of non-feeding at midday is probably beneficial in reducing the level of stress associated with feeding.

Rearing environment

There is no rule of thumb indicating what depths to be used in fish culture, but juvenile salmonids will feel stressed if the water level is too low. For example, the minimum depth that brown trout choose in natural running waters is approximately twice its length, that is, a fish with a length of 20 cm needs a minimum water depth of about 40 cm (Heggenes, 1989). From the fish point of view, the greater the depth the better. For practical reasons, large depths may be disadvantageous, so the depth of rearing units must be a compromise between the needs of the fish and the constraints of the farmer.

The suitable light level for fish to forage efficiently is about 100–200 lux for Atlantic salmon (Fraser and Metcalfe, 1997). Salmonids are probably efficient foragers at this light intensity, and are at lower risk of being killed by day-active predators, e.g. birds. However, 200 lux is a very low light intensity for farm personnel to work in and, again, it may be necessary for a compromise between fish and workers. Nevertheless, the light intensity should not exceed 500 lux.

The colour inside tanks should mimic the natural conditions as much as possible. Dark colours are preferred by most species, whereas bright ones will create a marked contrast between the fish and tank surroundings, which is associated with high predation risk.

As described earlier, many fish species have a need for hiding in different situations. This is mainly an anti-predator behaviour that is very important for fish at different times of the day or season, or when the environment in general signals high predation risk. It would be desirable if an environment that contains hiding areas was created, at least in tanks. The easiest way to do this for salmonids is to put a sheet on top of the tank that covers and shades about 30–50% of the area. In addition, various types of substrates that create cavities or substrate for them to burrowing in can also be used within tanks, but the problem with objects in the water is that they disturb cleaning routines.

Successful rearing methods in intensive systems really deal with making the fish apply a schooling behaviour. A schooling behaviour is achieved by taking advantage of the plasticity the fish have in their behaviour repertoire and rearing them in conditions where it does not pay to be territorial but where it is more advantageous to swim in groups. Examples of such conditions are given. It is common sense that fish species for which schooling 'comes naturally' are used in aquaculture instead of species that have a very high tendency for forming territories and thus spend all their energy in fighting and eating each other. A fish farmer who keeps continuous track of the social situations in each unit according to the given guidelines should be able to reduce growth variation, enhance the total production and reduce stressful conditions. Also, to select for friendlier fish that are more willing to school should of course be desirable for the aquaculture industry, although no breeding with this in mind exists.

References

Abbott, J.C. and Dill, L.M. (1989) The relative growth of dominant and subordinate juvenile steelhead trout (*Salmo gairdneri*) fed equal ratios. *Behaviour* 108, 104–113.

Abbott, J.C., Dunbrack, R.L. and Orr, C.D. (1985) The interaction of size and experience in dominance relationships of juvenile steelhead trout (*Salmo gairdneri*). *Behaviour* 92, 241–253.

Adams, C.E., Huntingford, F.A., Krapal, J., Jobling, M. and Brunett, S.J. (1995) Exercise, agonistic behaviour and food acquisition in Arctic char, *Salvelinus alpinus. Environmental Biology of Fishes* 43, 213–218.

Adams, S.M. (1990) Status and use of biological indicators for evaluating the effects of stress on fish. *American Fisheries Society Symposium* 8, 1–8.

Alanärä, A. (1992) Demand feeding as a self-regulating feeding system for rainbow trout (*Oncorhynchus mykiss*) in net pens. *Aquaculture* 108, 347–356.

Alanärä, A. and Brännäs, E. (1996) Dominance in demand-feeding behaviour in Arctic charr and rainbow trout: the effect of stocking density. *Journal of Fish Biology* 48, 242–254.

Alanärä, A., Winberg, S., Brännäs, E., Kiessling, K., Höglund, A. and Elofsson, U. (1998) Feeding behaviour, brain serotonergic activity and energy reserves of Arctic char (*Salvelinus alpinus*) within a dominance hierarchy. *Canadian Journal of Fisheries and Aquatic Sciences* 76, 212–220.

Anderson, D.P. (1990) Immunological indicators: effects of environmental stress on immune protection and disease outbreaks. *American Fisheries Society Symposium* 8, 38–50.

Bachman, R.A. (1984) Foraging behaviour of free-ranging wild and hatchery brown trout in a stream. *Transactions of the American Fisheries Society* 113, 1–32.

Baker, R.F. and Ayles, G.B. (1990) The effects of varying density and loading level on the growth of Arctic charr (*Salvelinus alpinus* L.) and rainbow trout (*Oncorhynchus mykiss*). *World Aquaculture* 21, 58–61.

Bégout, M.-L. and Lagardère, J.-P. (1995) An acoustic telemetry study of seabream (*Sparus aurata* L.): first results on activity rhythm, effects of environmental variables and space utilization. *Hydrobiologia* 300/301, 417–423.

Bosakowski, T. and Wagner, E.J. (1995) Experimental use of cobble substrates in concrete raceways for improving fin condition of cutthroat (*Oncorhynchus clarki*) and rainbow trout (*O. mykiss*). *Aquaculture* 130, 159–165.

Brännäs, E. and Alanärä, A. (1992) Monitoring the feeding activity of individual fish with a demand feeding system. *Journal of Fish Biology* 42, 209–215.

Brännäs, E. and Alanärä, A. (1997) Is diel dualism in feeding activity influenced by competition between individuals? *Canadian Journal of Zoology* 75, 661–669.

Brown, G.E. and Brown, J.A. (1996) Does kin-biased territorial behavior increase kin-biased foraging in juvenile salmonids? *Behavioral Ecology* 7, 24–29.

Brown, G.E., Brown, J.A. and Srivastava, R.K. (1992) The effect of stocking density on the behaviour of Arctic charr (*Salvelinus alpinus*). *Journal of Fish Biology* 41, 955–963.

Chilton, E.W. and Poarch, S.M. (1997) Distribution and movement behavior of radio-tagged grass carp in two Texas reservoirs. *Transactions of the American Fisheries Society* 126, 467–476.

Christiansen, J.S. and Jobling, M. (1990) The behaviour and the relationship between food intake and growth of juvenile Arctic charr, *Salvelinus alpinus* L. subjected to sustained exercise. *Canadian Journal of Zoology* 68, 2185–2191.

Daan, S. (1981) Adaptive daily strategies in behaviour. In: Aschoff, J. (ed.) *Handbook of Behavioural Neurobiology (Biological Rhythms)*, Vol. 4. Springer-Verlag, Berlin, pp. 275–298.

Davidson, W. and Goldspink, G. (1977) The effect of prolonged exercise on the lateral musculature of the brown trout *Salmo trutta* L. *Journal of Experimental Biology* 70, 1–12.

Dill, L.M. (1983) Adaptive flexibility in the foraging behavior of fishes. *Canadian Journal of Fishery and Aquatic Sciences* 40, 398–408.

Doyle, R.W. and Talbot, A.J. (1986) Artificial selection for growth and correlated selection on competetive behaviour in fish, *Canadian Journal of Fishery and Aquatic Sciences* 43, 1059–1064.

East, P. and Magnan, P. (1987) The effect of locomotor activity on the growth of brook charr *Salvelinus fontinalis*. *Canadian Journal of Zoology* 65, 843–846.

Elliott, J.M. (1994) *Quantitative Ecology and the Brown Trout*. Oxford University Press, Oxford, 286 pp.

Eriksson, L.-O. (1973) Spring inversion of the diel rhythm of locomotor activity in young sea-going brown trout, *Salmo trutta* L. and Atlantic salmon, *Salmo salar* L. *Aquilo, Ser. Zoologica* 14, 68–79.

Fausch, K.D. (1984) Profitable stream positions for salmonids: relating specific growth rate to net energy gain. *Canadian Journal of Zoology* 62, 441–451.

Fernnö, A., Furevik, D., Huse, I. and Bjordal, Å. (1988) A multiple approach to behaviour studies of salmon reared in marine net pens. *International Council of the Exploration of the Sea* C. M. F:15.

Fraser, N.H.C. and Metcalfe, N.B. (1997) The costs of becoming nocturnal: feeding efficiency in relation to light intensity in juvenile Atlantic salmon. *Functional Ecology* 11, 385–391.

Fraser, N.H.C., Metcalfe, N.B. and Thorpe, J.E. (1993) Temperature-dependent switch between diurnal and nocturnal foraging in salmon. *Proceedings of the Royal Society London, Series B, Biological Sciences* 252, 135–139.

Furevik, D.M., Bjordal, Å., Huse, I. and Fernö, A. (1993) Surface activity of Atlantic salmon (*Salmo salar* L.) in net pens. *Aquaculture* 110, 119–128.

Gerking, S.D. (1959) The restricted movements of fish populations. *Biological Reviews* 34, 221–242.

Gimfer-Artigas, E. (1998) Feeding strategies for bream and bass. In: *Aquaculture and Water: Fish Culture, Shellfish Culture and Water Usage*. Aquaculture Europe '98, Bordeaux, France, pp. 93–94.

Gjedrem, T. (1993) *Fiskeoppdrett* (in Norwegian). Landbruksforlaget, Oslo, Norway.

Godin, J.-G.J. (1982) Migrations of salmonid fishes during early life history phases: daily and annual timing. In: Brannon, E.L. and Salo, E. (eds) *Salmon and Trout Migratory Symposium*. University of Washington Press, Seattle, pp. 22–50.

Grant, J.W.A. (1997) Territoriality. In: Godin, J.-G.J. (ed.) *Behavioural Ecology of Teleost Fishes*. Oxford University Press, Oxford, pp. 81–103.

Grant, J.W.A. and Noakes, D.L.G. (1992) Feeding and social behaviour of brook and lake charr. In: Thorpe, J.E. and Huntingford, F.A. (eds) *The Importance of Feeding Behavior for the Efficient Culture of Salmonid Fishes*, Vol. 2. World Aquaculture Workshops, The World Aquaculture Society, pp. 13–20.

Hecht, T. and Appelbaum, S. (1988) Observations on intraspecific aggression and coeval sibling cannibalism by larval and juvenile *Clarias gariepinus* (Clariidae: Pisces) under controlled conditions. *Zoology* 214, 21–44.

Heggenes, J. (1989) Physical habitat selection by brown trout (*Salmo trutta*) in riverine systems. *Nordic Journal of Freshwater Research* 64, 74–90.

Helfman, G. (1993) Fish behaviour by day, night, and twilight. In: Pitcher, T.J. (ed.) *The Behaviour of Teleost Fishes*. Croom Helm, London, pp. 366–387.

Hemmer, H. (1991) *Domestication. The Decline of Environmental Appreciation*. Cambridge University Press, Cambridge.

Hoar, W.S. (1942) Diurnal variations in feeding activity of young salmon and trout. *Journal of the Fisheries Research Board of Canada* 6, 90–101.

Houlihan, D.F. and Laurent, P. (1987) Effects of exercise training on the performance, growth and protein turnover of rainbow trout. *Canadian Journal of Fisheries and Aquatic Sciences* 44, 1614–1621.

Huntingford, F.A., Metcalfe, N.B., Thorpe, J.E., Graham, W.D. and Adams, C.E. (1990) Social dominance and body size in Atlantic salmon parr (*Salmo salar*). *Journal of Fish Biology* 36, 877–881.

Hvidsten, N.A., Jensen, A.J., Vivaas, H., Bakke, O. and Heggberget, T.G. (1995) Downstream migration of Atlantic salmon smolts in relation to water flow, water temperature, moon phase and social interaction. *Nordic Journal of Freshwater Research* 70, 38–48.

Jenkins, T.M. (1969) Social structure, position choice and microdistribution of two trout species (*Salmo trutta* and *Salmo gairdneri*) resident in mountain streams. *Animal Behaviour Monograph* 2, 57–123.

Jobling, M. (1995) Physiological and social constraints on growth of fish with special reference to Arctic charr, *Salvelinus alpinus* L. *Aquaculture* 44, 83–90.

Johnsson, J.I. and Björnsson, B.T. (1994) Growth hormone increases growth rate, appetite and dominance in juvenile rainbow trout, *Oncorhynchus mykiss*. *Animal Behaviour* 48, 177–186.

Jørgensen, E.H., Christiansen, J.S. and Jobling, M. (1993) Effects of stocking density on food intake, growth performance and oxygen consumption in Arctic char (*Salvelinus alpinus*). *Aquaculture* 110, 191–204.

Juell, J.E. (1995) The behaviour of Atlantic salmon in relation to efficient cage-rearing. *Reviews in Fish Biology and Fisheries* 5, 320–335.

Kadri, S., Huntingford, F.A., Metcalfe, N.B. and Thorpe, J.E. (1996) Social interactions and the distribution of food among one-sea-winter Atlantic salmon (*Salmo salar*) in a sea-cage. *Aquaculture* 139, 1–10.

Kalleberg, H. (1958) Observations in a stream tank of territoriality and competition in juvenile salmon and trout (*Salmo salar* L. and *S. trutta* L.). *Reports from the Institute of Freshwater Research in Drottningholm* 39, 55–98.

Keeley, E.R. and Grant, J.W.A. (1995) Allometric and environmental correlates of territory size in juvenile Atlantic salmon (*Salmo salar*). *Canadian Journal of Fisheries and Aquatic Sciences* 52, 186–196.

Keeley, E.R. and McPhail, J.D. (1998) Food abundance, intruder pressure, and body size as determinants of territory size in juvenile steelhead trout (*Oncorhynchus mykiss*). *Behaviour* 135, 65–82.

Kils, U. (1989) Some aspects of schooling for aquaculture. *International Council of the Exploration of the Sea*. C. M. F:12.

Kinghorn, B.P. (1983) A review of quantitative genetics in fish breeding. *Aquaculture* 31, 283–304.

Magurran, A.E. (1993) Individual differences and alternative behaviours. In: Pitcher, T.J. (ed.) *Behaviour of Teleost Fishes*. Croom Helm, London, pp. 441–477.

McCarthy, I.D., Carter, C.G. and Houlihan, D.F. (1992) The effect of feeding hierarchy on individual variability in daily feeding of rainbow trout, *Oncorhynchus mykiss* (Walbaum). *Journal of Fish Biology* 41, 257–263.

Metcalfe, N.B. (1986) Intraspecific variation in competitive ability and food intake in salmonids: consequences for energy budgets and growth rates. *Journal of Fish Biology* 28, 525–531.

Metcalfe, N.B., Huntingford, F.A., Thorpe, J.E. and Adams, C.E. (1990) The effects of social status on life history variation in juvenile salmon. *Canadian Journal of Zoology* 68, 2630–2636.

Metcalfe, N.B., Taylor, A.C. and Thorpe, J.E. (1995) Metabolic rate, social status and life-history strategies in Atlantic salmon. *Animal Behaviour* 49, 431–436.

Mikheev, V.N., Adams, C.E., Huntingford, F.A. and Thorpe, J.E. (1996) Behavioural responses of benthic and pelagic Arctic charr to substratum heterogeneity. *Journal of Fish Biology* 49, 494–500.

Milinski, M. (1990) Information overload and food selection. In: Hughes, R.N. (ed.) *Behavioural Mechanisms of Food Selection*. Springer-Verlag, Berlin, pp. 721–736.

Morgan, M.J. (1988) The effect of hunger, shoal size and the presence of a predator on shoal cohesiveness in bluntnose minnows, *Pimephales notatus* Rafinesque. *Journal of Fish Biology* 32, 963–971.

Pawson, M.G., Kelley, D.F. and Picket, G.D. (1987) The distribution and migrations of bass, *Dicentrarchus labrax* L., in waters around England and Wales as shown by tagging. *Journal of the Marine Biology Association, UK* 67, 183–217.

Peuhkuri, N., Ranta, E. and Seppä, P. (1997) Size-assortative shoaling in three-spined sticklebacks. *Ethology* 108, 318–324.

Pickett, G.D. and Pawson, M.G. (1994) *Sea Bass – Biology, Exploitation and Conservation*. Chapman & Hall, London, 337 pp.

Pitcher, T.J. and Parrish, J.K. (1993) Functions of shoaling behaviour in teleosts. In: Pitcher, T.J. (ed.) *Behaviour of Teleost Fishes*. Chapman & Hall, London, pp. 363–427.

Puckett, K.J. and Dill, L.M. (1985) The energetics of feeding territoriality in juvenile coho salmon (*Oncorhynchus kisutch*). *Behaviour* 92, 97–111.

Reinhardt, U.G. and Healey, M.C. (1997) Size-dependent foraging behaviour and use of cover in juvenile coho salmon under predation risk. *Canadian Journal of Zoology* 75, 1642–1651.

Ruzzante, D. E. (1994) Domestication effects on aggressive and schooling behaviour in fish. *Aqaculture* 120, 1–24.

Ryer, C.H. and Olla, B.L. (1995) The influence of food distribution upon the development of aggressive and competitive behaviour in juvenile chum salmon, *Oncorhynchus keta*. *Journal of Fish Biology* 46, 264–272.

Sánchez-Vázquez, F.J., Azzaydi, M., Martinez, F.J., Zamora, S. and Madrid, J.A. (1998) Annual rhythms of demand-feeding activity in sea bass: evidence of a seasonal phase inversion of the diel feeding pattern. *Chronobiology International* (in press).

Schreck, C.B., Patino, C.K., Winton, J.R. and Holway, J.E. (1985) Effects of rearing density on indices of smoltification and performance of coho salmon, *Oncorhynchus kisutch*. *Aquaculture* 45, 345–358.

Shaw, E. (1961) Minimal light intensity and the dispersal of fish schools. *Bulletin from the Institute of Oceanography* No. 1213, 8 pp.

Shepherd, J. and Bromage, N. (1992) *Intensive Fish Farming*. Blackwell Science, 404 pp.

Smith, C. and Reay, P. (1991) Cannibalism in teleost fish. *Reviews in Fish Biology and Fisheries* 1, 41–64.

Swain, D.P. and Riddel, B.E. (1990) Variation in agonistic behaviour between newly emerged juveniles from hatchery and wild populations of coho salmon, *Oncorhynchus kisutch*. *Canadian Journal of Fisheries and Aquatic Science* 47, 566–571.

Symons, P.E.K. (1970) The possible role of social and territorial behaviour of Atlantic salmon parr in the production of smolts. *Technical Reports of the Fisheries Research Board of Canada* 206, 1–25.

Thompson, I. (1993) Nutrition and disease resistance in fish. PhD thesis, University of Aberdeen, Aberdeen.

Thorpe, J.E., Talbot, C., Miles, M.S., Rawlings, C. and Keay, D.S. (1990) Food consumption in 24 hours by Atlantic salmon (*Salmo salar* L.) in a sea cage. *Aquaculture* 90, 41–47.

Totland, G.K., Kryvi, H., Jødestøl, K.A., Christiansen, E.N., Tangerås, A. and Slinde, E. (1987) Growth and composition of the swimming muscle of adult Atlantic salmon (*Salmo salar* L.) during long-term sustained swimming. *Aquaculture* 66, 299–313.

Vijayan, M.M. and Leatherland, J.F. (1989) Cortisol-induced changes in plasma glucose, protein, and thyroid hormone levels, and liver glycogen content of coho salmon (*Oncorhynchus kisutch* Walbaum). *Canadian Journal of Zoology* 67, 2746–2750.

Wedemeyer, G.A. (1976) Physiological response of juvenile coho salmon (*Oncorhynchus kisutch*) and rainbow trout (*Salmo gairdneri*) to handling and crowding stress in intensive fish culture. *Journal of the Fisheries Research Board of Canada* 33, 2699–2702.

Winberg, S. and Nilsson, G.E. (1993) Roles of brain monoamine neuro-transmitters in agonistic behaviour and stress reactions, with particular reference to fish. *Comparative Biochemistry and Physiology* 3, 597–614.

Wootton, R.J. (1998) *Ecology of Teleost Fishes*, 2nd edn. Kluwer Academic Publishers, Dordrecht, 432 pp.

Yamamoto, T., Ueda, H. and Higashi, S. (1998) Correlation among dominance status, metabolic rate and otolith size in masu salmon. *Journal of Fish Biology* 52, 281–290.

Young, P.S. and Cech, J.J. (1994) Effects of different exercise conditioning velocities on the energy reserves and swimming stress responses in young-of-the-year striped bass (*Morone saxatilis*). *Canadian Journal of Fisheries and Aquatic Sciences* 51, 1528–1534.

Zayan, R. (1991) The specificity of social stress. *Behavioural Processes* 25, 81–93.

Contemporary Topics in Social Behaviour

In this final section of the book we focus on topics that have been mentioned in the previous chapters, but which have not been discussed in detail. For the most part they are relatively new areas in pure and applied ethology, but ones, nevertheless, we think are essential to include in a book on social behaviour in farm animals. Unlike the structured and consistent layout of the previous six chapters dealing with social behaviour in the different species, the authors in this section have been given free rein to speculate on these fledgling research areas.

Much in this book on social behaviour has dealt with bonds between animals and so it seems necessary to focus some attention on what happens when these bonds are broken. Of course, bonds such as that between mother and young are broken in nature, but this is usually a gradual process. In contrast, animals used in agriculture are often separated abruptly and at times when they themselves may not choose to have separated. It is inevitable that there are consequences of this separation. Separation or death of a loved one in humans is regarded as a severe stressor, having major consequences for human health and well-being. The first chapter in this section (Chapter 11) examines some of the consequences of breaking bonds on the health, production and welfare of farm animals.

It is accepted that humans react differently to different situations. That animals respond differently as well has generally been regarded as a nuisance by researchers, who have to use larger numbers of individuals in their experiments as a consequence. Chapter 12 takes the opposite view, proposing that individual differences present an interesting and worthwhile area of research in their own right. Even if each species has a typical behavioural repertoire, as presented in the previous section, within this there is still a large variation. It is suggested that this has implications for the way we manage our animals. For example, we should no longer design housing for the average animal but can, and perhaps should, take into consideration that animals have different strategies for coping with their environment and modify the design accordingly.

For a farmer, managing and dealing with individual differences in animals is the cornerstone of good stockmanship. Dog and cat owners are also well aware of the individual differences between their pets and have what could be described as a social relationship with them. Nevertheless, the scientific study of human–animal interaction is surprisingly new, only really gaining momentum in the past 20 years. Chapter 13 examines how farm animals respond to people and addresses the question of whether or not farm animals can be said to have social relationships with people. Considering that all our farm animals come into more or less regular contact with humans, it is necessary to understand how people fit into the social lives of animals.

Whether the social relationship is across species, such as between man and animal, or within a species, for it to become established there needs to be a certain level of recognition and probably also communication. In close relationships, there may even be transfer of information about emotional states, such as suffering or pleasure. All these topics come under the subject of cognition. As we try and understand more about the behaviour of animals, it is natural that we reach the point where we want to understand how animals perceive and respond to their environment. In a book on social behaviour, it seems inevitable that we should choose to have a chapter that tackles the subject of social cognition (Chapter 14).

Breaking Social Bonds

11

Ruth Newberry[1] and Janice Swanson[2]

[1]Center for the Study of Animal Well-being, Department of
Veterinary and Comparative Anatomy, Pharmacology and
Physiology and Department of Animal Sciences, Washington
State University, Pullman, WA 99164-6520, USA; [2]Department
of Animal Sciences and Industry, 134C Weber Hall, Kansas
State University, Manhattan, KS 66506, USA

*(Editors' comments: A social bond is a mutual social attachment that is
relatively long lasting and survives temporary separation. In humans,
breaking such a bond, by separation or death, results in a strong
emotional response with both behavioural and physiological symptoms.
In this chapter the authors address whether there could be similar
consequences in animals by referring to studies on routine management
practices in farm animals where there is the potential for bonds to be
broken, such as during weaning, moving animals between groups, selling
or slaughtering. While some results, such as the increased vocalization,
suggest distress when animals are separated, it is not clear how long this
lasts or whether the bond is re-established when the animals are reunited.
Results do indicate variation between species depending on their social
structure, for example, whether the male and female form lifelong
monogamous pairs, like geese, and whether the young hide and so are
used to prolonged periods of separation from their mother.)*

11.1. Introduction

In humans, the breaking of social bonds evokes an emotional response
to loss, referred to as grief, which is characterized by depression,
anxiety, guilt, anger, anhedonia and loneliness. Grief is manifested by
behavioural and physiological symptoms such as agitation, sleep disturbances, crying, loss of appetite, cognitive impairment and increased
susceptibility to disease (Stroebe and Stroebe, 1987). Given the
potentially serious consequences of breaking bonds for the health and

well-being of humans, it is not surprising that questions are raised about the extent to which breaking bonds could have similar consequences in other animal species.

In this chapter, we present characteristics of social bonds in livestock and poultry and describe processes by which these bonds are broken under natural and agricultural conditions. We examine the impact of separating bonded animals on their health, well-being and productivity and explore techniques used to minimize adverse consequences. There has been relatively little systematic research on this topic in livestock and poultry. Most of the available information concerns the mother–young bond (see Section 3.5) and the impact of early weaning. Given the potential economic and welfare consequences of breaking bonds prematurely, we emphasize the need for improved understanding of the degree of emotional attachment occurring between animals in managed groups and the persistence of distress following separation.

11.2   Social Bonds

11.2.1   Characteristics of social bonds

Before addressing the issue of breaking bonds, it is necessary to have an understanding of what constitutes a social bond and in which contexts bonds occur in farmed species. We define a social bond as a mutual, affectionate, emotional attachment between two individuals that is relatively long lasting and survives temporary separations (Hinde, 1974; Gubernick, 1981). This definition arbitrarily excludes temporary associations that form between males and females solely for mating although they may have some features of a bond. A social bond is characterized by affiliative behaviour directed by one or both parties towards the other, such as allogrooming, providing food and protection, resting in contact, playing, synchronizing activities, and maintaining proximity, reinstatement behaviour when separated, and greeting following temporary separations (Fig. 11.1). In some situations, the presence of one party has a calming effect and reduces fear in the other. For example, in some species a mother provides a base of security for her infant, who moves away from her to explore and returns to her when alarmed (Bowlby, 1979). Affiliative behaviour between bonded individuals is associated with physiological responses, such as reduced heart rate and elevated β-endorphin levels (Keverne et al., 1989; Feh and de Mazieres, 1993).

On a physiological level, oxytocin, vasopressin, prolactin and opioids have been implicated in the development and maintenance of mammalian social bonds (Keverne and Kendrick, 1992; Kalin et al.,

Fig. 11.1. Affiliative behaviour between a sow and her piglets in a semi-natural environment.

1995; Bridges, 1996; Insel, 1997; Panksepp, 1998). In some species, oxytocin appears to be more important for bonding in females whereas vasopressin is important in males (Insel, 1997). In females primed with oestrogen and progesterone, vaginocervical stimulation of females during mating and parturition can stimulate the release of oxytocin. Dwyer and Lawrence (1997) were able to elicit maternal behaviour in some hormonally primed non-pregnant sheep through mechanical vaginocervical stimulation providing that they had had previous maternal experience.

From an evolutionary perspective, social bonds should be dynamic, being maintained only as long as the potential benefits to both parties in terms of (inclusive) fitness exceed the costs. Thus, we expect to find parent–offspring bonds when the fitness of both the parent and young, under natural conditions, is enhanced by parental care of the young. These bonds seem to be aimed primarily at ensuring provision of care to offspring rather than the young of others when the young are mobile or otherwise likely to co-mingle (Gubernick, 1981). Monogamous mating systems have been favoured during evolution under conditions where one parent is unlikely to be successful rearing young alone. Bonds between monogamous mates persist over multiple breeding seasons under conditions where the development of a workable relationship takes time or when locating a suitable mate is costly. For example, long-term pair bonds are found in wild geese where successful hatching and rearing of goslings may be achieved only after several seasons together (Choudhury and Black, 1993). Parent–offspring bonds should be maintained beyond the infant care period when there is likely to be a fitness benefit to both parties if offspring remain in the natal group, for example, as helpers of younger siblings. Long-term social bonds between parents, offspring and siblings may also provide protection against incest and inbreeding depression.

The term 'bond' is somewhat misleading in that it implies that the level of attachment is equal for both parties. Asymmetry often exists in the level of attachment of two individuals and the degree of asymmetry can change over time and may reverse direction (Hinch *et al.*, 1987). Parent–offspring bonds weaken when provision of parental care is no longer in the best interests of both parties and there may be conflict between parent and offspring over the timing of withdrawing parental care (Trivers, 1974). In addition, when bonds break under natural conditions, the process is often gradual such that there is no clear moment in time when one can draw the line between a bond and a more benign social association.

Despite difficulties in deciding when a bond exists, we require more evidence than membership in the same group before classifying a relationship between two individuals as a bond. Group members can derive benefits from mutual recognition and cohesion without having a preferential association with a specific individual. For example, the removal of a group member may not evoke an emotional response from any remaining group member although the individual removed may be distressed and highly motivated to return to the group. The behavioural and physiological responses of specific individuals to separation and reunion provide important clues that help us to distinguish social bonds from more benign associations.

11.2.2 Responses to separation

When bonded individuals are separated, and are motivated to regain contact, bouts of reinstatement behaviour such as locomotion (searching) and vocal signalling are performed by both parties, interspersed or followed by periods of energy-conserving depression (Panksepp, 1998). Bowlby (1979) refers to these phases as 'protest' and 'despair', respectively. Reinstatement behaviour, although costly in terms of energy, time expenditure and increased predation risk, is thought to be adaptive under natural conditions by increasing the probability that the animals will be reunited (Panksepp, 1998). Reinstatement behaviour may be accompanied by altered feeding and sleep patterns, suspension of play, elevated corticosteroid levels and changes in heart rate and core body temperature (Cirulli *et al.*, 1996;

Koolhaas *et al.*, 1997). However, the context and duration of separation must be taken into account when interpreting responses to separation (Gubernick, 1981). In species in which the young remain in a nest during early development, such as pigs and rabbits, responses of young to departure of their mother are likely to be low unless the separation is prolonged or the young are removed from the nest. In contrast, even brief separations from their dam are likely to evoke reinstatement behaviour in follower species such as sheep and goats.

If bonded animals are separated for a long period and they are able to adapt to the separation, their responses to separation wane and they resume normal activities. Behaviour suggestive of prolonged grief following separation from an attachment figure, including unresponsiveness, listlessness, head hanging and sunken eyes, has been reported in some individuals of relatively long-lived, intelligent mammalian species such as chimpanzees, elephants and dolphins, and in monogamous birds such as ravens and geese (Lorenz, 1991; de Waal, 1996). Pronounced responses to separation have been associated with maladaptive outcomes including cognitive impairment and depressed immunity (Boccia *et al.*, 1997). To our knowledge, evidence for prolonged grief has not been reported in the common agricultural species.

At the physiological level, oxytocin and vasopressin appear to play important roles in mammalian responses to separation from attachment figures, as illustrated by comparison of two closely related rodent species (Insel, 1997). Prairie voles are monogamous and their offspring show a pronounced distress response when separated from their parents (ultrasonic vocalizations and elevated corticosterone). Montane voles, on the other hand, are polygynous and their young show less separation distress and disperse at a younger age. In addition, prairie voles, but not montane voles, continue to show a preference for their mate for weeks after separation. These differences in social behaviour are associated with differences in the distribution of oxytocin and vasopressin receptors in the brain and differences in the transient expression of oxytocin and vasopressin receptor binding during postnatal development. Such differences may contribute to species, breed and individual differences in degree of social attachment and timing of voluntary dispersal.

11.2.3 Mother–offspring bonds in farmed species

Most species used in agriculture (e.g. cattle, sheep, goats, pigs, horses, chickens, turkeys) have a polygynous mating system and relatively precocial young that are mobile soon after birth. Under these conditions, mothers are solely responsible for offspring care and form bonds with their young soon after parturition or hatching (Veissier *et al.*, 1998). The rhea is a farmed polyandrous species in which the father will care for his offspring if given the opportunity (Lábaque *et al.*, 1999). Pigeons and geese are examples of farmed monogamous species in which the male and female bond with each other and with their offspring if given the opportunity.

Genetic differences in maternal attachment to offspring have been associated with selection for different productivity traits (Dwyer *et al.*, 1998). For example, cows of dairy breeds are derived from generations of removal of calves within 24 hours post-partum. Dairy cows usually respond less to separation from their calf than cows of beef breeds, and are more tolerant of alien calves. In contrast, beef cows often raise their calves in herds under extensive conditions. Selection forces would dictate that strong maternal bonds are maintained and care-giving preferences lie with their own calves. Similarly, sheep of hill breeds, selected for successful rearing of lambs under harsh range conditions, tend to be more maternally responsive than sheep of lowland breeds.

Few studies have investigated the persistence of mother–young bonds in agricultural species following separation. One exception is the study by Lamb *et al.* (1997), who reported that beef cows spent little time in proximity to their calf after a 4-week separation starting at about 15 days of age, suggesting that the bond between them had been extinguished. In contrast, cows whose calves remained within head and neck contact behind a barrier during this period retained their bond to their calf. The same was true if the cow was hobbled so that an unrelated calf could suckle from her while her own calf was kept behind the barrier. When these animals were subsequently released on pasture after 4 weeks, the cow readily allowed her own calf to suckle and resisted efforts by the unrelated calf to suckle from her by kicking and moving away. The unrelated calves survived (with relatively low growth rates) by stealing milk from any cow that would tolerate it, either by suckling at the same time as the cow's own calf or by suckling from behind the cow. However, cows whose own calves had been completely removed at 15 days and who were hobbled to allow an unrelated calf to suckle developed a bond with the unrelated calf. They ignored their own calf when reunited after 4 weeks and continued to nurse the unrelated calf when subsequently released on pasture. Post-partum anovulation was prolonged in cows that were being suckled and had a continuously reinforced bond to either their own or an adopted calf.

Although not subject to systematic investigation, litter size appears to affect the strength of bonding between mother and young, with weaker attachments forming when litters are large, as found in domestic pigs. In multi-sow groups, some piglets are opportunistic and will seek a teat on any tolerant sow where they are able to displace an existing teat occupant (Newberry and Wood-Gush, 1985). Piglets that establish themselves in the teat order of a different sow associate more closely with their adopted sow and littermates than with their natural mother and littermates. There is no evidence that they retain a strong bond with their natural mother or littermates (Newberry and Wood-Gush, 1986).

In agricultural species living in relatively natural conditions, associations between mothers and offspring generally persist beyond weaning and, sometimes, beyond the birth of subsequent offspring (e.g. cattle: Reinhardt and Reinhardt, 1981; pigs: Newberry and Wood-Gush, 1985; sheep: Hinch *et al.*, 1990; horses: Kimura, 1998). Interestingly, Swanson and Stricklin (1985) observed that agonistic behaviour at the feed trough was halved in groups of related beef cows (composed of mothers and daughters) compared with groups of unrelated cows although the mothers and daughters had been separated for at least a year after weaning prior to being reunited. Further systematic investigation of the nature of responses to separation and encounters following reunion after varying periods is needed to evaluate the persistence of social bonds between mothers and young.

11.2.4   Other bonds in farmed species

The extent of social bonding between siblings, other relatives and unrelated group members in agricultural species is unclear. Porter *et al.* (1995) observed that lambs emitted fewer distress bleats when separated from the ewe at 3 weeks of age if they were paired with their twin than if paired with an unfamiliar lamb. They suggested that the twins were bonded. For a more definitive answer, it would be useful to assess the lambs' responses to separation from each other and subsequent reunion. In a semi-natural enviroment, suckling piglets associated more closely with littermates than with piglets from other litters (Newberry and Wood-Gush, 1986). However, they did not exhibit special associations with specific littermates and there was no behavioural indication that they were distressed by the disappearance (death) of specific littermates. On the contrary, young porcine littermates use their canine teeth to compete with each other for high-quality teats to the extent that weaker siblings obtain poorer teats, spend more time at the udder and are at a greater risk of being crushed by the sow (Fraser and Thompson, 1990). Litter size may affect the strength of bonds between littermates, with twins responding to separation more strongly than members of large litters. Increasing age may also strengthen bonds between specific pairs of littermates and peers, especially between same-sex pairs. From a production stand-point, it would be useful to determine the extent to which the presence of 'friends', as opposed to other familiar individuals, can buffer the depressive and immune suppressive effects of maternal separation, as has been reported in some primates (Boccia *et al.*, 1997). In free-ranging populations, it would also be of interest to determine whether 'friendships' between specific adult group members increase offspring survival (e.g. through communal nesting as in pigs, tolerating suckling by friends' offspring, and protecting friends' young from predatory and infanticidal attacks). The existence of clear benefits to both parties from close association would increase the likelihood that evolution has favoured mechanisms that would facilitate bonding between them.

It might be assumed that unrelated animals confined together develop social bonds. Heifers will form associations with unrelated peers but the strength of the associations is stronger the younger the group is formed (Veissier *et al.*, 1998). Older animals may take a long time to affiliate with one another and may never form emotional attachments (e.g. pigs: Newberry and Wood-Gush, 1986; dogs: Tuber *et al.*, 1996). Adult domestic fowl tend to fight with former group members after an absence of only 2 to 3 weeks (Chase, 1982). Among unrelated pigs kept together for several weeks, Ewbank and Meese (1971) reported that low-ranking individuals were attacked after a 3-day absence whereas dominant pigs were accepted back into the group without contest after an absence of up to 25 days. All pigs, regardless of social status, were 'groomed' when returned to the group whereas this behaviour was not directed towards complete strangers. However, the antagonism exhibited between the familiar pigs after temporary separations suggests that they were not socially bonded.

11.3   When are Bonds Broken?

11.3.1   Under natural conditions

Under natural conditions, social bonds may be broken through death of one member of the pair or natal dispersal. Death may result from predation, disease, accident, infanticide, siblicide, aggression or cannibalism. Occasionally, a youngster may become separated from the dam as a result of getting trapped behind an obstacle, lost or left behind. If not reunited quickly, the young animal is unlikely to survive. Individuals may become isolated from other group members when seriously ill or injured. If they recover, the duration of the absence and memory capacities of the animals will affect whether they are accepted back into the group or treat each other as strangers.

Weaning usually occurs gradually and at variable ages according to the food supply and timing of arrival of the mother's next offspring (e.g. horses: Khalil and Kaseda, 1997) (Fig. 11.2). In semi-natural

Fig. 11.2. The point at which a mother weans her foal is variable in horses. It often depends on the timing of arrival of her next offspring.

environments, weaning of domestic piglets has been reported to occur at ages ranging from 2 to 4 months (Newberry and Wood-Gush, 1985; Jensen, 1995). The suckling frequency gradually declines until the offspring no longer seek to suckle, or weaning may be accelerated by aggression directed by the mother towards her young. The sow may also make it increasingly difficult for her young to suckle by walking away from them, ignoring their begging calls and resting in sternal rather than lateral recumbency (see Section 6.1.6). In cattle (see Section 5.1.4), natural weaning generally occurs around 6 months of age but may not occur until 1 year. As in pigs, cows can be quite individual about whether and when they choose to deny suckling privileges to their growing offspring. Although weaning is usually associated with reduced contact between mother and young, cows still show a preference for their own yearling calves over other calves (Veissier *et al.*, 1998) and sheep may maintain associations with their lambs for 1–2.5 years (Hinch *et al.*, 1990; Lawrence, 1990).

Natal dispersal refers to the movement of animals from their birthplace to their first breeding site or reproductive home range. Depending on the distance travelled and the degree of subsequent interaction with members of the natal group, dispersal can also be viewed as a time of breaking or weakening bonds with former group members. Emigration is risky. Dispersing animals may have difficulty finding and competing for food and other resources. Lack of familiarity with their surroundings can increase the risk of predation. By staying in the natal group, or delaying dispersal, animals may be able to inherit a breeding position within the group, gain physical condition and social experience and increase their inclusive fitness by caring for younger kin. On the other hand, by leaving, they may benefit from reduced competition for resources, including food and mates, and avoidance of inbreeding. Thus, the timing of dispersal can be variable.

In feral horses, most females leave the natal group voluntarily to join another stallion while in oestrus and most males leave voluntarily during a period of food shortage or when a sibling is born (Monard *et al.*, 1996; Khalil and Kaseda, 1997). Departure usually occurs between 1 and 4 years of age and there may be no evidence of prior weakening of social bonds with natal group members. In wild boars, dispersal appears to be a gradual process. Weaned juveniles spend some time associating with their mother, some time in their natal home range and some time on exploratory forays away from both the sow and the natal home range (Cousse *et al.*, 1994). In some cases, attempts at dispersal may fail and dispersers may return to the natal group for varying periods before dispersing permanently (McNutt, 1996; Khalil and Kaseda, 1997).

11.3.2   Under farm conditions

Feral animals are able to make choices about their social associations according to their perceived costs and benefits (Vehrencamp, 1983). In contrast, animals used in agriculture are grouped according to human interests and bonded individuals are often separated at times when they would not choose to separate themselves. Whereas separation in nature often results from gradual weakening of bonds, separation of animals on farms is typically abrupt and permanent. Weaning is often performed earlier than would occur in nature although the timing of weaning varies between production systems. For example, in dairy cattle, where milk is the desired commodity, calves are usually removed from the cow shortly after birth and before any lasting bond is formed between them. Calves may be allowed a short period to suckle colostrum from the cow or colostrum may be harvested and supplied artificially to the calf. In contrast, beef cows and calves are usually separated between 4 and 6 months of age, approaching the age range in which natural weaning occurs. In intensive meat production systems, there is economic pressure to advance the age at weaning to shorten the interval between births and increase the number of offspring produced per dam per unit of time.

Social groups of farm animals are rarely stable for extended periods because animals are removed and replaced, or kept in large groups with continual mingling of strangers. As animals grow and take up space within a confined area, it becomes necessary to divide groups or move them to larger enclosures. At this time, animals may be regrouped on the basis of body weight, sex, productivity or other traits of human interest. Animals may be separated as a result of sale or slaughter for human consumption. Sick or disabled animals may be abruptly removed from their group and placed in isolation or grouped with other animals (e.g. in the bulling pen or cull cow milking string). Pregnant females may be physically isolated during their pregnancy or prior to parturition. Temporary separations are also common, for example, when animals are removed from their group for breeding, showing, health checks, foot care, grooming or performance of minor surgical procedures. With the exception of early weaning, there has been a lack of systematic studies investigating the impact of these separations on the removed animals and on those remaining in the group.

11.4 Consequences of Breaking Bonds

11.4.1 Temporary absences and changes in group membership

Separation distress produced by periodically removing neonates from their mothers may produce permanent alteration of corticotrophin-releasing hormone (CRH) gene expression, resulting in the elevation of central CRH and the proliferation of CRH receptors (Nemeroff, 1998). These changes may sensitize the hypothalamic–pituitary–adrenal axis and produce changes in behaviour and physiology indicative of depression (Matthews *et al.*, 1996; Koolhaas *et al.*, 1997). In dogs, repeated absences of the human caretaker to whom the dog is bonded can lead to separation anxiety (Lund and Jørgensen, 1999). Young livestock being hand-reared by a human may be susceptible to this condition (see Section 13.2.1). Sensitization to separation may explain why male beef calves briefly separated from their dams for 'processing' (ear-marking, vaccination, branding, dehorning and castration) 6 weeks prior to weaning had poorer weight gains after weaning than those that were 'processed' at the time of weaning (Holroyd and Petherwick, 1997). It may also explain why lambs separated daily from their dams starting at 3.5 weeks of age were more sensitive to parasite infection than lambs weaned suddenly at 3 months of age (Orgeur *et al.*, 1998). Whether bonded animals are temporarily or permanently separated by humans, they have no way of knowing how long the separation will last. Nevertheless, if separations are routine, brief and untraumatic, and if the animals have the cognitive capacity, they may be able to develop an expectation for the timing of reunion and experience no adverse effects from separation.

In livestock and poultry kept in small groups, disruption of social relationships through removal of animals from one group and introduction to another group containing unfamiliar individuals results in a physiological stress response and changes in immune function. High levels of agonistic behaviour occur as the unfamiliar animals fight to establish dominance relationships (Gross and Siegel, 1981; Fraser and Rushen, 1987; Hasegawa *et al.*, 1997). Evidence from primates suggests that repeated changes in group membership can result in a long-term effect on central serotonergic function (Fontenot *et al.*, 1995). However, when group composition is changed during routine animal husbandry, it is often unclear to what extent behavioural and physiological responses result from: (i) introduction to unfamiliar individuals or social isolation; (ii) changes in group size; (iii) movement to a new location; (iv) changes in diet, temperature, pathogenic challenge, human caretakers and other environmental factors that differ between locations; and (v) the breaking of social bonds, if any, through separation from former group members.

11.4.2   Sudden, early weaning

Sudden, early weaning applies to all livestock that are weaned prior to the time when weaning would occur if they were left together (see Section 3.11). It involves: (i) abrupt severance of the mother–young bond; (ii) an abrupt change in diet; and (iii) depriving the offspring of the opportunity to perform suckling behaviour at an age when they are highly motivated to perform this comforting behaviour. Usually, both the mother and offspring are moved to new locations after weaning. Interestingly, relocation of piglets following weaning can have a greater impact on growth than the aggression associated with mixing with strangers (Csermely and Ballarini, 1988).

Much of the focus of research into early weaning has targeted the control of mortality in neonates. For example, segregated early weaning of piglets, in which piglets are weaned at 5–21 days of age and transferred to a different facility, was originally developed to exploit health benefits from the passive disease resistance acquired by the piglets from suckling the colostrum (Robert *et al.*, 1999). Selman *et al.* (1971) examined the impact of removing dairy calves from their dams at birth. Calves that were allowed to suckle from their dam, or kept with their dam but fed colostrum from a bucket and not allowed to suckle, had higher immunoglobulin levels than calves separated from their mothers at birth and fed colostrum from a bucket. Hence, it was the absence of the mother rather than natural suckling that appeared to reduce colostrum absorption and increase the risk of neonatal mortality.

A useful measure of the extent to which psychological well-being is impaired by separation of animals may be obtained from studying their signalling behaviour. The frequency of high-pitched vocalizations given by piglets following separation from the sow appears to represent an honest signal of their need to be reunited with the sow. Hungrier, colder and slower-growing piglets give more of these calls than well-fed, warm and rapidly growing piglets (Weary and Fraser, 1995; Weary *et al.*, 1997, 1999). The sow responds to these calls by vocalizing and approaching rapidly (Weary *et al.*, 1996). Even when they are not hungry or cold, thriving 2-week-old piglets vocalize more than thriving 4-week-old piglets when separated from the sow (Weary *et al.*, 1999). This result suggests that the piglets derive comfort from the presence of the sow in addition to the milk and warmth that she provides and that the bond between sow and piglets is stronger in the younger piglets.

The continued motivation to perform suckling behaviour poses problems when neonates are weaned. In young calves kept in groups, cross-suckling of body parts of penmates produces soreness and infection and may lead to milk sucking in adulthood. Urine drinking is a manifestation of preputial suckling. In piglets, early weaning is associated with high levels of belly nosing, a stereotyped behaviour that resembles massaging of the sow's udder (Robert *et al.*, 1999). This behaviour is more prevalent when piglets are weaned at 2 than at 4 weeks of age (Weary *et al.*, 1999). Mason (1996) also observed higher levels of stereotyped behaviour in mink following weaning when the mink were weaned at a younger age. In piglets, ear and tail chewing of penmates may be derived from suckling behaviour. Alternatively, this destructive behaviour may be stimulated by hunger associated with difficulty adapting to a solid diet following weaning. Digestive upsets and a check or reduction in weight immediately following weaning are common although the provision of high-quality, easily digested diets containing blood plasma minimizes this problem in pigs (Robert *et al.*, 1999).

Separation of young from their mother and other adults reduces opportunities for cultural transmission of information about predators (Mateo and Holmes, 1997) and food (Thorhallsdottir *et al.*, 1990). When reared in the absence of their mother or other adults, turkey poults may fail to find and consume appropriate food (Lewis and Hurnik, 1979). Turkeys are a semi-precocial species unlike domestic fowl. Chicks of the latter species are fully precocial and are less likely to have problems finding food. Separation from adults may also release young from a policing influence of adults resulting in reduced group cohesion (Frank, 1996).

In the absence of the mother, offspring may become attached to mother substitutes such as a foster mother, human caretaker, imprinting object, heater or artificial udder. We are not aware of any studies comparing the relative costs to young livestock and poultry of separating them from different types of mother figures. However, rearing them in the absence of the mother may have unexpected costs. For example, turkey poults reared in the absence of the hen may have inappropriate activity cycles and use up their energy reserves more quickly than those that respond to the brooding cycles set by the hen.

Relatively little attention has been paid to the costs of early weaning on the mothers. When offspring are removed suddenly from the dam, the dam may experience discomfort from a distended udder. Distress may also be induced by the sudden withdrawal of pleasurable experiences associated with physical contact with the offspring (e.g. grooming, huddling, nursing). Anecdotal reports suggest that cows recover from the separation more quickly if they are pregnant than if they are open.

11.4.3   Death of a group member

Concerns have been expressed that agricultural species may be distressed by observing or otherwise sensing the deaths of conspecifics to which they are bonded. It is not known whether these species have a concept of death or associate the death or injury of others with their own mortality. At a slaughterhouse, animals are exposed to many novel and startling stimuli that elicit fear-related responses independently of any comprehension of death. Animals may communicate their fear to conspecifics through postures, vocalizations and alarm odours. In sheep and pigs, Anil *et al.* (1996, 1997) were unable to detect any additional response specifically associated with witnessing the stunning and sticking of conspecifics. However, they did not investigate the impact of bonding between the slaughtered animals and the witnesses.

Even if rendered unconscious prior to death, slaughter is accompanied by bleeding, raising the question of whether the blood of an attachment figure can cause a specific aversive response in the animals. Aversion to conspecific blood has been documented in chickens (Jones and Black, 1979). In contrast, Yngvesson and Keeling (1998) reported that hens pecked more at blood-stained than at clean feathers. Pigs preferred to chew on blood-soaked rather than on clean tail models (Fraser, 1987). Pigs and poultry species can be cannibalistic, indicating that aversion to conspecific blood is not a fixed trait in these species and that some individuals may actively seek out blood. Anil *et al.* (1997) reported that some pigs ingested the blood of slaughtered penmates. Cattle showed increased sniffing in the air and stretched locomotion when exposed to the odours of conspecific blood and the urine of stressed conspecifics but the presence of these odours did not alter feeding behaviour (Terlouw *et al.*, 1998). None of these studies investigated responses to the blood or other body fluids of attachment figures as opposed to other familiar and unfamiliar conspecifics. Learning is likely to be involved in associating blood and other characteristics of dying individuals with food or danger.

Behaviour suggestive of grieving has been documented in highly social, long-lived mammals such as elephants and various primate species (de Waal, 1996). 'Mourning' behaviour can include remaining with a dead individual after the group has moved away, returning to the site of the death and investigating the bones. There are anecdotal reports of mourning behaviour in dogs, with symptoms described as anorexia, apathy, lack of playfulness and apparent searching for the missing animal. However, there are no systematic accounts investigating whether responses to the disappearance of an attachment figure are altered if the death is witnessed or the dead body is investigated. Dog behaviour may be affected by a change in the behaviour of a human caretaker grieving over the lost animal and by social isolation resulting from the absence of a conspecific. Chickens and pigs may prey on dead group mates. It has not been reported whether they avoid preying upon individuals with whom they have had strong bonds. In farmed species, there is presumably selection against individuals that become anorexic or fail to ovulate for an extended period following breaking of a bond. On the other hand, given the investment that a mother has in her offspring, it would be disadvantagous to abandon an apparently unresponsive offspring too soon since it may only be sleeping or otherwise temporarily indisposed.

11.5 Lessening the Psychological Cost of Breaking Bonds

11.5.1 Substituting stimuli provided by attachment figures

When human interests dictate separating bonded animals, replacing stimuli previously obtained from an attachment figure may have a calming effect. For example, providing a teat to newly weaned animals provides a safe outlet for the performance of non-nutritive suckling and promotes digestion through the release of digestive hormones (de Passillé *et al.*, 1993). Grooming by a human may also be positively reinforcing (Taira and Rolls, 1996). Gentle contact with a human immediately following separation from the mother may assist the young to cope with subsequent temporary separations from their social group. For example, goats weaned at 1 week, kept in a social group and handled twice daily for 10 days after weaning were less distressed when isolated at 5 months of age than those that were not handled or handled in a similar manner starting at 6 weeks of age (Boivin and Braastad, 1996). However, presence of a 'trainer' cow in pens of newly weaned beef calves appears to have no benefit (J. Stookey, personal communication).

The benefit of providing auditory stimuli is equivocal. Playback of sow reinstatement vocalizations elicited increased vocalizing by isolated piglets and could not be viewed as a source of comfort (Weary *et al.*, 1997). Silence was more effective in quieting piglets separated from the sow singly or in littermate pairs than playback of meditation music, white noise or vocalizations of unfamiliar piglets (Cloutier *et al.*, 2000). These findings correspond to the observation of Lund and Jørgensen (1999) that behaviour associated with separation anxiety in dogs gradually diminished with time after departure of their human attachment figure but noises and other disturbances stimulated new episodes of this behaviour. In contrast, music was effective in quieting separation vocalizations of chicks (Panksepp, 1998). Studies of responses to playback of different types of vocalization performed by an individual to whom the subject is bonded would be of interest.

Isolation of agricultural species from herd or flockmates typically produces a fear response characterized by elevated heart rate, increased vocalization and potentially injurious escape attempts (Veissier and LeNeindre, 1992). Mirrors were as effective in calming chicks as the presence of a social companion (Montevecchi and Noel, 1978; Panksepp, 1998) whereas, in sheep, the sight of a mirror image was less effective than the presence of a companion (Parrott *et al.*, 1988). Isolated beef heifers had a lower heart rate when exposed to a mirror than in the absence of a mirror, especially when the animals viewed the mirror from the front rather than the side. The reflected side-view may have been perceived as a threat posture, suggesting that the animals viewed their reflection as a representation of a conspecific (Piller *et al.*, 1999). There is no evidence that they recognized themselves in the mirror or responded to their mirror image as if it was a specific individual to whom they were bonded. The latter would be problematic since the mirror image portrays their own visual features and not those specific to another individual. Sheep are able to distinguish between pictures and odours of different familiar individuals, although the neural pathways for the identification of specific faces may take a month or more to become fully functional (Kendrick, 1998). Perhaps providing a picture or video image of an attachment figure could alleviate separation distress in older animals. For younger animals, the (unstressed) odour of the attachment figure may be more effective.

11.5.2 Weaning methods

Ideally, we could devise economical methods of managing livestock and poultry that would not require the premature breaking of social bonds. There are some production systems that do not rely on breaking bonds. For example, sows in the family pen system are kept together throughout their lives and their piglets are weaned naturally and remain in the family pen until they are sold for slaughter (Stolba, 1981).

If young must be separated from the dam at an early age, it is usually assumed that the separation will be less traumatic if made prior to development of a bond. However, Lidfors (1996) found that dairy calves removed from the dam after 4 days and placed 5 m away and in sight of the cow performed very little vocalization. Although specific call types were not analysed, there did not appear to be a strong difference in the amount of vocalizing by the cow after removal of the calf immediately post-partum or after 4 days. Similar results were obtained by Hopster *et al.* (1995) when dairy calves were removed from their dam at 3 days post-partum. Providing that calves remaining with the cow are observed to ensure that they are suckling colostrum, there may be benefits to remaining with the cow for several days resulting from activation and grooming of the calf by the cow, and improved absorption of colostrum (Selman *et al.*, 1971).

Techniques that accustom animals to brief separations appear to reduce the impact of separation at weaning although care is needed to avoid sensitization of the stress response to separation. Pajor *et al.* (1999) compared piglet behaviour and growth in 'confined' pens where the sow and piglets were continually together versus 'get-away' pens where the sow could leave her piglets temporarily. In the 'get-away' pens, piglets gained experience of short periods of physical separation from their mother. Although they did not show differences in levels of vocalization, rooting or belly nosing in the 2 weeks after weaning at 5 weeks of age, these piglets ate more solid food and gained more weight than piglets reared in 'confined' pens. The results suggest that they adjusted more easily to long-term separation from the sow.

Evidence from beef cattle, elk and horses suggests that separating mothers and young partially by allowing them to make contact across a fence is less traumatic than isolating them completely. Stookey *et al.* (1997) observed that beef calves had elevated levels of standing, walking and vocalizing during the first 3 days after weaning. Calves placed in pens adjacent to their mothers vocalized and walked less on the day of weaning than calves placed 1 km away from their mothers. By day 3, behavioural differences between the two treatment groups had disappeared. However, over the 3 days, the calves remaining in contact with their mothers gained more weight. Although economic advantages may not be gleaned from changing to a contact weaning system, it seems promising with regard to enhancing the well-being of beef calves during initial weaning stress. A similar experiment was conducted with farmed wapiti (Haigh *et al.*, 1997). Cows and calves were placed in adjacent paddocks or separated by 50 m (no visual contact). The 'contact-weaned' calves vocalized less and spent less time standing, walking, running and fence pacing than their 'remote-weaned' counterparts.

Foals are typically weaned at around 6 months of age. They tend to react to weaning with vigorous attempts to rejoin the mare, resulting in a high risk of injury (Apter and Householder, 1996). Apart from welfare concerns, scarring is of economic concern to breeders since horses are valued for their aesthetics. When completely isolated from each other at weaning, both mares and foals had elevated cortisol levels and housing the foals in pairs in stalls provided no apparent advantage over housing them singly (Malinowski *et al.*, 1990; Hoffman *et al.*, 1995; but see Houpt *et al.*, 1984). Foals that were allowed to maintain contact with their mothers across a fence performed less reinstatement behaviour (Table 11.1; McCall *et al.*, 1985) and had lower cortisol levels than foals that were isolated from their mothers (McCall *et al.*, 1987). Providing a concentrate supplement pre-weaning had some benefit in reducing distress at weaning (Table 11.1; McCall *et al.*, 1985; Hoffman *et al.*, 1995). For mares and foals on pasture, removing one or two mares at a time every 2 days reduced foal vocalization and locomotion and increased grazing compared with removing all mares at once (Holland *et al.*, 1996).

Data on factors reducing separation calling by piglets indicate that piglets adjust to weaning more rapidly when fed easily digested solid food and placed in a warm and comfortable environment (Fraser *et al.*, 1994; Weary and Fraser, 1995; Weary *et al.*, 1997). Ingestion of creep feed prior to weaning improved the weaning weight of some piglets although it had little effect on post-weaning growth of piglets weaned at 4 weeks of age (Fraser *et al.*, 1994). Keeping weaned piglets in a

Table 11.1. Behaviour (least-square means) of foals during the first 5 h after weaning. (Adapted from McCall *et al.*, 1985.)

Separation at weaning[1]	Creep feed before weaning	Lying down (min)	Walking (min)	Trotting[2] (min)	Vocalizations (no.)	
					Hour 3	Hour 5
Remote	No	0.0	182.1[a]	22.9	97.0[a]	50.3[a]
Remote	Yes	2.5	118.8[a,b]	16.3	66.8[b]	39.8[a]
Contact	No	8.3	108.3[b]	6.3	7.2[c]	2.7[b]
Contact	Yes	0.0	27.5[c]	0.0	8.3[c]	2.3[b]
Not weaned	Yes	12.5	3.8[c]	0.0	2.0[c]	0.0[b]
SE		3.1	9.1	3.9		2.8

[a,b,c]Column means with different superscripts differ (*P* < 0.05).
[1]Remote: mare out of visual, auditory and olfactory contact; contact: fenceline contact with mare; not weaned: mare kept with foal.
[2]Treatment differences approach significance (*P* < 0.06).

familiar environment allows them to adjust to the separation more easily (Csermely and Ballarini, 1988).

Studies of the neurobiological control of attachment are needed to improve our understanding of the welfare implications of separating animals at different stages in the course of commercial animal production. Pharmacological interventions may alleviate some symptoms of separation distress although, at the present time, they are only likely to be economically viable for highly valued animals such as racehorses.

11.6 Conclusions

In animals used in agriculture, there is often a lack of evidence upon which to base a decision about whether or not to label a specific relationship a social bond. This confusion presents difficulties when discussing the breaking of bonds. Ideally, bonds would be recognized based on information about behavioural and physiological responses of one animal towards another when together, when separated and when reunited after varying lengths of time. It is not fully established how long members of the different livestock and poultry species remain distressed following separation and how long they could be separated and still retain an emotional attachment evident upon reunion. Previous studies have tended to look at their responses following separation for relatively short periods of time (e.g. 10 min: Weary and Fraser, 1995; up to 24 h: Ramirez *et al.*, 1996).

The time is ripe for further study of the implications of breaking social bonds on the behaviour, health and well-being of livestock and poultry. Experimental approaches are ethically challenging given that, if an emotional attachment exists, breaking that attachment will cause distress. This distress is expected to vary in duration and severity according to the strength of the attachment and the conditions under which it is broken. Based on the great variability in timing of dispersal under natural conditions, we can predict considerable individual variation in responses to abrupt separation at a specific age. Guidelines are needed for deciding when to intervene to alleviate separation distress and how this may best be accomplished.

References

Anil, M.H., Preston, J., McKinstry, J.L., Rodway, R.G. and Brown, S.N. (1996) An assessment of stress caused in sheep by watching slaughter of other sheep. *Animal Welfare* 5, 435–441.

Anil, M.H., McKinstry, J.L., Field, M. and Rodway, R.G. (1997) Lack of evidence for stress being caused to pigs by witnessing the slaughter of conspecifics. *Animal Welfare* 6, 3–8.

Apter, R.C. and Householder, D.D. (1996) Weaning and weaning management of foals: a review and some recommendations. *Journal of Equine Veterinary Science* 16, 428–435.

Boccia, M.L., Scanlan, J.M., Laudenslager, M.L., Berger, C.L., Hijazi, A.S. and Reite, M.L. (1997) Juvenile friends, behavior, and immune responses to separation in bonnet macaque infants. *Physiology and Behavior* 61, 191–198.

Boivin, X. and Braastad, B.O. (1996) Effects of handling during temporary isolation after early weaning on goat kids' later responses to humans. *Applied Animal Behaviour Science* 48, 61–71.

Bowlby, J. (1979) *The Making and Breaking of Affectional Bonds.* Tavistock Publications, London, 184 pp.

Bridges, R.S. (1996) Biochemical basis of parental behavior in the rat. *Advances in the Study of Behavior* 25, 215–242.

Chase, I.D. (1982) Dynamics of hierarchy formation: the sequential development of dominance relationships. *Behaviour* 80, 218–240.

Choudhury, S. and Black, J.M. (1993) Mate-selection and sampling strategies in geese. *Animal Behaviour* 46, 747–757.

Cirulli, F., Terranova, M.L. and Laviola, G. (1996) Affiliation in periadolescent rats: behavioral and corticosterone response to social reunion with familiar and unfamiliar partners. *Pharmacology, Biochemistry and Behavior* 54, 99–105.

Cloutier, S., Weary, D.M. and Fraser D. (2000) Sound enrichment: can ambient sound reduce distress among piglets during weaning and restraint? *Journal of Applied Animal Welfare Science* (in press).

Cousse, S., Spitz, F., Hewison, M. and Janeau, G. (1994) Use of space by juveniles in relation to their postnatal range, mother, and siblings: an example in the wild boar, *Sus scrofa* L. *Canadian Journal of Zoology* 72, 1691–1694.

Csermely, D. and Ballarini, G. (1988) Some aspects of weight gain and social behaviour in caged weaned pigs. *Monitore Zoologico Italiano* 22, 1–16.

de Passillé, A.M.B., Christopherson, R.J. and Rushen, J. (1993) Nonnutritive suckling and postprandial secretion of insulin, CCK and gastrin by the calf. *Physiology and Behaviour* 54, 1069–1073.

de Waal, F. (1996) *Good Natured: the Origins of Right and Wrong in Humans and Other Animals.* Harvard University Press, Cambridge, Massachusetts, 296 pp.

Dwyer, C.M. and Lawrence, A.B. (1997) Induction of maternal behaviour in non-pregnant, hormone primed ewes. *Animal Science* 65, 403–408.

Dwyer, C.M., McLean, K.A., Deans, L.A., Chirnside, J., Calvert, S.K. and Lawrence, A.B. (1998) Vocalisations between mother and young in sheep: effects of breed and maternal experience. *Applied Animal Behaviour Science* 58, 105–119.

Ewbank, R. and Meese, G.B. (1971) Aggressive behaviour amongst groups of domesticated pigs on removal and return of individuals. *Animal Production* 13, 685–694.

Feh, C. and de Mazieres, J. (1993) Grooming at a preferred grooming site reduces heart rate in horses. *Animal Behaviour* 46, 1191–1194.

Fontenot, M.B., Kaplan, J.R., Manuck, S.B., Arango, V. and Mann, J.J. (1995) Long-term effects of chronic social stress on serotonergic indices in the prefrontal cortex of adult male cynomologus macaques. *Brain Research* 705, 105–108.

Frank, S.A. (1996) Policing and group cohesion when resources vary. *Animal Behaviour* 52, 1163–1169.

Fraser, D. (1987) Mineral-deficient diets and the pig's attraction to blood: implications for tail biting. *Canadian Journal of Animal Science* 67, 909–918.

Fraser, D. and Rushen, J. (1987) Aggressive behavior. In: Price, E.O. (ed.) *Farm Animal Behavior. Veterinary Clinics of North America: Food Animal Practice*, Vol. 3, pp. 285–305.

Fraser, D. and Thompson, B.K. (1990) Armed sibling rivalry among piglets. *Behavioural Ecology and Sociobiology* 29, 9–15.

Fraser, D., Feddes, J.J.R. and Pajor, E.A. (1994) The relationship between creep feeding behavior of piglets and adaptation to weaning: effect of diet quality. *Canadian Journal of Animal Science* 74, 1–6.

Gross, W.B. and Siegel, P.B. (1981) Long term exposure of chickens to three levels of social stress. *Avian Diseases* 25, 312–325.

Gubernick, D.J. (1981) Parent–infant attachment in mammals. In: Gubernick, D.J. and Klopher, P.H. (eds) *Parental Care in Mammals*. Plenum Press, New York, pp. 243–305.

Haigh, J.C., Stookey, J.M., Bowman, P. and Waltz, C. (1997) A comparison of weaning techniques in farmed Wapiti (*Cervus elaphus*). *Animal Welfare* 6, 255–264.

Hasegawa, N., Nishiwaki, A., Sugawara, K. and Ito, I. (1997) The effects of social exchange between two groups of lactating primiparous heifers on milk production, dominance order, behavior and adrenocortical response. *Applied Animal Behaviour Science* 51, 15–27.

Hinch, G.N., Lecrivain, E., Lynch, J.J. and Elwin, R.L. (1987) Changes in maternal–young associations with increasing age of lambs. *Applied Animal Behaviour Science* 17, 305–318.

Hinch, G.N, Lynch, J.J., Elwin, R.L. and Green, G.C. (1990) Long term associations between Merino ewes and their offspring. *Applied Animal Behaviour Science* 27, 93–103.

Hinde, R.A. (1974) *Biological Basis of Human Social Behaviour*. McGraw-Hill, New York, pp. 7–20.

Hoffman, R.M., Kronfeld, D.S., Holland, J.L. and Greiwe-Crandell, K.M. (1995) Preweaning diet and stall weaning method influences on stress response in foals. *Journal of Animal Science* 73, 2922–2930.

Holland, J.L., Kronfeld, D.S., Hoffman, K.M., Greiwe-Crandell, K.M, Boyd, T.L., Cooper, W.L. and Harris, P.A. (1996) Weaning stress is affected by nutrition and weaning methods. *Pferdeheilkunde* 12, 257–260.

Holroyd, R. and Petherwick, C. (1997) The impact of weaning and processing on the health and performance of beef weaners. In: Hemsworth, P.H., Spinka, M. and Kostal, L. (eds) *Proceedings of the 31st International Congress of the ISAE*. Institute of Animal Biochemistry and Genetics, Slovak Academy of Sciences, Ivanka pri Dunaji, Slovakia, p. 159.

Hopster, H., O'Connell, J. and Blokhuis, H.J. (1995) Acute effects of cow–calf separation on heart rate, plasma cortisol and behaviour in multiparous dairy cows. *Applied Animal Behaviour Science* 44, 1–8.

Houpt, K.A., Hintz, H.F. and Butler, W.R. (1984) A preliminary study of two methods of weaning foals. *Applied Animal Behaviour Science* 12, 177–181.

Insel, T.R. (1997) A neurobiological basis of social attachment. *American Journal of Psychiatry* 154, 726–735.

Jensen, P. (1995) The weaning process in free-ranging domestic pigs: within- and between-litter variations. *Ethology* 100, 14–25.

Jones, R.B. and Black, A.J. (1979) Behavioral responses of the domestic chick to blood. *Behavioral and Neural Biology* 27, 319–329.

Kalin, N.H., Shelton, S.E. and Lynn, D.E. (1995) Opiate systems in mother and infant primates coordinate intimate contact during reunion. *Psychoneuroendocrinology* 20, 735–742.

Kendrick, K.M. (1998) Intelligent perception. *Applied Animal Behaviour Science* 57, 213–231.

Keverne, E.B. and Kendrick, K.M. (1992) Oxytocin facilitation of maternal behavior in sheep. *Annals of the New York Academy of Science* 652, 102–121.

Keverne, E.B., Martensz, N.D. and Tuite, B. (1989) Beta-endorphin concentrations in cerebrospinal fluid of monkeys are influenced by grooming relationships. *Psychoneuroendocrinology* 14, 155–161.

Khalil, A.M. and Kaseda, Y. (1997) Behavioral patterns and proximate reason of young male separation in Misaki feral horses. *Applied Animal Behaviour Science* 54, 281–289.

Kimura, R. (1998) Mutual grooming and preferred associate relationships in a band of free-ranging horses. *Applied Animal Behaviour Science* 59, 265–276.

Koolhaas, J.M., Meerlo, P., De Boer, S.F., Strubbe, J.H. and Bohus, B. (1997) The temporal dynamics of the stress response. *Neuroscience and Biobehavioral Reviews* 21, 775–782.

Lábaque, M.C., Navarro, J.L. and Martella, M.B. (1999) A note on chick adoption: a complementary strategy for rearing rheas. *Applied Animal Behaviour Science* 63, 165–170.

Lamb, G.C., Lynch, J.M., Grieger, D.M., Minton, J.E. and Stevenson, J.S. (1997) *Ad libitum* suckling by an unrelated calf in the presence or absence of a cow's own calf prolongs postpartum anovulation. *Journal of Animal Science* 75, 2762–2769.

Lawrence, A.B. (1990) Mother–daughter and peer relationships of Scottish hill sheep. *Animal Behaviour* 39, 481–486.

Lewis, N.J. and Hurnik, J.F. (1979) Stimulation of feeding in neonatal turkeys by flashing lights. *Applied Animal Ethology* 5, 161–171.

Lidfors, L. (1996) Behavioural effects of separating the dairy calf immediately or 4 days post-partum. *Applied Animal Behaviour Science* 49, 269–283.

Lorenz, K. (1991) *Here Am I – Where Are You? The Behavior of the Greylag Goose.* Harcourt Brace Jovanovich, New York, 267 pp.

Lund, J.D. and Jørgensen, M.C. (1999) Behaviour patterns and time course of activity in dogs with separation problems. *Applied Animal Behaviour Science* 63, 219–236.

Malinowski, K., Hallquist, N.A., Helyar, N.A., Sherman, A.R. and Scanes, C.G. (1990) Effect of different separation protocols between mares and foals on plasma cortisol and cell-mediated immune response. *Journal of Equine Veterinary Science* 5, 363–368.

Mason, G. (1996) Early weaning enhances the later development of stereotypy in mink. In: Duncan, I.J.H., Widowski, T.M. and Haley, D.B. (eds) *Proceedings of the 30th International Congress of the International Society for Applied Ethology.* Col. K.L. Campbell Centre for the Study of Animal Welfare, Guelph, Ontario, p. 16.

Mateo, J.M. and Holmes, W.G. (1997) Development of alarm-call responses in Belding's ground squirrels: the role of dams. *Animal Behaviour* 54, 509–524.

Matthews, K., Wilkinson, L.S. and Robbins, T.W. (1996) Repeated maternal separation of preweanling rats attenuates behavioral responses to primary and conditioned incentives in adulthood. *Physiology and Behavior* 59, 99–107.

McCall, C.A., Potter, G.D. and Kreider, J.L. (1985) Locomotor, vocal and other behavioral responses to varying methods of weaning foals. *Applied Animal Behaviour Science* 14, 27–35.

McCall, C.A., Potter, G.D., Kreider, J.L. and Jenkins, W.L. (1987) Physiological responses in foals weaned by abrupt or gradual methods. *Journal of Equine Veterinary Science* 7, 368–374.

McNutt, J.W. (1996) Sex-biased dispersal in African wild dogs, *Lycaon pictus. Animal Behaviour* 52, 1067–1077.

Monard, A.M., Duncan, P. and Boy, V. (1996) The proximate mechanisms of natal dispersal in female horses. *Behaviour* 133, 1095–1124.

Montevecchi, W.A. and Noel, P.E. (1978) Temporal effects of mirror-image stimulation on pecking and peeping in isolate, pair- and group-reared domestic chicks. *Behavioural Biology* 23, 531–535.

Nemeroff, C.B. (1998) The neurobiology of depression. *Scientific American* 278, 42–49.

Newberry, R.C. and Wood-Gush, D.G.M. (1985) The suckling behaviour of domestic pigs in a semi-natural environment. *Behaviour* 95, 11–25.

Newberry, R.C. and Wood-Gush, D.G.M. (1986) Social relationships of piglets in a semi-natural environment. *Animal Behaviour* 34, 1311–1318.

Orgeur, P., Mavric, N., Yvore, P., Bernard, S., Nowak, R., Schaal, B. and Lévy, F. (1998) Artificial weaning in sheep: consequences on behavioural, hormonal and immunopathological indicators of welfare. *Applied Animal Behaviour Science* 58, 87–103.

Pajor, E.A., Weary, D.M., Fraser, D. and Kramer, D.L. (1999) Alternative housing for sows and litters 1. Effects of sow-controlled housing on responses to weaning. *Applied Animal Behaviour Science* 65, 105–121.

Panksepp, J. (1998) *Affective Neuroscience: the Foundations of Human and Animal Emotions.* Oxford University Press, New York, 466 pp.

Parrott, R.F., Houpt, K.A. and Misson, B.H. (1988) Modification of the responses of sheep to isolation stress by the use of mirror panels. *Applied Animal Behaviour Science* 19, 331–338.

Piller, C.A.K., Stookey, J.M. and Watts, J.M. (1999) Effects of mirror-image exposure on heart rate and movement of isolated heifers. *Applied Animal Behaviour Science* 63, 93–102.

Porter, R.H., Nowak, R. and Orgeur, P. (1995) Influence of a conspecific agemate on distress bleating by lambs. *Applied Animal Behaviour Science* 45, 239–244.

Ramirez, A., Quiles, A., Hevia, M.L., Sotillo, F. and Ramirez, M.C. (1996) Effects of immediate and early post-partum separation on maintenance of maternal responsiveness in parturient multiparous goats. *Applied Animal Behaviour Science* 48, 215–224.

Reinhardt, V. and Reinhardt, A. (1981) Cohesive relationships in a cattle herd (*Bos indicus*). *Behaviour* 77, 121–151.

Robert, S., Weary, D.M. and Gonyou, H. (1999) Segregated early weaning and welfare of piglets. *Journal of Applied Animal Welfare Science* 2, 31–40.

Selman, I.E., McEwan, A.D. and Fisher, E.W. (1971) Studies on dairy calves allowed to suckle their dams at fixed times post partum. *Research in Veterinary Science* 12, 1–6.

Stolba A. (1981) A family system in enriched pens as a novel method of pig housing. In: *Alternatives to Intensive Housing Systems*. Universities Federation for Animal Welfare, Potters Bar, UK, pp. 52–67.

Stookey, J.M., Schwartzkopf-Genswein, C.S. and Watts, J.M. (1997) Effects of remote and contact weaning on behaviour and weight gain of beef calves. *Journal of Animal Science* 75 (suppl. 1), 157.

Stroebe, W. and Stroebe, M.S. (1987) *Bereavement and Health: the Psychological and Physical Consequences of Partner Loss*. Cambridge University Press, Cambridge, UK, 288 pp.

Swanson, J.C. and Stricklin, W.R. (1985) Kinship affects displacement and agonistic activity among beef cows at the feed trough. *Journal of Animal Science* 61 (suppl. 1), 211.

Taira, K. and Rolls, E.T. (1996) Receiving grooming as a reinforcer for the monkey. *Physiology and Behavior* 59, 1189–1192.

Terlouw, E.M.C., Boissy, A. and Blinet, P. (1998) Behavioural responses of cattle to the odours of blood and urine from conspecifics and to the odour of faeces from carnivores. *Applied Animal Behaviour Science* 57, 9–21.

Thorhallsdottir, A.G., Provenza, F.D. and Balph, D.F. (1990) The role of the mother in the intake of harmful foods by lambs. *Applied Animal Behaviour Science* 25, 35–44.

Trivers, R. (1974) Parent–offspring conflict. *American Zoologist* 14, 249–264.

Tuber, D.S., Hennessy, M.B., Sanders, S. and Miller, J.A. (1996) Behavioral and glucocorticoid responses of adult dogs (*Canis familiaris*) to companionship and social separation. *Journal of Comparative Psychology* 110, 103–108.

Vehrencamp, S.L. (1983) A model for the evolution of despotic versus egalitarian societies. *Animal Behaviour* 31, 667–682.

Veissier, I. and LeNeindre, P. (1992) Reactivity of Aubrac heifers exposed to a novel environment alone or in groups of four. *Applied Animal Behaviour Science* 33, 11–15.

Veissier, I., Boissy, A., Nowak, R., Orgeur, P. and Poindron, P. (1998) Ontogeny of social awareness in domestic herbivores. *Applied Animal Behaviour Science* 57, 233–245.

Weary, D.M. and Fraser, D. (1995) Calling by domestic piglets: reliable signals of need? *Animal Behaviour* 50, 1047–1055.

Weary, D.M., Lawson, G.L. and Thompson, B.K. (1996) Sows show stronger responses to isolation calls of piglets associated with greater levels of piglet need. *Animal Behaviour* 52, 1247–1253.

Weary, D.M., Ross, S. and Fraser, D. (1997) Vocalizations by isolated piglets: a reliable indicator of piglet need directed towards the sow. *Applied Animal Behaviour Science* 53, 249–257.

Weary, D.M., Appleby, M.C. and Fraser, D. (1999) Responses of piglets to early separation from the sow. *Applied Animal Behaviour Science* 63, 289–300.

Yngvesson, J. and Keeling, L.J. (1998) Are cannibalistic laying hens more attracted to blood in a test situation than other hens? In: Veissier, I. and Boissy, A. (eds) *Proceedings of the 32nd Congress of the International Society for Applied Ethology.* INRA, Clermont-Ferrand, France, p. 173.

Individual Differences and Personality

12

Hans W. Erhard[1] and Willem G.P. Schouten[2]

[1]Macaulay Land Use Research Institute, Craigiebuckler, Aberdeen AB15 8QH, UK; [2]Department of Animal Husbandry, Wageningen University, PO Box 338, 6700 AH Wageningen, The Netherlands

(Editors' comments: Throughout this book we have been making generalizations about the behaviour of farm animals. This chapter takes another approach and focuses on the differences rather than the similarities between individuals.

For these individual differences to be open to study, there has to be some consistency, that is to say, the variation between individuals cannot just be random. The systematic differences between high- and low-ranking individuals are well known and are discussed here. But Erhard and Schouten also take up differences in the extent to which individuals seek out social contact, in maternal abilities and success in rearing offspring, and in mating behaviour. Both these last criteria, maternal and sexual behaviour, are of importance in farm animals since they can potentially be used in genetic selection and breed improvements.

As in all areas of science, it is important that terms are defined. This is particularly true for words that are already used in everyday speech. For the term 'personality' to be used, for example, it has to be consistent across time. Another term is 'coping' and this is used in situations that lie outside the normal range of situations for an individual. As Erhard and Schouten say, if a problem cannot be solved it has to be dealt with, and the fact that individuals may cope with situations in different but predictable ways has led to the use of the term 'coping strategies'.

The final part of this chapter deals with the practical relevance and implications of research on individual variation. Here selection for undesirable personality traits in agricultural animals, such as rejection of young or aggression towards humans, is discussed, as is the fact that, if we could select individuals for our experiments more appropriately, we could reduce the number of animals that were necessary for statistical significance.)

12.1 Introduction

The behaviour of animals is usually described on a species level, e.g. 'the behaviour of sheep, pigs', etc. Buildings for housing animals are designed for 'the average cow, pig or sheep'. Sometimes the approach to animal behaviour is problem-based, as in optimal foraging theory, which uses the solution to a problem as the reference point for the behaviour of individual animals. In the science of animal nutrition, the average weight gain, and composition of this gain, is used to calculate the dietary requirements and to predict feeding behaviour. All these approaches assume or rely on a population of animals which is uniform, with very little variation between individuals. Nevertheless, given the same situation and the same stimulus, individual animals may show considerable differences in what they do and how they do it. Some of these differences appear to be random and unpredictable, whereas others have an element of consistency. The latter are the topic of this chapter. Differences between individuals have been the basis of genetic selection since ancient times (see Chapter 4). Domestication has been based to a large extent on the selection of the tamest, most docile individuals, at least initially. This selection of animals for behavioural traits is particularly evident in different breeds of dogs, which are the result of a selection for very specific aspects of behaviour or personality, e.g. pointers, setters, retrievers, terriers and so on. Sheep which have traditionally been herded by a shepherd have been selected for a high level of flocking behaviour, whereas sheep which are kept in an environment with a patchy distribution of food, such as the Scottish Blackface, show a larger tendency to disperse (see Section 8.1.4).

Individual variation has also played an important role in the design of buildings and machinery. A fence which will hold back the average deer, sheep, cow, bull or horse will prove ineffective for half of the population, and will therefore prove useless. Fences are designed with the extremes of the population in mind. The same should be the case for handling facilities and pen design in general. For example, the comfortable distance which allows animals to pass one another in a passageway depends on the personal space of the two individuals concerned. Passageways should therefore be wide enough to take the largest 'personal space' in the herd into account, so that all animals can freely pass each other and have access to the entire building.

These are just a few examples of the importance of individual variation as opposed to population averages.

12.2 Definition

There is no universally accepted definition of 'individual variation' or 'individual differences'. Sometimes these terms are used for variation which cannot be explained by the experimental treatment, or which cannot be labelled in other ways. Other similar labels might be 'breed differences' and 'sex differences'. This approach, however, makes the structure of research and the description of 'individual differences' sometimes unnecessarily complicated. In this chapter, we will use a slightly different approach.

If an animal shows fear-related behaviour, we may say that this animal experiences fear, it is afraid. If this individual is reliably likely to be afraid, then we can say that this animal is fearful, it has a disposition to be afraid. While the actual fear is a result of the inter-action between the situation and the individual, the disposition to be afraid is a property of the animal. Following the structure of personality suggested by Eysenck (1967), fear is the 'state' the animal is in at a given moment (response) , and 'fearfulness' is the corresponding personality 'trait'. Zuckerman (1983) illustrates the difference between state and trait by contrasting the sentences 'I am nervous right now' (state of being nervous) and 'I am a nervous person' (trait 'nervousness').

Within a personality trait, individuals may be found in different categories (e.g. active or passive (Benus, 1988)) or on a continuum (e.g. the shy–bold continuum (Wilson *et al.*, 1994)). The trait can be regarded as a dimension, in which the variation is distributed.

It has to be emphasized that this approach is merely descriptive, and that it does not make any assumptions about the underlying cause of the individual's personality. It describes an animal's propensity to behave in certain ways in certain situations. This propensity will be affected by genetic factors and by the environment, either in the short term (some situations are more likely to be related to certain personality traits than others, e.g. if an animal is forced into a novel environment, it may show its fearfulness instead of its inquisitiveness) or in the longer term (early experience may affect an individual's personality).

If an individual animal has a low propensity to react strongly to being handled by humans, then this can be regarded as evidence that it is docile. This docility is an attribute of the individual, and not of the breed, or group, or sex, or litter. Specific breeds may be characterized by consisting of a high proportion of docile individuals, but it is still these individuals who are docile, not the breed or the group.

It is on the basis of this point of view that we include 'breed differences' in this chapter on individual differences.

12.3 Agonistic Behaviour in a Group

Group-living animals form dominance relationships, which may or may not be organized in hierarchies (see Section 2.2.1). Often hierarchies are linear or near-linear, which results in the presence of high-, medium- and low-ranking animals in a group. This distinction between dominant and subordinate or high- and low-ranking does not describe the individual, but rather its position in relation to other individuals (Drews, 1993). A high rank is associated with priority of access to valued resources, so that low-ranking individuals may suffer if they have to compete for resources such as food and shelter. While each hierarchy will have high- and low-ranking animals, how they behave is, at least to a certain extent, dependent on their personality.

Stability in a social hierarchy is often considered to be beneficial, since it reduces the number of fights. Whether individuals at the bottom of the hierarchy consider stability a good thing, however, probably depends on how they are treated by the top-ranking individuals. With dominance relationships, therefore, there are two separate issues to consider: how they are established and, once established, how they are reinforced.

Many studies on mixing in pigs have reported considerable individual variation in the expression of aggressive behaviour (Kelley *et al.*, 1980; McGlone and Morrow, 1988; Mount and Seabrook, 1993; Hessing *et al.*, 1994). Erhard and Mendl (1997) described a behavioural test which can be used to assess individual aggressiveness in growing pigs. They used a resident–intruder situation, in which they confronted a resident test pig with an unfamiliar intruder pig who was approximately 60% of the test pig's body weight and 2 to 3 weeks younger. The interval between the time when the resident pig first made contact with the intruder to when it attacked was relatively stable across 2 days, as well as across a 4-week period, and was not affected by the sex or body weight of the intruder pigs. Only if the intruder pigs were less than 50% of the resident's body weight were they less likely to be attacked. In a later study (Erhard *et al.*, 1997), eight pigs from two litters (4 + 4) were mixed, in the combinations of short with short, short with long, and long with long attack latency pigs. This study showed that the attack latency in the aggression test described above predicted the behaviour of the pigs immediately after mixing and for the following week. Pigs with short attack latency (SAL) were more aggressive after mixing than pigs with long attack latency (LAL). The type of aggression displayed depended largely on the opponents. SAL pigs mixed with other SAL pigs spent more time fighting and fought more vigorously. When they were mixed with LAL pigs, they fought less (because the LAL pigs withdrew), but performed more bites and chasing behaviour. Thus, when an aggressive pig bites (attacks) another pig, then the opponent either bites back, leading to a fight, or it withdraws, in which case the attack remains a bite or, if the attack persists, turns into a chase. Hence, an absence of fights can be the result of a lack of attacks, or of the withdrawal of the individual who is attacked, and should therefore not be used as an indicator of the general level of aggression. Erhard *et al.* (1997) suggested that the presence of aggressive individuals in a group and, therefore, the performance of high levels of aggressive behaviour were more important in determining the effect that mixing had on the group than which type of aggression was performed (i.e. whether fighting occurred or not). Based on these results and on faster group integration when only low-aggressive pigs were mixed, they concluded that a reduction in the number of high-aggressive pigs in a population would be beneficial for individual pigs' welfare at mixing (Erhard *et al.*, 1997).

Mendl *et al.* (1992) described the interaction between personality and position in the social hierarchy for group-housed primiparous pigs (gilts). They categorized pigs into three groups, according to their ability to displace others (success in agonistic interactions), and defined them as having 'high success', 'low success' and 'no success' (similar to high-, medium- and low-ranking). Both high- and low-success pigs showed high levels of aggressive behaviour, while no-success pigs were least aggressive. The pigs could also be categorized as high-aggressive (high and low success; high and medium rank) and low-aggressive (no success; low rank). Since the pigs had not been tested for their individual aggressiveness before entering the group, it is possible that the low aggressiveness of the no-success pigs was a result of their low rank in the hierarchy. Results from Erhard *et al.* (1997), however, which showed high-aggressive litters always winning when mixed with low-aggressive litters, suggest that the low rank is probably a result of the low aggressiveness rather than its cause. High-success gilts had the highest gain in body weight after they joined the group and the highest total weight of piglets born alive at the first farrowing, while low-success gilts had the lowest total weight of piglets born alive, lower than no-success pigs. The authors suggested that the strategy the pigs used to cope with their social environment (performing high or low levels of aggressive behaviour) was more important than the actual position in the hierarchy (Mendl *et al.*, 1992). Alternatively, one might say that a low level of aggressiveness is of benefit to animals at the bottom of the hierarchy, while a high level of aggressiveness is only of benefit to those at the top (Mendl and Deag, 1995). Since hierarchies by definition cannot consist of top-ranking animals alone, a group consisting of low-aggressive animals only may fare better than a mixed group.

This conclusion is in disagreement with that of Hessing *et al.* (1994), who suggested that a combination of high- and low-aggressive pigs in a group would lead to a more stable social structure, which would be beneficial for the welfare of the individuals concerned. One possible reason for this disagreement is that Hessing *et al.* (1994) investigated the pigs over a longer period, while Erhard *et al.* (1997) only looked at the first week after mixing. Also, Hessing *et al.* (1994) did not mix pigs according to aggressiveness, but according to their resistance to manual restraint (the back test), which they found to be predictive of aggressiveness (Hessing *et al.*, 1993). To date, the jury is still out on this question.

This interaction between individual aggressiveness and rank may be a general phenomenon, and not only restricted to pigs. Cook *et al.* (1996) divided sheep into three categories (high-aggressive, moderately aggressive, non-aggressive), similar to the categories used by Mendl *et al.* (1992). They found significant differences between categories in the response to pain. Their data do not allow for the testing of how behaviour of different types of high-ranking animals (high- or low-aggressive) affects low-ranking ones, but it would be interesting to find out how the level of aggressiveness in high-ranking animals affects the behaviour and welfare of low-ranking animals.

Individual differences in aggressive behaviour have also been found to relate to experience. Mendl and Paul (1991), for instance, reported higher levels of nursing and of general maternal care in mice selected for short attack latency than in the control line. The agonistic behaviour of pigs who were raised in a barren environment is less well developed than that of pigs raised in a stimulus-rich environment (de Jonge *et al.*, 1996). Investigating the effect of social experience, specifically of meeting unfamiliar pigs, van Putten and Buré (1997) found that experience of mixing increased the pigs' 'social skills' and helped reduce the number of agonistic interactions, fighting time and the number of lesions of the integument. Whether this effect is long-lasting is not known at the moment.

12.4   Sociality and Sociability

Individual animals differ not only in their propensity to perform agonistic behaviour, but also in non-agonistic elements of social behaviour. They may differ in the extent to which they need social companionship (sociality), in how close they want to be to other group members (sociability) or in how frequently they interact with other group members. It is not known how closely these three aspects of social motivation are linked. Sociality has been assessed by recording vocalizations when in social isolation (e.g. Syme, 1981; Faure *et al.*, 1983) and in the 'treadmill test', which measures the distance Japanese quail chicks run towards conspecifics (Mills and Faure, 1990). When they created selection lines of Japanese quail based on this 'social reinstatement' behaviour, Mills and Faure (1991) found that selection for low levels of social reinstatement behaviour led to a lower tendency to move towards conspecifics, but not to a higher tendency to move away. Syme (1981) found consistent differences in the reactions of individual sheep to social isolation. It is not yet completely clear to what extent the behaviour of animals in social isolation reflects their sociality (motivation to end the social isolation) or their particular reaction pattern when in a stressful situation ('temperament'; reactivity), or a combination of the two. In sheep, sociability, the tendency to be close to other group members, is usually assessed using nearest neighbour (NN) distances. These distances differ considerably between breeds, hill sheep, such as Scottish Blackface and Welsh Mountain, having larger NN distances than lowland breeds, such as Suffolks (Lynch *et al.*, 1992, citing Arnold, 1985; Dwyer and Lawrence, 1999). Sibbald *et al.* (1998) describe a method which uses nearest neighbour identity in a group of sheep to test for consistent differences in sociability between sheep. These differences are important for extensive grazing systems, where individuals have to spread out in order to fully utilize the pasture (Fig. 12.1).

Similar individual propensities to stay with or near other group members are expressed in differences in flocking behaviour. When in perceived danger, such as encountering a sheepdog, sheep of most

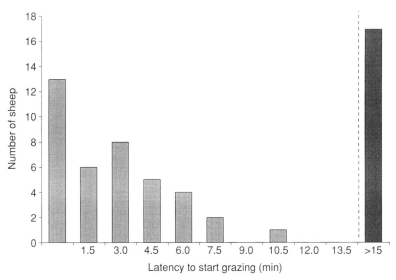

Fig. 12.1. The relationship between sociability and foraging behaviour when individual sheep had to move away from the rest of the group in order to graze. This figure shows the frequency distribution of the latency to start grazing of 56 Scottish Blackface sheep. The bar on the right-hand side represents 17 sheep who did not graze within the 15 min of the test. (A.M. Sibbald, unpublished data.)

breeds (including hill breeds) will flock together, whereas Soay sheep are more likely to disperse (Lynch *et al.*, 1992). Herding systems which rely on flocks of sheep being moved over large distances depend on a strong group cohesion, i.e. a high degree of sociability (see Section 8.2.2).

The frequency of non-agonistic social interactions is another area of social behaviour in which animals show a degree of variation. The Salers breed of cattle, for instance, was found to have a greater frequency of non-agonistic social interactions than Friesians (Le Neindre and Sourd, 1984; Le Neindre, 1989).

12.5   Maternal Behaviour

A special case of a social interaction is that between mother and offspring (see Chapter 3). Maternal behaviour plays an important role in extensive husbandry systems, where the opportunity for intervention by humans to compensate for insufficient maternal skills is reduced. Within-breed differences in maternal behaviour are important in animal breeding, and between-breed differences are important for choosing the right breed for the conditions in which they are to be kept.

Generally, individuals belonging to breeds which traditionally are kept under extensive conditions with little human supervision are likely to perform at least adequate maternal behaviour and rear their young with little help from humans, since dams with inadequate maternal skills will not reproduce sufficiently, and hence will have been excluded from passing on their genes to the next generation. Individuals kept under intensive conditions (giving birth under close supervision) are more likely to show poor maternal behaviour, since the human carer can step in to ensure the survival of the young. Dwyer and Lawrence (1999) studied differences between two breeds of sheep: a lowland breed (Suffolk), typically lambed under close human supervision, and a hill breed (Scottish Blackface), typically lambed in extensive conditions, with little human supervision. They found that the maternal behaviour of primiparous ewes showed considerable differences between breeds. Compared with Scottish Blackface ewes, Suffolk ewes were more likely to abandon a lamb and to be aggressive towards it. They spent less time grooming their lambs in the first 2 h after the lamb's birth, and displayed less cooperative behaviour in response to attempts by the lamb to suck. Also, more Suffolk lambs died in the postnatal period than Scottish Blackface lambs. The strength of the bond between ewe and lamb was assessed relative to the ewes' motivation to stay with their penmates. In this test, a lamb was removed from the pen in which the dam and her penmates were housed, and the reaction of the dam, which ranged from ignoring to following, was recorded. Scottish Blackface ewes were more likely to follow their lambs when they were taken away by a human, and, although this could be a result of a lower motivation to stay with their penmates, it still shows that the relative motivation of Scottish Blackface ewes to stay with their lambs as opposed to their penmates is greater than that of Suffolk ewes.

These results are not specific to sheep, as Le Neindre (1989) showed in a study comparing the maternal behaviour of two breeds of cattle, a dairy breed (Friesian) and a beef breed (Salers). He suggested that the motivation to maintain close mother–young contact was more important for Salers cows than for Friesians. A strong mother–young bond is undesirable in dairying, where it is important that the cow lets down her milk for humans, in the absence of her calves.

12.6 Sexual Behaviour

Individual differences in sexual behaviour have been reported in the sexual orientation of male animals, particularly sheep (e.g. Zenchak and Anderson, 1980; Zenchak *et al.*, 1981), and in quantitative aspects of their behaviour, for instance their sexual motivation (Price *et al.*, 1998) and their serving capacity (Price *et al.*, 1992). Serving capacity, i.e. the number of ejaculations a ram achieves during a 30-min exposure to four oestrous ewes, has been suggested as a predictor of sexual performance of rams in the field (Lynch *et al.*, 1992). This serving capacity can reveal a considerable variation between individuals. For example, Price *et al.* (1992) reported an average of 0–8.5 ejaculations in three 30-min serving capacity tests. Perkins *et al.* (1992) showed that rams with high serving capacity (HP) were superior in sexual performance and reproductive success when exposed to groups of 30 oestrus-synchronized ewes over a 9-day period. The number of lambs born alive per ewe was 1.65 ± 0.09 for HP rams as opposed to 0.52 ± 0.09 for rams with a low serving capacity.

The ewe : ram ratios recommended in the literature range from 15 to 25 ewes per ram for young, inexperienced rams (Smith *et al.*, 1983; Lynch *et al.*, 1992) up to 100 ewes per ram for experienced rams (Lynch *et al.*, 1992, citing Allison and Davis, 1976). The considerable differences in serving capacity between rams indicate that it might be possible to be more specific in which ewe : ram ratio should be used in the field by basing it on the ram's own serving capacity.

12.7 Personality: Discussion of Term and Concept

We have used the term 'personality' several times in this chapter, because we think that it is particularly well suited to describe and structure 'individual variation' in animal behaviour. Based on the way Eysenck (1967) described the dispositional approach to personality, we can structure individual differences in a way which provides us with a guide for the design and validation of behavioural tests and for the study of personality. This approach organizes the different aspects of personality into three levels: the state, the trait and the type. The term 'state' is used for the behaviour an individual performs or the 'mood' it is in at a specific moment in time in a specific situation (e.g. 'fear-related behaviour' or 'afraid' in a novel environment). If an individual is repeatedly found to be in similar 'states' in similar situations, we can make assumptions about the underlying personality 'trait' (e.g. fearfulness). The next level up from traits is the 'supertrait', or 'type'. If trait dimensions are linked in such a way that an individual's position in one dimension predicts its position in another dimension, then individuals can be categorized by their position in a 'type' dimension. The classic example from human psychology is the 'Big Five': neuroticism/stability, extroversion/introversion, openness, agreeableness/antagonism and conscientiousness/undirectedness (McCrae and Costa, 1987).

This approach brings the study of individual differences in non-human animals closer to what is done in humans. By structuring the different levels of personality, it also clarifies what evidence is required at the different levels in the model. The evidence required to attribute a specific behaviour to a state is clearly different from that required for a trait or type. For a state, we need standardized behavioural tests which give reliable information about the state an individual is in at the moment of testing. The time scale is the moment of testing. For a trait, we need to show that individuals are repeatedly found in a similar state within a specific context. For a type, we need to show that different traits are related (Mendl and Deag, 1995).

12.8 Behavioural Tests

Individual variation can be assessed by observing animals in their normal environment (e.g. Gosling, 1998), or by using behavioural tests. It is not always clear what the requirements are for these tests to provide valuable, reliable information. If we call the behaviour shown in a behavioural test a state, then the requirement for the validity of a test is that it clearly identifies which state an animal is in. It is important to know how many false positives or false negatives we get in a test, if we want to use the information correctly. The tests need not be perfect, however. A test for fear, for instance, in which all animals who are afraid show a specific behaviour, but not all animals who do not show the behaviour are free of fear, can still be used to assess presence of fear, even though it cannot be used to assess absence of fear. The test results need not be consistent across time, they only need to be consistent within the state. In other words, an individual does not have to show fear-related behaviour every time it is in the test situation. It does, however, have to show fear-related behaviour every time it is afraid. Examples for how behavioural tests can be evaluated are given by Perkins *et al.* (1992), who showed that serving capacity in a test situation predicts sexual activity of rams in the field, and by Erhard *et al.* (1997), who showed that latency to attack in a resident–intruder situation predicted aggressive behaviour of growing pigs when unfamiliar pigs were mixed.

Consistency across time is required before states can be used as indicators of a personality trait. In order to label an individual as being of a fearful disposition, for instance, it has to be shown that it has a propensity to experience fear (or to perform fear-related behaviour) when exposed to fear-inducing situations. It may be more appropriate to show a consistency between different test situations within the same context (fear) than to show that the very same test results in the same behaviour (Romeyer and Bouissou, 1992).

A question which is often asked is 'If I carry out the same test with the same individual twice, will I get the same result?' The answer to this question will depend on the extent to which the animals adapt their behaviour according to the experience they had in the test situation (Erhard and Mendl, 1999). An animal, through habituation, may cease to find a particular situation threatening, and therefore show less and less fear-related behaviour when repeatedly exposed to the same test situation. This information is not required for the validation of a test, but is important for its application and interpretation. If the test is used to investigate the reaction to novelty, for example, it cannot be repeated. However, if we want to use experimental animals as their own controls, by testing them before and after a certain treatment, we need to know whether we can use exactly the same test before and after.

12.9 Coping Strategies

A very influential idea in the field of individual differences is that of 'coping strategies', first described in rodents (Benus, 1988; see Fig. 12.2). Similar phenomena have since been reported in spiders (Riechert, 1993) and great tits (Verbeek, 1998). However, before we describe these examples, we will discuss the terminology and the background of these examples of 'individual variation', since they have led to so much controversy and confusion in the past. To begin with, the term 'coping' is used in human psychology for behaviour shown in situations which

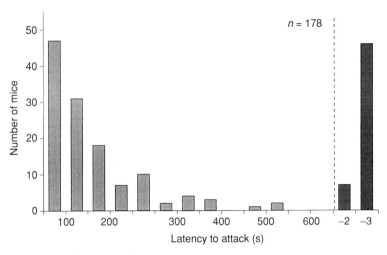

Fig. 12.2. A frequency distribution of attack latencies of 178 male house mice. The bars on the right-hand side represent mice who did not attack in two (−2) or all three (−3) of the tests conducted. (Based on Benus, 1988, p. 5, Figure 2.)

lie outside an individual's competency. If a problem cannot be solved, it has to be coped with (e.g. an incurable or terminal illness). In animals it is used in general for behaviour shown in difficult (challenging) situations, sometimes implying that the behaviour is successful, sometimes only labelling the situation as difficult (see Wechsler, 1995). The term 'strategy' led to the discussion as to whether the underlying distribution was continuous (e.g. the shy–bold continuum) or categorical (e.g. active or passive coping) (Forkman *et al.*, 1995; Jensen, 1995).

Another source of confusion is that terms such as 'strategy', 'style', 'tactic', 'pattern' are used to label very different types of variation. The variation in behaviour we observe may be caused by the circumstances (e.g. high or low predation risk, known or unknown territory, food scarcity or abundance, etc.); it may be due to a tendency of an individual to behave in a specific way, independent of the situation (either genetically determined or acquired, but consistent across time); or it may be random (or due to causes as yet unknown). It would be helpful if the terminology reflected these differences. We suggest that 'strategy' and 'tactic' may be more appropriate for differences which are situation dependent, whereas 'style' or 'personality' is less dependent on the situation, but rather a property of the individual.

The examples given below follow the structure of personality outlined above. The behaviours in the different test situations are equivalent to 'states', the different contexts (e.g. flexibility, aggressiveness) are traits, and the 'active and passive coping' strategies link the traits at the level of the type.

12.9.1 Coping strategies in mice

The model of active/passive coping is based on studies on lines of wild house mice, divergently selected for short and long latencies to attack (Benus *et al.*, 1989). In extensive studies on male mice from these two genetic lines, Benus (1988) found fundamental differences between the behaviour of the two strains. In a defeat test, aggressive mice ('SAL' for short attack latency) were more likely to show flight or attack behaviour (the latter when there was no opportunity to escape), whereas non-aggressive mice ('LAL' for long attack latency) were more likely to show immobility. In an active shock-avoidance test, SAL mice performed well, in that they escaped from the shock, whereas there was a clear dichotomy within the LAL mice into high- and low-avoidance individuals (Benus *et al.*, 1989). When faced with an inescapable shock, SAL mice did not change their level of activity, whereas the activity of LAL mice was suppressed (Benus, 1988). When the mice were trained to run a maze and, subsequently, a change was introduced, SAL mice, unlike the LAL mice, did not react to this change, which was interpreted as their forming behavioural routines. When the maze was changed continuously, so that it was not possible to form a routine, SAL mice did worse in the maze than LAL mice (Benus, 1988). Based on these results, Benus hypothesized that the behaviour of LAL mice was more controlled by external influences, whereas the behaviour of SAL mice was more intrinsically controlled. This hypothesis was tested in an experiment, in which the adaptation of mice to changes in the light/dark cycle was investigated. In agreement with the hypothesis, LAL mice adapted faster to the change than SAL mice (Benus, 1988). The neurochemical background of these differences between the two selection lines was confirmed in an experiment investigating the response to apomorphine (Benus *et al.*, 1991). SAL mice showed a greater increase of stereotypic behaviour than LAL mice, and it was suggested that there was a link between the dopaminergic system and the flexibility of behaviour. Koolhaas *et al.* (1997) reviewed and discussed the behavioural, neuroendocrinological and central-nervous differences between aggressive and non-aggressive mice and rats in more detail.

Based on the differences in the level of locomotion between the two lines of mice when they were confronted with a challenge, Benus (1988) suggested the terms 'active and passive coping strategies'.

12.9.2 Coping strategies in great tits (*Parus major*)

Verbeek (1998) described consistent individual differences in behaviour of great tits. From a natural population she collected young male great tits, hand reared them and tested the birds during the first 18 weeks of their life. The birds could be characterized by their exploratory and aggressive behaviour. The birds showed consistent reactions to different unfamiliar objects in their home cage over a period of several weeks. Some birds approached the objects quickly while others were very slow. Birds that approached unfamiliar objects quickly were also faster in exploring an unfamiliar aviary than birds who approached an unfamiliar object slowly. Most birds either approached within the first 10 s (39%) or failed to approach within the 2 min of the test (25%). These behavioural differences were also found in a food-search test. In this test the birds were trained to find food that was always offered at the same location. After changing the location of food, the fast explorers (FE) persisted in visiting the place where the food used to be. The slow explorers (SE) soon changed their behaviour and stopped visiting the usual feeding place. These birds were more alert and explorative than the fast explorers (Verbeek *et al.*, 1994).

In dyadic interactions FE birds started more fights and won more fights than SE birds (Verbeek *et al.*, 1996). Verbeek also observed agonistic interactions in groups of great tits. In all groups a stable dominance hierarchy was established after a few days. During the first day the results were similar to the observations in the dyadic interactions: FE birds initiated and won more fights than SE birds. However, once the hierarchy had been established, SE birds had higher average ranks than FE birds. The familiarity or unfamiliarity with the environment explained this counter-intuitive result. The birds in the groups were unfamiliar with the aviary they were tested in. When the birds were separately familiarized with the aviary before grouping, FE birds won more often than SE birds on the first day and had higher ranks once the hierarchy was established. SE birds explored more thoroughly an unfamiliar environment while FE birds were more focused on fights. The better and more detailed knowledge of the environment might have increased the chance of winning fights in SE birds (Sandell and Smith, 1991). In a familiar aviary both FE and SE birds had a good knowledge of the environment and the advantage of the SE birds vanished.

The results show that SE great tits are better adapted to new and changing environments, while FE great tits perform better in familiar and stable environments. The behavioural differences found between SE and FE great tits are in agreement with the differences found in mice, rats and pigs. The behaviour of FE birds corresponds to the behaviour found in rodents and pigs that adopt an active behavioural strategy (active copers) to gain and maintain control over the social and physical environment when challenged (Benus, 1988; Benus *et al.*, 1989; Hessing *et al.*, 1993, 1994). The behaviour of the SE birds agrees with the behaviour found in rodents and pigs adopting a passive behavioural strategy (passive copers).

12.9.3   Coping strategies in pigs

Hessing *et al.* (1993) suggested the existence of coping strategies in pigs, when they found a link between the response to manual restraint (the back test) and other behaviours (Fig. 12.3). With the restraint test piglets could be divided into resistant (R) and non-resistant (NR) individuals. R pigs mostly appeared to be the dominant animals and showed more aggression after mixing than NR pigs. NR pigs were mostly found in the middle and lower ranks of the hierarchy. In a novel environment R pigs showed escape behaviour and explored a novel object placed in this environment rapidly but superficially. NR pigs, however, hardly tried to escape and after some hesitation explored the novel object intensively (Hessing *et al.*, 1994). When challenged, R pigs showed a strong increase in heart rate while in the NR pigs a small increase or a decrease in heart rate was found (Hessing *et al.*, 1994). The immune reactivity also differed between R and NR pigs. R pigs showed a higher cell-mediated response to non-specific and specific antigens than NR pigs. Post-mortem examination at slaughter showed a higher incidence of heart muscle alterations in R than in NR pigs and more stomach wall damage in NR pigs. Recently, striking differences in apomorphine susceptibility between R and NR pigs have been found (Bolhuis *et al.*, 1998). A low dose of apomorphine (0.2 mg kg^{-1} body weight, s.c.) provoked unusual activities, like jumping while rotating around the hind legs, walking on the wrists, walking forward with the

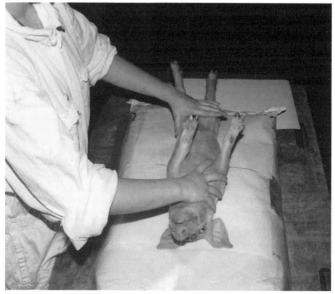

Fig. 12.3.   Pig response to manual restraint (the back test).

hind legs while the forehoof rests on the floor. The NR pigs showed twice as many of these activities as the R pigs (Bolhuis *et al.*, 1998). This difference in apo-susceptibility indicates a difference in the dopaminergic system between R and NR types of pigs. In summary, the above-described differences between R and NR pigs resemble findings in the rat, and indicate that R pigs are 'active copers' whereas NR pigs use a more 'passive' coping strategy. Forkman *et al.* (1995) reported that they did not find this dichotomy of active/passive pigs. Instead they identified three personality traits related to aggression, sociability and exploration, without finding a link between them.

12.10   Questions Still Open

The discussion about the distinction between traits is still going on, since there is not a generally agreed definition of 'trait', of when exactly a number of states can be said to represent a trait, and what distinguishes one trait from another.

The question about the shape of the distribution applies to state, trait and type. On all three levels, we face the question as to whether the distribution is on a continuous scale or on a categorical one and much time has been spent discussing this issue. Eysenck (1967) suggested normal distributions, while the coping style theory suggests distinct categories. Whether the distribution is continuous or categorical may depend upon the specific behaviour or on the trait, and a decision should be made based on the distribution found, rather than based on a pre-formed theory. It is also possible that an underlying continuous distribution of, for example, a hormone will lead to a categorical distribution in the resulting behaviour. Trying to transform every distribution into a normal one is as suspect as forcing data into artificial categories.

Once specific personality traits have been described and tests developed, we can start investigating which aspects are more determined by the genetic make-up of the individuals and which aspects are more affected by experience or the environment.

12.11   Application

Knowledge of animals' personality can be used in experimental designs to reduce the within-treatment variation, by standardizing experimental animals (Erhard *et al.*, 1997). This would reduce the number of animals required in experiments, and could be of significant benefit to the animals, and the research budget. It also allows for the use of animals as their own control. Price *et al.* (1998), for instance, tested ewes for their acceptance of unfamiliar lambs before they used them in a fostering study. They subsequently only used ewes who had rejected the alien lamb, and in doing so eliminated the need for a control group. Lambs in the control group would have had a very high rejection rate, and as a result would have been subject to high rates of aggressive behaviour by the ewe. Thus, the knowledge of the test sheep's personality not only resulted in a reduction of the number of test animals, but also avoided suffering.

If aspects of personality or entire personality traits are identified as undesirable, as is the case with aggressiveness in pigs, possible solutions can be identified. It may be possible to influence the development of personality traits by changing rearing conditions (de Jonge *et al.*, 1996), or to alleviate their consequences by providing additional experience, for instance to improve their social skills (van Putten and Buré, 1997). Alternatively, if the variation has a strong genetic component, genetic selection for or against specific personality traits may be an option, provided care is taken that the selection in favour of or against one trait does not result in undesirable changes in another trait. Also, a specific personality trait may affect the behaviour in different situations in different ways. Aggression against unfamiliar animals, for instance, may be highly undesirable. However, if this aggression is the result of a strong motivation to bond with group members, then a selection against it may result in a weakening of these social bonds. Le Neindre (1989), for example, found a link between aggression and the cow–calf bond. A selection against aggressiveness might therefore be accompanied by a deterioration of maternal behaviour.

Furthermore, personality in non-human animals may be a useful model for personality research in human animals, particularly in the context of therapy (Gosling, 1998).

12.12 Conclusion

Individual differences, rather than being a mere nuisance to behavioural scientists, play an important role in animal husbandry. They are the basis for animal breeding, they are important for designing buildings, for choosing the right animals for specific environments and for learning about backgrounds of specific behaviours. Also, in the behavioural sciences, they can be incorporated in designs as a tool, for example by using animals as their own control or by standardizing animals across treatments.

References

Benus, R.F. (1988) *Aggression and coping. Differences in behavioural strategies between aggressive and non-aggressive male mice*. PhD thesis, University of Groningen.

Benus, R.F., Bohus, B., Koolhaas, J.M. and van Oortmerssen, G.A. (1989) Behavioural strategies of aggressive and non-aggressive male mice in active shock avoidance. *Behavioural Processes* 20, 1–12.

Benus, R.F., den Daas, S., Koolhaas, J.M. and van Oortmerssen, G.A. (1991) Behavioural differences between artificially selected aggressive and non-aggressive mice: response to apomorphine. *Behaviour and Brain Research* 43, 203–208.

Bolhuis, J.E., Schouten, W.G.P., Wiegant, V.M. and de Jong, I.C. (1998) Individual differences in pigs: behavioural responses to apomorphine. In: *Proceedings of the 32nd Congress of the International Society for Applied Ethology*, 21–25 July 1998. Clermont-Ferrand, France.

Cook, C.J., Maasland, S.A. and Devine, C.E. (1996) Social behaviour in sheep relates to behaviour and neurotransmitter responses to nociceptive stimuli. *Physiology and Behavior* 60, 741–751.

de Jonge, F.H., Bokkers, E.A.M., Schouten, W.G.P. and Helmond, F.A. (1996) Rearing piglets in a poor environment: development aspects of social stress in pigs. *Physiology and Behavior* 60, 389–396.

Drews, C. (1993) The concept and definition of dominance in animal behaviour. *Behaviour* 125, 283–313.

Dwyer, C.M. and Lawrence, A.B. (1999) Variability in the expression of maternal behaviour in primiparous sheep: effects of genotype and litter size. *Applied Animal Behaviour Science* 58, 311–330.

Erhard, H.W. and Mendl, M. (1997) Measuring aggressiveness in growing pigs in a resident–intruder situation. *Applied Animal Behaviour Science* 54, 123–136.

Erhard, H.W. and Mendl, M. (1999) Tonic immobility and emergence time in pigs – more evidence for behavioural strategies. *Applied Animal Behaviour Science* 61, 227–237.

Erhard, H.W., Mendl, M. and Ashley, D.D. (1997) Individual aggressiveness of pigs can be measured and used to reduce aggression after mixing. *Applied Animal Behaviour Science* 54, 137–151.

Eysenck, H.J. (1967) The structure of personality. In: Eysenck, H.J. (ed.) *The Biological Basis of Personality*, Vol. 2. Charles C. Thomas American Lecture Series, Springfield, Illinois, pp. 34–74.

Faure, J.-M., Jones, R.B. and Bessei, W. (1983) Fear and social motivation as factors in open-field behaviour of the domestic chick. A theoretical consideration. *Biology of Behaviour* 8, 103–116.

Forkman, B., Furuhaug, I.L. and Jensen, P. (1995) Personality, coping patterns, and aggression in piglets. *Applied Animal Behaviour Science* 45, 31–42.

Gosling, S.D. (1998) Personality dimensions in spotted hyenas (*Crocuta crocuta*). *Journal of Comparative Psychology* 112, 107–118.

Hessing, M.J.C., Wiepkema, P.R., van Beek, J.A.M., Schouten, W.G.P. and Krukow, R. (1993) Individual behavioural characteristics in pigs. *Applied Animal Behaviour Science* 37, 285–295.

Hessing, M.J.C., Schouten, W.G.P., Wiepkema, P.R. and Tielen, M.J.M. (1994) Implications of individual behavioural characteristics on performance in pigs. *Livestock Production Science* 40, 187–196.

Jensen, P. (1995) Individual variation in the behaviour of pigs – noise or functional coping pattern? *Applied Animal Behaviour Science* 44, 245–255.

Kelley, K.W., McGlone, J.J. and Gaskins, C.T. (1980) Porcine aggression: measurement and effects of crowding and fasting. *Journal of Animal Science* 50, 336–341.

Koolhaas, J.M., de Boer, S.F. and Bohus, B. (1997) Motivational systems or motivational states: behavioural and physiological evidence. *Applied Animal Behaviour Science* 53, 131–143.

Le Neindre, P. (1989) Influence of cattle rearing conditions and breed on social relationships of mother and young. *Applied Animal Behaviour Science* 23, 117–127.

Le Neindre, P. and Sourd, C. (1984) Influence of rearing conditions on subsequent social behaviour of Friesian and Salers heifers from birth to six months of age. *Applied Animal Behaviour Science* 12, 43–52.

Lynch, J.J., Hinch, G.N. and Adams, D.B. (1992) *The Behaviour of Sheep*. CAB International, Wallingford, UK.

McCrae, R.R. and Costa, P.T. (1987) Validation of the five-factor model of personality across instruments and observers. *Journal of Personality and Social Psychology* 52, 81–90.

McGlone, J.J. and Morrow, J. (1988) Reduction of pig agonistic behavior by androstenone. *Journal of Animal Science* 66, 880–884.

Mendl, M. and Deag, J.M. (1995) How useful are the concepts of alternative strategy and coping strategy in applied studies of social behaviour? *Applied Animal Behaviour Science* 44, 119–137.

Mendl, M. and Paul, E.S. (1991) Parental care, sibling relationships and the development of aggressive behaviour in two lines of wild house mice. *Behaviour* 116, 11–41.

Mendl, M., Zanella, A.J. and Broom, D.M. (1992) Physiological and reproductive correlates of behavioural strategies in female domestic pigs. *Animal Behaviour* 44, 1107–1121.

Mills, A.D. and Faure, J.M. (1990) The treadmill test for measurement of social motivation in *Phasianidae* chicks. *Medical Science Research* 18, 179–180.

Mills, A.D. and Faure, J.M. (1991) Divergent selection for duration of tonic immobility and social reinstatement behaviour in Japanese quail (*Coturnix coturnix japonica*) chicks. *Journal of Comparative Psychology* 105, 25–38.

Mount, N.C. and Seabrook, M.F. (1993) A study of aggression when group housed sows are mixed. *Applied Animal Behaviour Science* 36, 377–383.

Perkins, A., Fitzgerald, J.A. and Price, E.O. (1992) Sexual performance of rams in serving capacity tests predicts success in pen breeding. *Journal of Animal Science* 70, 2722–2725.

Price, E.O., Erhard, H.W., Borgwardt, R. and Dally, M.R. (1992) Measures of libido and their relation to serving capacity in the ram. *Journal of Animal Science* 70, 3376–3380.

Price, E.O., Dally, M.R., Erhard, H.W., Gerzevske, M., Kelly, M., Moore, N., Schultze, A. and Topper, C. (1998) Manipulating odor cues facilitates add-on fostering in sheep. *Journal of Animal Science* 76, 961–964.

Riechert, S.E. (1993) The evolution of behavioural phenotypes: lessons learned from divergent spider populations. *Advances in the Study of Behavior* 22, 103–134.

Romeyer, A. and Bouissou, M.F. (1992). Assessment of fear reactions in domestic sheep, and influence of breed and rearing conditions. *Applied Animal Behaviour Science* 34, 93–119.

Sandell, M. and Smith, H.G. (1991) Dominance, prior occupancy, and winter residency in the great tit (*Parus major*). *Behavioural Ecology and Sociobiology* 29, 147–152.

Sibbald, A.M., Smart, T.S. and Shellard, L.J.F. (1998) A method for measuring the social behaviour of individuals in a group: an example with sheep. In: Noldus, L.P.J.J. (ed.) *Measuring Behavior '98. Proceedings of the 2nd International Conference on Methods and Techniques in Behavioral Research*.

Smith, B., Wickersham, T. and Miller, K. (1983) *Beginning Shepherd's Manual*. Iowa State University Press, Ames, Iowa

Syme, L.A. (1981) Social disruption and forced movement orders in sheep. *Animal Behaviour* 29, 283–288.

van Putten, G. and Buré, R.G. (1997) Preparing gilts for group housing by increasing their social skills. *Applied Animal Behaviour Science* 54, 173–183.

Verbeek, M.E. (1998) Bold or cautious: behavioural characteristics and dominance in great tits. PhD thesis, University of Wageningen.

Verbeek, M.E., Drent, P.J. and Wiepkema, P.R. (1994). Consistent individual differences in early exploratory behaviour of male great tits. *Animal Behaviour* 48, 1113–1121.

Verbeek, M.E., Boon, A. and Drent, P.J. (1996) Exploration, aggressive behaviour and dominance in pair-wise confrontations of juvenile male great tits. *Behaviour* 133, 945–963.

Wechsler, B. (1995) Coping and coping strategies: a behavioural view. *Applied Animal Behaviour Science* 43, 123–134.

Wilson, D.S., Clark, A.B., Coleman, K. and Dearstyne, T. (1994) Shyness and boldness in humans and other animals. *TREE* 9, 442–446.

Zenchak, J.J. and Anderson, G.C. (1980) Sexual performance levels of rams (*Ovis aries*) as affected by social experiences during rearing. *Journal of Animal Science* 50, 167–174.

Zenchak, J.J., Anderson, G.C. and Schein, M.W. (1981) Sexual partner preference of adult rams (*Ovis aries*) as affected by social experiences during rearing. *Applied Animal Ethology* 7, 157–167.

Zuckerman, M. (1983) The distinction between trait and state scales is not arbitrary: comment on Allen and Potkay's 'On the arbitrary distinction between traits and states'. *Journal of Personality and Social Psychology* 44, 1083–1086.

People as Social Actors in the World of Farm Animals

13

Jeffrey Rushen,[1] Anne Marie de Passillé,[1] Lene Munksgaard[2] and Hajima Tanida[3]

[1]Dairy and Swine Research and Development Centre, Agriculture and Agri-Food Canada, Lennoxville, Quebec J1M 1Z3, Canada; [2]Department of Health and Welfare, Danish Institute of Agricultural Science, Research Centre Foulum, PO Box 50, Tjele, Denmark; [3]Department of Animal Science, Faculty of Applied Biological Sciences, Hiroshima University, Higashi-Hiroshima 724, Japan

---

(Editors' comments: Domestic animals, by definition, are managed animals and this usually necessitates that animals and humans interact. But why include a chapter on human–animal interactions in a book on social behaviour in farm animals? The reason is the growing evidence that this relationship between species is a social one, similar in many respects to the relationship between conspecifics. This is most obvious in the Fulani cattle herdsmen, who control their cattle in Africa by inserting themselves into the social system of the cattle, asserting their dominance over the cattle on some occasions, but also moving among the herd scratching and rubbing them in affiliative behaviour at other times.

In recent years there has been a surge in the amount of research on human–animal interactions. This research falls into several categories, but among the most interesting is the question of whether animals can recognize individual people and what mechanism they use. That dogs can distinguish between different people from an object bearing only that person's odour is well known. Studies with cattle, pigs and sheep, however, seem to imply that visual rather than olfactory cues are most important.

Through the process of domestication there has probably been selection for ease of handling, and so a genetic basis influencing how animals react to humans, but learning, especially during the early period of life, is also important. Certainly handling animals reduces their fear of humans. If it is carried out early in the animal's life, then sometimes this effect is disproportionately large, suggesting that there may be a form of imprinting on humans, although evidence for this varies between species.

©CAB *International* 2001. *Social Behaviour in Farm Animals*
(eds L.J. Keeling and H.W. Gonyou)

353

The chapter concludes that humans are an important social influence in the lives of farm animals.)

13.1   Introduction

It is generally accepted that the behaviour of companion animals and that of their owners is closely connected, but this is much less appreciated for farm animals. Recently, the marked effect that farmers or stock people can have upon the behaviour and the productivity of farm animals has become apparent (Seabrook and Bartle, 1992; Hemsworth and Coleman, 1998). An essential part of the process of domestication involves animals becoming tamer (Price, 1998) through a reduction in their fear of people. However, there are more subtle aspects to the relationship between people and domestic animals. The relationship between people and companion animals is often described as a social relationship: Topal *et al.* (1998) consider that the attachment of a dog to its owner is analogous to the attachment of a child to a parent. In a very interesting review, Tennessen and Hudson (1981) examined the different species which have been successfully domesticated, and found several similarities in their basic social structure. Domesticated species tended to have wild ancestors that lived in groups rather than being solitary, with dominance hierarchies rather than a territorial system. Studies of traditional herding societies (Lott and Hart, 1979) suggest that these traits may have aided the process of domestication by allowing people to enter into social relationships with the animals and so exploit their natural social behaviour.

In this chapter, we examine some of the factors that can influence how farm animals respond to people. Since this book is about the social behaviour of animals, we try to address the question as to whether or not farm animals can be said to have 'social' relationships with people. We will examine the evidence that people can establish dominance over animals, the extent that people and farm animals can communicate and the extent that farm animals imprint on people.

13.2   The Nature of the Relationship Between People and Farm Animals

13.2.1   Social behaviour and social communication between people and farm animals

The relationships between people and animals have been described in terms of predator–prey relations and there is some evidence that some animals may indeed 'see' people as predators: Kendrick (1991) noted that the same neurons were active when sheep were looking at the face of a person as when looking at a dog. These were distinct from the neurons that were active when looking at faces of other sheep.

However, the type of relationship that can develop between people and animals that are in constant contact, such as companion animals, can obviously be much more subtle than this, and it has been claimed that such relationships are genuine social relationships, similar to what would be seen between conspecifics. Estep and Hetts (1992) consider that one of the most interesting cases of relationships that can exist between people and animals is where each responds to the other as though it were a conspecific. This would be apparent in the use of affiliative behaviour between people and animals or in the ability to develop dominance relationships, or the ability to use species-specific communication (see Chapter 2).

It is often claimed that people can, and indeed should, establish dominance over their animals. For example, Grandin (2000) emphasizes the importance of establishing social dominance over animals in order to make handling safe and easy. Much of the aggression that domestic animals, particularly cattle, show towards people, has been interpreted in terms of the animal trying to establish its dominance over people. Whether people and farm animals can establish true dominance relationships, however, has not been looked at in any detail. One of the few detailed studies of how people appear to enter into genuine social relationships, including dominance relationships, with animals was Lott and Hart's (1979) study of the Fulani cattle herdsmen of sub-Saharan Nigeria. According to Lott and Hart, while 'western' farmers tend to rely primarily on physical means to control animals, e.g. fences or restraints, the Fulani control their animals primarily through their knowledge and exploitation of the cattle's natural social behaviour and by inserting themselves into the social system of the cattle. Interestingly, the Fulani appear quite aggressive in establishing dominance over their cattle, especially the bulls in the herd. They respond to any threats from the bull by attacking the bull or threatening it by yelling and waving a stick. The herdsmen continually reinforce their dominance over the other cattle by occasionally hitting them for no obvious reason, and by breaking up fights between other cattle. This routine use of physical aggression seems contrary to what is often argued: that good stockmen should not use physical force on their livestock. However, while, in recent years, the emphasis has been put on the negative effects of animals' fear of people, we need to remember that it might be equally important for the safety of the handlers to ensure some degree of dominance, especially over the larger farm animals, and, in some cases, a degree of physical force may be the most effective way of achieving this (see Grandin, 2000).

However, establishing social dominance is not the only form of social relationships that the Fulani have with their cattle. The Fulani herdsmen establish amicable relationships with their animals by spending considerable time walking among them, and scratching them on the head and neck, places where the cattle often groom themselves or each other. Grooming is a common affiliative behaviour of cattle so these affiliative behaviours that the Fulani herdsmen use are the same sort of social behaviours that the cattle use with each other.

Apart from anecdotes, there are relatively few reports of animals directing social behaviour towards people. Human-reared dogs evidently direct towards people a variety of social behaviours normally shown to other dogs and so can be considered to have established some social bonds. Unfortunately, this aspect of human–animal relationships has not been systematically studied in any detail for farm animals. Certainly young lambs reared by people will readily follow people to keep in proximity (e.g. Markowitz *et al.*, 1998). Sambraus and Sambraus (1975) reported that young goats, lambs and pigs that were reared by people would direct their sexual behaviour preferentially towards people rather than conspecifics. In some studies, it is reported that young calves will play with their human handler (Boivin *et al.*, 1992a; Jago *et al.*, 1999).

A social relationship between people and animals might also be apparent in the ability of people to provide social support for animals, especially animals in social isolation. For example, dam-reared lambs separated from their mothers vocalize frequently and Boivin *et al.* (1997) and Korff and Dyckhoff (1997) have recently shown that human-reared lambs also vocalize less in the presence of people than when alone and this effect was more apparent for the shepherd that had raised the lambs than for a stranger (Boivin *et al.*, 1997). Similar results have also been reported for human-reared goats (Boivin and Braastad, 1996). Thus it seems that for these animals the presence of people provides some form of social comfort.

Estep and Hetts (1992) suggest that another criterion that could be used to describe the relationship between people and animals as 'social' would be in the use of species-specific communicative behaviour, either by the animal attempting to 'communicate' with people or by the people attempting to communicate with the animal. There are clear cases when animals do attempt to communicate with people using species-specific communicative behaviours: a dog threatening a person looks very much like a dog threatening another dog. Furthermore, there is little doubt that people can use vocal commands to alter the behaviour of companion animals such as dogs or horses, although this seems a relatively unexploited way of influencing the behaviour of farm animals. For example, Murphey and Moura Duarte (1983) report that cattle herdsmen in Brazil are able to attract individual calves by calling their name, although Hemsworth *et al.* (1986) were not able to influence the approach behaviour of pigs by calling to them.

However, in most cases, the ability to 'communicate' with animals is the result of deliberate training of the animal with rewards to respond to certain vocal or gestural signals with the appropriate behaviour. Moreover, the signals used by people are arbitrary from the point of view of the animal, that is, it is probably as easy to teach a dog to sit in response to the word 'stand' as it is to the word 'sit'. This type of communication bears little relationship to the dog's own communicative system.

A series of studies of sheepdogs (McConnell and Baylis, 1985; McConnell, 1990) has found some evidence that communication between animals and people involves more than simply training animals to respond appropriately to arbitrary signals. McConnell and Baylis (1985) recorded the types of calls shepherds use to control the behaviour of sheepdogs and examined the acoustic structure of the calls. The calls that were used to indicate which direction the dog should move appeared to be arbitrary, in that there was no consistent acoustic structure to the calls, which differed widely between the shepherds. However, there were marked similarities in the calls that different shepherds used to make the dog stop or to move. To make the dogs move, all shepherds tended to use rapidly repeated calls of short duration and of rising frequency, while, to inhibit the dog's motion, all shepherds used single, prolonged notes of a descending frequency. Furthermore, McConnell (1990) found that the use of the appropriate call type greatly facilitated training dogs to respond appropriately to the call. McConnell suggested that the structure of the call may affect the dogs' level of arousal and reported some anecdotal evidence for similar effects for horses and cats. This suggests that these may reflect general attributes of mammalian communication, and that the shepherds were using these basic properties of mammalian communicative systems to control their dogs. This research suggests that animals' response to calls given by people reflects not just training, but that there is some fundamental association between the acoustic structure of the calls and the meaning, i.e. its effect upon the animals' behaviour. Further knowledge of such associations would greatly increase the ease of communicating with and handling animals.

13.2.2 Recognition by animals of individual people

To be able to have a genuine social relationship with people, animals would need to be able to recognize individual people. There is good evidence that many (if not all) species of mammals and birds are able to recognize individuals of their own species (see Section 14.1). Most people would probably assume that companion animals, such as dogs, can tell people apart, and it would not seem surprising to find this capacity in farm animals. Farm animals do not always respond differently to different people. For example, Hemsworth *et al.* (1994, 1996) found that pigs do not differentiate between different people, and that they do not behave differently with familiar and unfamiliar people. Despite this, many studies suggest that animals can tell different people apart. For example, dogs' ability to distinguish people by odour is well known (Settle *et al.*, 1994) and many species of animals have been reported to react differently to different people (Rushen *et al.*, 1999b). Tanida *et al.* (1995) subjected newly weaned piglets to gentle handling from one person and then tested their responses to familiar and unfamiliar humans. In addition to showing a general reduction in their fear of humans, handled pigs also preferred to interact with the familiar human over the stranger. Thus, the animals generalized their positive experiences with the familiar handler to some extent, yet still responded very differently to the two humans, showing their ability to discriminate. The ability of pigs to differentiate people has been subsequently confirmed (Tanida and Nagano, 1996). De Passillé *et al.* (1996) noted that, when a person entered a calf's pen, periods of contact were shorter and more frequent if the person was unfamiliar. When calves were handled repeatedly by different people, one of whom treated them positively while the other handled them aversively, they contacted the positive handler significantly more than the aversive handler, showing that they could distinguish between them. Munksgaard *et al.* (1997) found similar results for adult dairy cows, and subsequent work has confirmed the ability of both cattle (Boivin *et al.*, 1998; Taylor and Davis, 1998; Munksgaard *et al.*, 1999; Rushen *et al.*, 1999a) and sheep (Boivin *et al.*, 1997; Davis *et al.*, 1998) to distinguish individual people.

These studies show that farm animals can clearly tell people apart. What cues, however, might the animals be using? This has been investigated in cattle, sheep and pigs.

Cattle have a reasonable degree of visual acuity and are capable of colour vision (Albright and Arave, 1997). However, among cattle, the sense of smell rather than sight is the means of individual recognition in dominance relationships although visual as well as olfactory cues are required to determine which of a pair of animals will control a food source (Bouissou, 1971) and maternal recognition in *Bos indicus* calves appears to be based on visual cues (Murphey *et al.*, 1990). Visual cues, especially those associated with the clothing worn, also seem important in recognizing people. De Passillé *et al.* (1996) showed that dairy calves can distinguish between people wearing different colours of clothing. Rybarczyk *et al.* (1999) showed that very young calves could easily distinguish between people wearing different colour clothes, although most calves could not discriminate between people wearing the same colour clothes. Munksgaard *et al.* (1997) and Rushen *et al.* (1999a) found that adult dairy cows could similarly distinguish between people based on the treatment received, at least when the two people wore different colour clothes. However, the cows did not discriminate between the same handlers when they both wore the same colour, suggesting that the colour worn is an important cue that cows use to recognize people. Despite this, the cows did not generalize their behaviour to unknown people who wore the same colour clothes, suggesting that clothing colour is not the only cue that the cows use. More recently, Munksgaard *et al.* (1999) found that cattle had greater difficulty recognizing people who wore the same colour clothes compared with people who wore different colour clothes, although Taylor and Davis (1998) showed that cattle in fact could learn to distinguish people who wore the same colour clothes. Boivin *et al.* (1998) showed that changes of clothing did not affect beef cattle's ability to distinguish familiar from unfamiliar people. Together, these results suggest that, while clothing colour is used by cattle to distinguish people, it is by no means the only cue that they use.

Bouissou *et al.* (1996) found that ewes displayed lower fear reactions to the slides of individuals of their own breed than to slides of individuals of another breed. Kendrick *et al.* (1995, 1996) showed that sheep can discriminate between breeds of sheep and individual sheep within a breed on the basis of facial cues and that sheep have neural circuits that respond preferentially to faces (Kendrick, 1991; Kendrick *et al.*, 1995, 1996). Although they did not test the ability of sheep to distinguish between the faces of different humans, they believe that sheep use facial discrimination for distinguishing between humans as they do conspecifics. Given that recognition of sheep appears to be primarily visually based, it seems reasonable that recognition of humans by sheep will also be based largely on visual features. Korff and Dyckhoff (1997) reported that bottle-fed lambs would vocalize less in the presence of a familiar person than when alone. Reduced vocalization also occurred in the presence of a visual image of that person, but not in the presence of a recording of the person's voice. This indicates the importance of visual cues.

Pigs are thought to be preferentially olfactory and auditory animals and not to be highly dependent on vision (see Section 6.13). It is known, however, that pigs can discriminate among wavelengths of light and therefore presumably among colour hues (Tanida *et al.*, 1991). Pigs appear to use a combination of visual, olfactory and acoustic stimuli to recognize each other (Ewbank *et al.*, 1974; Meese and Baldwin, 1975; Shillito Walser, 1986; Horrell and Hodgson, 1992). Using operant conditioning techniques and a Y-maze, Tanida and Nagano (1998) showed that miniature pigs can discriminate between a stranger and their familiar handler based on visual, auditory and/or olfactory cues. Pigs handled gently and offered food treats were trained to choose their regular handler in a Y-maze. All pigs correctly chose their handler over a stranger with greater than 80% accuracy within four sessions. Further testing, wherein the humans' voice, odour and figure were systematically masked, revealed a poorer level of performance, suggesting that pigs rely on more than one modality to recognize people. Further work (Tanida and Koba, 1997) showed that pigs that had learned to approach one person in order to receive a food reward would approach a second person who wore the same colour clothes. However, this did not occur if the second person wore the same perfume. When exposed for a sufficient period of time to people wearing the same colour and style of clothing, miniature pigs can discriminate between people based on their face and body size (Koba and Tanida, 1999). This suggests the dominance of visual cues over olfactory cues in pigs' recognition of individual people, which is surprising given their reliance on olfactory and auditory cues to recognize each other.

Thus, it is clear that a number of farm animal species can tell different people apart. The particular cues that they use are just beginning to be investigated. Some animals, such as dogs, are clearly very good at using olfactory cues to identify people (Settle *et al.*, 1994) but, contrary to what is most popularly believed, cattle, pigs and sheep do not rely mainly on olfactory cues. Rather, visual cues seem to be the most important.

13.3　Factors that Influence the Response of Farm Animals to People

13.3.1　Genetic basis

Differences among animals in their responses to people and in other behavioural characteristics may be influenced by genetics. Genetic differences in the behaviour of domestic animals, including their responses to people, have recently been reviewed (Grandin, 1998).

For wild animals, we could expect some natural selection to favour some degree of fear of people, who are potential predators. As animals and humans became co-dependent during the process of domestication, natural selection pressure to maintain animals' fear of human beings would have been relaxed (see Chapter 4). In fact, during the domestication process, animals most probably were selected for docility and tameness by farmers seeking ease of handling (e.g. Price, 1998). Genetic selection to increase tameness is now being systematically applied to the newly domesticated animals such as deer (Pollard *et al.*, 1994), mink (Hansen, 1996) and foxes (e.g. Belyaev *et al.*, 1985), and such deliberate attempts to use selective breeding to favour increased tameness generally report success within a few generations (Belyaev *et al.*, 1985; Hansen, 1996).

The process of genetic selection underlying tameness is likely to be a continuing one even for animals that have long been domesticated since animals that are difficult to handle continue to be culled. However, with the increased mechanization and intensification of animal agriculture the target of artificial selection has gradually shifted to productivity traits with greater emphasis on efficiency of production, rather than on handling difficulties *per se*. However, this may have undesirable effects; Grandin and Deesing (1998) suggest that intense selection for productivity traits, especially rapid growth and lean meat, has resulted in nervous and aggressive animals that are difficult to handle.

Many studies have found breed differences in docility in cattle. These have recently been reviewed by Burrow (1997). Less research has examined genetic differences in other farm animal species. Breed differences in responses to people have also been noted for sheep (Le Neindre *et al.*, 1993; Lankin, 1997). In poultry, the degree of tonic immobility and flightiness (which does often involve responses to people) has a strong genetic component (reviewed in Craig and Muir, 1998; Faure and Mills, 1998). Hens of a 'flighty' line, which typically withdrew from humans, and those of a docile line, which typically stood still or actively approached people, have been successfully created by divergent selection (Murphy and Duncan, 1977, 1978). Mink have been successfully selected for increased or reduced fear of people (Hansen, 1996). In goats, individual differences in responses to people appear stable and there is a strong correlation between fraternal twins, even when these are reared separately (Lyons *et al.*, 1988). In pigs, heritability of reactivity to humans has been estimated at 0.38 (Hemsworth *et al.*, 1990).

Interestingly, some of the genes that favour tameness appear to have pleiotropic effects, influencing other characteristics of the animals, for example coat colour (Price, 1998). More recently, Grandin *et al.* (1995) found that the temperament of cattle is correlated with the position of their hair whorls, a surprising finding that has been replicated (Randle, 1998).

This indicates that a considerable degree of genetic variance remains for farm animals' responses to people and suggests that animals' responses to people could be moderated by judicious consideration of objective measures of fearfulness of humans in mating systems, such as cross-breeding, and in genetic selection programmes. However, while it is clear that genetic differences underlie some of the variance between animals in tameness, the mechanisms that may underlie these genetic differences have not been investigated in detail.

Belyaev *et al.* (1985) suggest that selection for tameness in silver foxes has led to an extended sensitive period for social bonds, and suggests that the neoteny that is often said to characterize domestic animals may have resulted from selection for more easily socialized animals. This is discussed in some detail by Price (1998), who finds little evidence for such neoteny except in the case of domestic dogs.

13.3.2   Learning and the effect of handling on animals' productivity and level of fear

Despite the obvious effects of genetics on animals' fear of people, there is much evidence that the responses of farm animals to people are affected by the animals' experience of people, especially the way they are handled (Hemsworth and Coleman, 1998). For example, animals' fear of people can be reduced by increased contact with people. This has been demonstrated many times, e.g. with poultry (Jones, 1994), pigs (Gonyou *et al.*, 1986), sheep (Mateo *et al.*, 1991), cattle (Hemsworth *et al.*, 1986; Boivin *et al.*, 1992a, b), goats (Boivin and Braastad, 1996), horses (Mal and McCall, 1996), rabbits (Podberscek *et al.*, 1991) and foxes (Pedersen and Jeppesen, 1990). Such effects appear to be particularly marked for young animals. There is clear evidence that increased handling or contact with people early in life leads to prolonged reduction in their fear of people. This has been shown for domestic poultry (Jones and Waddington, 1992; Jones, 1994), cattle (Boissy and Bouissou, 1988; Boivin *et al.*, 1992a, b, 1994; Jago *et al.*, 1999), pigs (Hemsworth *et al.*, 1986), goats (Lyons *et al.*, 1988; Lyons, 1989; Boivin and Braastad, 1996), sheep (Markowitz *et al.*, 1998), horses (Mal and McCall, 1996), partridges (Csermely *et al.*, 1983) and foxes (Pedersen and Jeppesen, 1990; Pedersen, 1993).

A number of mechanisms have been proposed to explain this change (Hemsworth and Coleman, 1998). First, many animals are neophobic, and the appearance of unfamiliar people may elicit orienting responses and fear. Habituation of these responses would occur where the animals are repeatedly exposed to people, and where this has no negative consequences for the animals. In such cases, we might expect some dishabituation to occur, that is, the initial fear responses to people should be reinstated if the animals do not have further exposure to people over a long period of time. Although it seems most likely that habituation is responsible for some of the reduced fear, this has not been systematically examined in any detail. For poultry fear of people does not have to be learned but is present at hatching (Murphy and Duncan, 1978) and simple exposure to people is sufficient to reduce fearfulness in poultry (Murphy and Duncan, 1978; Jones, 1994), as it is also in rabbits (Podberscek *et al.*, 1991). However, there have not been any studies as to the extent that dishabituation occurs.

Hemsworth and Coleman (1998) suggest that animals' responses to people can also be altered through classical conditioning or associative learning, that is, the animals can learn to associate people either with rewards or punishments. Taming animals with food rewards is an obvious example, and many farm animals do learn to approach people who feed them. In the case of associative learning, the change in the way that the animals will respond to people will depend on how the people treat the animals. Positive rewards should increase the animals' tendency to approach people, while aversive treatments ('punishments') would increase the animals' tendency to avoid people. This has been most clearly demonstrated for pigs (Hemsworth *et al.*, 1981; Gonyou *et al.*, 1986). Indeed, pigs (Tanida *et al.*, 1994, 1995), cattle (de Passillé *et al.*, 1996; Munksgaard *et al.*, 1997, 1999; Taylor and Davis, 1998; Rushen *et al.*, 1999a) and sheep (Davis *et al.*, 1998) can learn to approach or avoid individual people according to the type of treatment they receive. However, such associative learning cannot always be relied upon to reduce animals' fear of people. Murphy and Duncan (1978) were not able to reduce the avoidance of people by giving food rewards to poultry of a particularly flighty line.

An interesting question is: what type of rewards can people use to attract animals? That animals will approach people who feed them is not surprising, but can people also give animals 'social rewards'? That gentle handling alone without additional feeding can function to reduce an animal's fearfulness of people has been shown for cattle (Boissy and Bouissou, 1988; Boivin *et al.*, 1992a), pigs (Hemsworth *et al.*, 1996), goats (Boivin and Braastad, 1996), horses (Mal and McCall, 1996), poultry (Jones, 1994) and foxes (Pedersen, 1993), suggesting that people can be a source of social rewards.

13.3.3 Imprinting

A third mechanism that has been proposed to explain the effect of increased contact between people and animals, especially young animals, is imprinting. Lorenz (1935) used the term imprinting to refer to the process by which young birds would learn to recognize their mother and begin to follow her.

According to Lorenz this process differed in a number of ways from what was then known about normal learning. Principally, imprinting appeared to occur rapidly during a sensitive period of the animal's life, which was soon after hatching. Imprinting did not require the traditional reinforcement by food rewards to occur, and was irreversible; once a young bird had imprinted it was difficult to undo this and the birds rarely imprinted to other objects afterwards. Lorenz also noted that hand-reared birds such as ducklings would imprint to a person, following that person about as if he/she were their mother. Following Lorenz's lead, some have used the term imprinting to refer to the way that young animals with much exposure to people, or which are reared by people, can develop an attraction towards people (Sambraus and Sambraus, 1975; Belyaev *et al.*, 1985; Albright and Arave, 1997). For example, Sambraus and Sambraus (1975) reported that young goats, lambs and pigs that were separated from conspecifics until sexual maturity and reared by people would direct their sexual behaviour preferentially towards people rather than conspecifics, which they claim was a clear case of sexual imprinting. Despite the common use of the term, the evidence that farm animals 'imprint' on people is not very strong.

First, more recent research has shown that, even in precocial birds, 'imprinting' does not have the properties suggested by Lorenz. This is not the place for a detailed review on imprinting, and the reader is referred to the excellent reviews by Rogers (1995) and Bolhuis (1991). In summary, imprinting is not so different from other forms of learning as once thought, sharing many similarities with associative learning.

Secondly, studies on early handling of farm animals have found some evidence that the effect of this handling does show some, but not all, of the properties suggested by Lorenz to be typical of imprinting.

The reduction in fear that occurs as a result of early contact between animals and people does appear very persistent. Reduced fearfulness has usually been noted several months later, and in some cases years later, without apparently needing to be reinforced by further handling (e.g. Lyons *et al.*, 1988; Boivin *et al.*, 1992b). Where studies have examined the persistence of the effect by repeatedly testing animals as they age, there is often little decrease in the effect with age (e.g. Hemsworth *et al.*, 1986; Lyons *et al.*, 1988; Pedersen and Jeppesen, 1990; Pedersen, 1993; Boivin *et al.*, 1992b, 1994; Boivin and Braastad, 1996; Markowitz *et al.*, 1998), suggesting that the effect may be permanent. Such a decrease would be expected if the reduction in fear was due to habituation (a phenomenon known as dishabituation).

Research has also examined whether or not there is a sensitive period for this learning to occur, that is, whether extra handling is more effective at some ages (especially earlier in the animal's life) than at other ages. Clear evidence for such a sensitive period has been found for partridges (Csermely *et al.*, 1983), sheep (Markowitz *et al.*, 1998), horses (Mal and McCall, 1996) and goats (Boivin and Braastad, 1996). In some cases, this sensitive period is surprisingly short. For example, Markowitz *et al.* (1998) found that extra handling of lambs during the 2 days immediately following birth was far more effective than was handling during the next 10 days of life. In fact, handling during the remaining 10 days appeared to be effective only if the lambs were tested a few days later and was not effective when the lambs were tested 25 days later. This is the clearest evidence yet of a sensitive period in a mammal for attachment to people. In horses, a sensitive period may exist but which is substantially longer than that reported for sheep. Extra handling during the first 7 days after birth does not appear to affect handling ease later (Mal *et al.*, 1994), whereas extra handling during the first 42 days of life is more effective than the same amount of handling from 43 to 84 days of life (Mal and McCall, 1996).

Although it is often inferred that a sensitive period exists for cattle, the attempts to find such a sensitive period have not been successful (Boissy and Bouissou, 1988; Boivin *et al.*, 1992a, b). Boissy and Bouissou (1988) handled heifers either at 0–3 months, 6–9 months or 0–9 months of age. Reduced fearfulness and increased ease of handling were found for the group handled during 0–9 months of age, with less of an effect for the group handled during 6–9 months. The least effective period for handling was during the 0–3 months following birth. The authors concluded that extended prepubertal handling is most effective rather than short-term handling and that there appears not to be any critical period for this effect.

Interestingly, for domestic poultry, where there appears much evidence in favour of a sensitive period for filial imprinting (Rogers, 1995), there is no evidence for a sensitive period for the effects of handling by people; chicks handled between days 1 and 9 after hatching showed the same level of reduced fear as did chicks handled between days 10 and 18 (Jones and Waddington, 1992). This is so despite the fact that the reduction in fear due to handling is rapid, occurring within the first 4 days after hatching (Jones, 1994).

There is much less evidence that prior imprinting to conspecifics reduces the effect of early handling by people. Sambraus and Sambraus (1975) claimed that long-term isolation from conspecifics was necessary for animals (sheep, goats and pigs) to imprint to people, although this was not systematically examined. Cattle (Boivin *et al.*, 1994) and goats (Lyons, 1989) that are reared by their mothers are often more fearful of people than those reared by hand, but this probably reflects a difference in the amount of contact with people. Some studies appear to assume that a period of social isolation or prior weaning is necessary to aid young animals' socialization to people (Boivin and Braastad, 1996; Markowitz *et al.*, 1998). However, many studies have shown that extra handling of young animals is effective even if the animals are reared by their natural mothers (Hemsworth *et al.*, 1986; Markowitz *et al.*, 1988; Boivin *et al.*, 1992a), or if the animals are kept grouped with conspecifics (Pedersen, 1993; Hemsworth *et al.*, 1986; Jones, 1994; Boivin *et al.*, 1992b, 1994; Boivin and Braastad, 1996). Creel and Albright (1988) hypothesized that calves weaned from their mothers and reared in isolation from other calves would 'imprint' on people and be less fearful than calves reared in groups of other calves. However, they found no evidence that isolated calves approached people more readily or were easier to handle. Although, there was some evidence of reduced flight distance of isolated calves, the authors concluded that social isolation from other calves does not make animals less fearful of people. Close contact between people and bulls that are kept isolated from other bulls can substantially increase the aggression that the bulls show (Price and Wallach, 1990). However, whether this is because the bulls are 'imprinting' on people, or whether they lack social restraints on their aggression as a result of their lack of social companions (suggested by Price and Wallach, 1990) is not clear.

Hemsworth and Barnett (1992) conducted one of the few studies that has directly examined whether weaning from the mother aids the process of social isolation to people. Young pigs were subjected to extra handling and were either kept with their mother or separated soon after birth and raised artificially. There was no indication that the weaned pigs were less fearful of people or that prior weaning increased the effectiveness of the extra handling. More recently, Krohn (1996) found that calves reared with their mothers had longer flight distances from people than calves reared individually, even though the amount of contact between the calves and people was similar. Whether these differences reflect genuine species differences or procedural differences is not clear.

Whether increased early socialization to people interferes with the animal forming social bonds to conspecifics has not been looked at systematically. However, Markowitz *et al.* (1998) report that handling lambs soon after birth did not disrupt the formation of social bonds either with conspecifics or with the natural mother.

In summary, while there is some evidence for sensitive periods in some species, there is only limited evidence that socialization to conspecifics (either to a mother or to peers) interferes with socialization to people or that it reduces the effectiveness of extra handling in reducing animals' fear of people.

13.4 Conclusions

Many studies have now shown the importance of the response of farm animals to their caretaker in affecting their welfare and productivity. In this chapter, we have reviewed some of the evidence that the relationship between farm animals and people is much more than an instance of predator–prey relationships, as some suggest. In some ways, farm animals can be said to form genuine social relationships with people, similar in some respects to what they would normally form with conspecifics. This is shown by: (i) the ability of people to provide what normally would be considered as social rewards; (ii) the ability of people to provide social support for animals when they are under stress; (iii) the ability of herdsmen to insert themselves into the social system of animals; and (iv) the capacity of people to utilize some fundamental aspects of mammalian communication to communicate with animals. The capacity to recognize and respond differently to different people, which is an essential component of social relationships, has been amply demonstrated for most farm animals. The nature of the relationship between farm animals and people is dependent upon the genetics of the animals, but is plastic, being influenced by the type of handling the animals receive and by the amount and type of contact they have with people when they are developing.

References

Albright, J.L. and Arave, C.W. (1997) *The Behaviour of Cattle.* CAB International, Wallingford, UK.

Belyaev, D.K., Plyusnina, I.Z. and Trut, L.N. (1985). Domestication in the silver fox (*Vulpes fulvus* Desm): Changes in physiological boundaries of the sensitive period. *Applied Animal Behaviour Science* 13, 359–370.

Boissy, A. and Bouissou, M.-F. (1988) Effects of early handling on heifers' subsequent reactivity to humans and to unfamiliar situations. *Applied Animal Behaviour Science* 20, 259–273.

Boivin, X. and Braastad, B.O. (1996) Effects of handling during temporary isolation after early weaning on goat kids' later response to humans. *Applied Animal Behaviour Science* 48, 61–71.

Boivin, X., Le Neindre, P. and Chupin, J.M. (1992a) Establishment of cattle–human relationships. *Applied Animal Behaviour Science* 32, 325–335.

Boivin, X., Le Neindre, P., Chupin, J.M., Garel, J.P.and Trillat, G. (1992b) Influence of breed and early management on ease of handling and open-field behaviour of cattle. *Applied Animal Behaviour Science* 32, 313–323.

Boivin, X., Le Neindre, P., Garel, J.P. and Chupin, J.M. (1994) Influence of breed and rearing management on cattle reactions during human handling. *Applied Animal Behaviour Science* 39, 115–122.

Boivin, X., Nowak, R., Desprès, G., Tournadre, H. and Le Neindre, P. (1997) Discrimination between shepherds by lambs reared under artificial conditions. *Journal of Animal Science* 75, 2892–2898.

Boivin, X., Garel, J.P., Mante, A., and Le Neindre, P. (1998) Beef cattle react differently to different handlers according to the test situation and their previous interactions with their caretaker. *Applied Animal Behaviour Science* 55, 245–257.

Bolhuis, J.J. (1991) Mechanisms of avian imprinting: a review. *Biological Reviews* 66, 303–345.

Bouissou, M.F. (1971) Effets de l'absence d'information optiques et de contact physique sur la manifestation des relations hiérarchiques chez les bovins domestiques. *Annales de Biologie Animale Biochimie Biophysique* 11, 191–198.

Bouissou, M.F., Porter, R.H., Boyle, L. and Ferreira, G. (1996) Influence of conspecific image of own vs. different breed on fear reactions of ewes. *Behaviour Processes* 38, 37–44.

Burrow, H.M. (1997) Measurements of temperament and their relationships with performance traits of beef cattle. *Animal Breeding Abstracts* 65, 477–495.

Candland, D.S. (1969) Discriminability of facial regions used by the domestic chicken in maintaining the social dominance order. *Journal of Comparative and Physiological Psychology* 69, 281–285.

Craig, J.V. and Muir, W.M. (1998) Genetics and behavior of chickens: welfare and productivity. In: Grandin, T. (ed.) *Genetics and the Behavior of Domestic Animals.* Academic Press, San Diego, pp. 265–298.

Creel, S.R. and Albright, J.L. (1988) The effects of neonatal social isolation on the behavior and endocrine function of Holstein calves. *Applied Animal Behaviour Science* 21, 293–306.

Csermely, D., Mainardi, D. and Spano, S. (1983) Escape-reaction of captive young red-legged partridges (*Alectoris rufa*) reared with or without visual contact with man. *Applied Animal Ethology* 11, 177–182.

Davis, H., Norris, C. and Taylor, A. (1998) Wether ewe know me or not: the discrimination of individual humans by sheep. *Behavioural Processes* 43, 27–32.

de Passillé, A.M.B., Rushen, J., Ladewig, J. and Petherick, C. (1996) Dairy calves' discrimination of people based on previous handling. *Journal of Animal Science* 74, 969–974.

Estep, D.Q. and Hetts, S. (1992) Interactions, relationships, and bonds: the conceptual basis for scientist–animal interactions. In: Davis, H. and Balfour, A.D. (eds) *The Inevitable Bond – Examining Scientist–Animal Interactions.* Cambridge University Press, Cambridge, pp. 6–26.

Ewbank, R. Meese, G.B. and Cox, J.E. (1974) Individual recognition and the dominance hierarchy in the domesticated pig. The role of sight. *Animal Behaviour* 22, 473–480

Faure, J.M. and Mills, A. (1998) Improving the adaptability of animals by selection. In: Grandin, T. (ed.), *Genetics and the Behavior of Domestic Animals.* Academic Press, San Diego, pp. 235–264.

Gonyou, H.W., Hemsworth, P.H. and Barnett, J.L. (1986) Effects of frequent interactions with humans on growing pigs. *Applied Animal Behaviour Science* 16, 269–278.

Grandin, T. (1998) *Genetics and the Behavior of Domestic Animals.* CAB International, Wallingford, UK.

Grandin, T. (2000) Behavioural principles of handling cattle and other grazing animals under extensive conditions. In: Grandin, T. (ed.) *Livestock Handling and Transport*, 2nd edn. CAB International, Wallingford, UK, pp. 63–85.

Grandin, T. and Deesing, M.J. (1998) Genetics and behavior during handling, restraint, and herding. In: Grandin T. (ed.) *Genetics and the Behavior of Domestic Animals.* CAB International, Wallingford, UK, pp. 113–144.

Grandin, T., Deesing, M.J., Struthers, J.J. and Swinker, A.M. (1995) Cattle with hair whorl patterns above the eyes are more behaviorally agitated during restraint. *Applied Animal Behaviour Science* 46, 117–124.

Hansen, S.W. (1996) Selection for behavioural traits in farm mink. *Applied Animal Behaviour Science* 49, 137–148.

Hemsworth, P. and Barnett, J.L. (1992) The effects of early contact with humans on the subsequent level of fear of humans in pigs. *Applied Animal Behaviour Science* 35, 83–90.

Hemsworth, P.H. and Coleman, G.J. (1998) *Human–Livestock Interactions.* CAB International, Wallingford, UK.

Hemsworth, P.H., Barnett, J.L. and Hansen, C. (1981) The influence of handling by humans on the behavior, growth and corticosteroids in the juvenile female pig. *Hormones and Behavior* 15, 396–406.

Hemsworth, P.H., Gonyou, H. and Dzuik, P.J. (1986) Human communication with pigs: the behavioural response of pigs to specific human signals. *Applied Animal Behaviour Science* 15, 45–54.

Hemsworth, P.H., Barnett, J.L., Treacy, D. and Madgwick, P. (1990) The heritability of the trait fear of humans and the association between this trait and the subsequent reproductive performance of gilts. *Applied Animal Behaviour Science* 25, 85–95.

Hemsworth, P.H., Coleman, G.J., Cox, M. and Barnett, J.L. (1994) Stimulus generalization: the inability of pigs to discriminate between humans on the basis of their previous handling experience. *Applied Animal Behaviour Science* 40, 129–142.

Hemsworth, P.H., Price, E.O. and Borgwardt, R. (1996) Behavioural responses of domestic pigs and cattle to humans and novel stimuli. *Applied Animal Behaviour Science* 50, 43–56.

Horrell, I. and Hodgson, J. (1992) The bases of sow–piglet identification. 1. The identification by sows of their own piglets and the presence of intruders. *Applied Animal Behaviour Science* 33, 319–327.

Jago, J., Krohn, C.C. and Matthews, L.R. (1999) The influence of feeding and handling on the development of the human–animal interactions in young cattle. *Applied Animal Behaviour Science* 62, 137–152.

Jones, R.B. (1994) Regular handling and the domestic chicken's fear of human beings: generalisation of response. *Applied Animal Behaviour Science* 42, 129–143.

Jones, R.B. and Waddington, D. (1992) Modification of fear in domestic chicks, *Gallus gallus domesticus*, via regular handling and early environmental enrichment. *Animal Behaviour* 43, 1021–1033.

Kendrick, K.M. (1991) How the sheep's brain controls the visual recognition of animals and humans. *Journal of Animal Science* 69, 5008–5016.

Kendrick, K.M., Atkins, K., Hinton, M.R., Broad, K.D., Fabre-Nys, C. and Keverne, B. (1995) Facial and vocal discrimination in sheep. *Animal Behaviour* 49, 1665–1676.

Kendrick, K.M., Atkins, K., Hinton, M.R., Heavens, P. and Keverne, B. (1996) Are faces special for sheep? Evidence from facial and object discrimination learning tests showing effects of inversion and social familiarity. *Behaviour Processes* 38, 19–35.

Koba, Y. and Tanida, H. (1999) How do miniature pigs discriminate between people? The effect of exchanging cues between a non-handler and their familiar handler on discrimination. *Applied Animal Behaviour Science* 61, 239–252.

Korff, J. and Dyckhoff, B. (1997) Analysis of the human animal interaction demonstrated in sheep by using the model of 'social support'. In: Hemsworth, P., Spinka, M. and Kostal, L. (eds) *Proceedings of the 31st International Congress of the International Society for Applied Ethology 13–16 August 1997, Prague*. Research Institute of Animal Production, Prague, pp. 87–88.

Krohn, C.C. (1996) The effects of early housing and handling of dairy calves on the subsequent man–animal relationship. In: Duncan, I.J.H., Widowski, T. and Hale, D.B. (eds) *Proceedings of the 30th International Congress of the International Society for Applied Ethology*. Colonel K.L. Campbell Centre for the Study of Animal Welfare, University of Guelph, Guelph, Ontario, p. 109.

Lankin, V. (1997) Factors of diversity of domestic behaviour in sheep. *Genetics, Selection and Evolution* 29, 73–92.

Le Neindre, P., Poindron, P., Trillat, G. and Orgeur, P. (1993) Influence of breed on reactivity of sheep to humans. *Genetics, Selection and Evolution* 25, 447–458.

Lorenz, K. (1935) Der Kumpan in der Umwelt des Vogels. *Journal Fur Ornithologie* 83, 137–213 and 289–413.

Lott, D.F. and Hart, B.J. (1979) Applied ethology in cattle cultures. *Applied Animal Ethology* 5, 309–319.

Lyons, D.M. (1989) Individual differences in temperament of dairy goats and the inhibition of milk ejection. *Applied Animal Behaviour Science* 22, 269–282.

Lyons, D.M., Price, E.O. and Moberg, G.P. (1988) Individual differences in temperament of dairy goats: consistency and change. *Animal Behaviour* 36, 1323–1333.

Mal, M.E. and McCall, C.A. (1996) The influence of handling during different ages on a halter training test in foals. *Applied Animal Behaviour Science* 50, 115–120.

Mal, M.E., McCall, C.A., Cummins, K.A. and Newland, M.C. (1994) Influence of preweaning handling methods on post-weaning learning ability and manageability of foals. *Applied Animal Behaviour Science* 40, 187–195.

Markowitz, T.M., Dally, M.R., Gursky, K. and Price, E.O. (1998) Early handling increases lamb affinity for humans. *Animal Behaviour* 55, 573–587.

Mateo, J.M., Estep, D.Q. and McCann, J.S. (1991) Effects of differential handling on the behaviour of domestic ewes (*Ovis aries*). *Applied Animal Behaviour Science* 32, 45–54.

McConnell, P.B. (1990) Acoustic structure and receiver response in domestic dogs, *Canis familiaris*. *Animal Behaviour* 39, 897–904.

McConnell, P.B. and Baylis, J.R. (1985) Interspecific communication in cooperative herding: acoustic and visual signals from human shepherds and herding dogs. *Zeitschrift für Tierpsychologie* 67, 302–328.

Meese, G.B. and Baldwin, B.A. (1975) The effects of ablation of the olfactory bulbs on aggressive behaviour in pigs. *Applied Animal Ethology* 1, 251–262.

Munksgaard, L., de Passillé., A.M.B., Rushen, J., Thodberg, K. and Jensen, M.B. (1997) Discrimination of people by dairy cows based on handling. *Journal of Dairy Science* 80, 1106–1112.

Munksgaard, L., de Passillé, A.M.B., Rushen, J. and Ladewig, J. (1999) Dairy cows' use of colour cues to discriminate between people. *Applied Animal Behaviour Science* 65, 1–11.

Murphey, M.R. and Moura Duarte A.F. (1983) Calf control by voice command in a Brazilian dairy. *Applied Animal Ethology* 11, 7–18.

Murphey, R.M., Ruiz-Marinda, C.R. and Moura Duarte, F.A. (1990) Maternal recognition in Gyr (*Bos indicus*) calves. *Applied Animal Behaviour Science* 27, 183–191.

Murphy, L.B. and Duncan, I.J.H. (1977) Attempts to modify the responses of domestic fowl towards human beings. I. The association of human contact with a food reward. *Applied Animal Ethology* 3, 321–334.

Murphy, L.B. and Duncan, I.J.H. (1978) Attempts to modify the responses of domestic fowl towards human beings. II. The effect of early experience. *Applied Animal Ethology* 4, 5–12.

Pedersen, V. (1993) Effects of different post-weaning handling procedures on the later behaviour of silver foxes. *Applied Animal Behaviour Science* 37, 239–250.

Pedersen, V. and Jeppesen, L.L. (1990) Effects of early handling on later behaviour and stress responses in the silver fox (*Vulpes vulpes*). *Applied Animal Behaviour Science* 26, 383–393.

Podberscek, A.L., Blackshaw, J.K. and Beattie, A.W. (1991) The effects of repeated handling by familiar and unfamiliar people on rabbits in individual cages and group pens. *Applied Animal Behaviour Science* 28, 365–373.

Pollard, J.C., Littlejohn, R.P. and Webster, J.R. (1994) Quantification of temperament in weaned deer calves of two genotypes (*Cervus elaphus* and *Cervus elaphus × Elaphurus davidianus* hybrids). *Applied Animal Behaviour Science* 41, 229–242.

Price, E.O. (1998) Behavioral genetics and the process of animal domestication. In: Grandin T. (ed.) *Genetics and the Behavior of Domestic Animals*. CAB International, Wallingford, UK, pp. 31–65.

Price, E.O. and Wallach, S.J.R. (1990) Physical isolation of hand-reared Hereford bulls increases their aggressiveness toward humans. *Applied Animal Behaviour Science* 27, 263–267.

Randle, H.D. (1998) Facial hair whorl position and temperament in cattle. *Applied Animal Behaviour Science* 56, 139–147.

Rogers, L.J. (1995) *The Development of Brain and Behaviour in the Chicken*. CAB International, Wallingford, UK.

Rushen, J., de Passillé, A.M.B. and Munksgaard, L. (1999a) Fear of people by cows and effects on milk yield, behavior and heart rate at milking. *Journal of Dairy Science* 82, 720–727.

Rushen, J., Taylor, A.A. and de Passillé, A.M.B. (1999b) Domestic animals' fear of humans and its effect on their welfare. *Applied Animal Behaviour Science* 65, 285–303.

Rybarczyk, P., Rushen, J. and de Passillé, A.M.B. (1999) Recognition of people by dairy calves. In: Boe, K., Bakken, M. and Braastad, B. (eds) *Proceedings of the 33rd International Congress of the International Society for Applied Ethology.* Agricultural University of Norway, As, Norway, p. 167.

Sambraus, H.H. and Sambraus, D. (1975) Pragung von Nutztleren auf Menschen. *Zeitschrift für Tierpsychologie* 38, 1–17.

Seabrook, M.F. and Bartle, N.C. (1992) Environmental factors influencing the production and welfare of farm animals. In: Phillips, C.J.C. and Piggins, D. (eds) *Farm Animals and the Environment.* CAB International, Wallingford, UK, pp. 111–130.

Settle, R.H., Sommerville, B.A., McCormick, J. and Broom, D.M. (1994) Human scent matching using specially trained dogs. *Animal Behaviour* 48, 1443–1448.

Shillito Walser, E.E. (1986) How early can piglets recognize their sows voice? *Applied Animal Behaviour Science* 15, 177 (abstract).

Tanida, H. and Koba, Y. (1997) How do miniature pigs discriminate between people? The effect of exchanging cues between a stranger and their familiar handler on discrimination. In: Hemsworth, P., Spinka, M. and Kostal, L. (eds) *Proceedings of the 31st International Congress of the International Society for Applied Ethology 13–16 August 1997, Prague.* Research Institute of Animal Production, Prague, p. 225.

Tanida, H. and Nagano, Y. (1996) The ability of miniature pigs to distinguish between people based on previous handling. In: Duncan, I.J.H., Widowski, T. and Hale, D.B. (eds) *Proceedings of the 30th International Congress of the International Society for Applied Ethology.* Colonel K.L. Campbell Centre for the Study of Animal Welfare, University of Guelph, Guelph, Ontario, p. 143.

Tanida, H. and Nagano, Y. (1998) The ability of miniature pigs to discriminate between a stranger and their familiar handler. *Applied Animal Behaviour Science* 56, 149–159.

Tanida, H., Senda, K., Suzuki, S., Tanaka, T. and Yoshimoto, T. (1991) Color discrimination in weanling pigs. *Animal Science and Technology (Japan)* 62, 1029–1034.

Tanida, H., Miura, A., Tanaka, T. and Yoshimoto, T. (1994) The role of handling in communication between humans and weanling pigs. *Applied Animal Behaviour Science* 40, 219–228.

Tanida, H., Miura, A., Tanaka, T. and Yishimoto, T. (1995) Behavioural response to humans in individually handled weanling pigs. *Applied Animal Behaviour Science* 42, 249–260.

Taylor, A. and Davis, H. (1998) Individual humans as discriminative stimuli for cattle (*Bos taurus*). *Applied Animal Behaviour Science* 58, 13–21.

Tennessen, T. and Hudson, R.J. (1981) Traits relevant to the domestication of herbivores. *Applied Animal Ethology* 7, 87–102.

Topal, J., Miklosi, A., Csanyi, V. and Doka, A. (1998) Attachment behavior in dogs (*Canis familiaris*): a new application of Ainsworth's (1969) strange situation test. *Journal of Comparative Psychology* 112, 219–229.

Social Cognition of Farm Animals

14

Suzanne T. Millman[1] and Ian J.H. Duncan[2]

[1]Humane Society of the United States, 2100 L Street NW, Washington, DC, USA; [2]Department of Animal and Poultry Science, University of Guelph, Guelph, Ontario N1G 2W1, Canada

(Editors' comments: This book deals with the social environment of animals, be that the natural environment of their ancestors or the commercial conditions under which we now keep them. The trend throughout this book is how this environment affects the animal's behaviour. Millman and Duncan take up the question of how that environment is perceived by the animal and this takes them into the area of animal cognition. Cognition can be defined as the process by which an animal perceives, stores and processes information.

The chapter takes up in more detail many of the issues mentioned previously, such as recognition, learning and communication. Recognition is important in the formation and maintenance of dominance hierarchies within groups and in the mother–young bond. Learning in animals is beyond the scope of this book, but social learning and the fact that animals may learn differently depending on whom they are learning from are included. Communication has been taken up as a separate section in each species chapter, but whether animals can communicate emotional states such as fear and frustration is an important aspect of the welfare of farm animals and is discussed here. Most of the work in this area is relatively recent and researchers are still developing and refining the methods. Indeed, some may argue that we can never know another human's thoughts, let alone the thoughts of an animal. Yet attempts have to be made if we are to have any appreciation of how that individual experiences what we do to it in the course of its lifetime.)

In response to societal concerns regarding the welfare of farm animals, applied ethologists are attempting to answer questions about animal cognition and consciousness. Understanding how farm animals

©CAB *International* 2001. *Social Behaviour in Farm Animals*
(eds L.J. Keeling and H.W. Gonyou)

perceive and respond to their environments is necessary to speculate intelligently on the extent of their suffering or pleasure. Cognition is the area of study that straddles an animal's behaviour and its experience, involving mechanisms by which an animal perceives, stores and processes information. These processes need not be complex, and may be accomplished consciously or unconsciously. Since it is impossible to observe animal thoughts directly, some researchers consider topics of cognition and consciousness to be beyond the scope of scientific enquiry. However, others have devoted their careers to developing methods through which such information can be gathered indirectly. In this chapter, we will explore the topic of social cognition of farm animals and, to illustrate these issues, we will examine the topics of recognition, social learning and communication. First, to what extent do farm animals perceive, store and process information about their social environments? Second, to what extent are animals aware or conscious of their social environment? Finally, what are the implications of social cognition for the management of farm animals?

14.1 Recognition

14.1.1 Categories of recognition

Many people familiar with farm animals contend that they are capable of individual recognition. However, there are other categories of recognition that could explain some of their contentions without imputing individual recognition. For example, the intense fighting that frequently occurs when groups of unfamiliar animals are mixed at abattoirs and stockyards suggests that these animals can recognize others as being familiar or unfamiliar, one of the simpler categories of social recognition. The fact that, in a herd or flock, subordinate animals will generally avoid more dominant animals suggests their ability to recognize equally familiar animals as having previously been associated with positive or negative experiences, a slightly more difficult ability. The ability of individual recognition is suggested by the fact that individuals within a herd or flock have preferred social partners, towards whom they maintain close proximity and direct social grooming. Similarly, the fact that dam and offspring form strong social bonds and are able to recognize each other even within a large herd or flock indicates individual recognition. Individual recognition, or the ability to identify another animal as being a particular, unique individual, is considered to be the most complicated category of recognition requiring the most cognitive ability. Hence, it would appear that farm animals have the ability of social recognition, including individual recognition.

Recognition is the process whereby memory is revealed. An object or an event previously encountered by an animal is treated differently because of the previous encounter. Recognition involves perception of an object or an event and formation of a memory. Memory involves the strengthening or weakening of synaptic connectivities. According to Bindra (1976), 'remembering' does not consist of storage, but seems to involve the fresh production or reconstruction of neural pathways representing an object or an event.

The topics of perception and sensation are beyond the scope of this chapter, and will be outlined only briefly here (for a review, see Piggins and Phillips, 1998). An animal is aware of only a fraction of the information bombarding its sensory organs at any one time, and only a fraction of that information is memorized and is thus available for recognition. Information filtering occurs at the level of the receptor organs and centrally, in the cortical regions of the brain. Some aspects of information filtering are 'hard-wired' or resistant to change, some are 'plastic' or flexible to change, and others may be initially plastic, but become hard-wired during development. From studies with humans using functional brain imaging techniques, such as magnetic resonance imaging (MRI), many areas of the brain that are active when a person perceives an object are also active when the person recalls that object (for a review, see Kendrick, 1998). Attempts are being made to determine whether animals are similar to humans with respect to the anatomical areas of the brain that respond to objects. Should similarities exist, it is possible that animals also possess the ability to form a mental image. Neurobiological research has provided some evidence of recognition in animals. For example, cells in the infero-temporal cortex respond with differing activity when sheep are presented with visual images of faces (Kendrick, 1998). However, the inability of animals to communicate their mental images, using language or drawings, limits inferences that may be drawn from such studies. Measuring brain activity and other techniques of neurobiology can, in isolation, provide only limited information about what an animal retains from its environment and how it is interpreted.

A second approach to the study of cognition is the utilization of an animal's behaviour, whereby researchers examine how animals respond to social information presented to them. For example, operant conditioning techniques have been used to test farm animals on their ability to discriminate and categorize individuals (Fig. 14.1). When sheep were rewarded with food for distinguishing between two adult ewe faces, they were able to accomplish this task easily, and were able to distinguish at least ten individual sheep (Kendrick *et al.*, 1996). Similarly, the ability of animals to recognize familiar individuals has been examined by comparing their behaviour towards them with behaviour towards unfamiliar individuals.

14.1.2   Mechanisms of recognition

One must be cautious in interpreting behavioural responses to draw conclusions about their cognitive abilities, since responses by animals are dependent on the choices available to them. For example, some research indicated that hens were unable to discriminate between live individuals when they were presented in the arms of a Y-maze (Bradshaw, 1991) or a T-maze (Lindberg and Nicol, 1996) (Fig. 14.1). These results were surprising, since chickens form social hierarchies and appear to identify dominant and subordinate individuals within a flock. However, results of these experiments may have been affected by the methods with which hens were asked to discriminate. Dawkins and Woodington (1997) found that the ability of hens to discriminate between two objects was strongly dependent on the distance at which the objects were presented. Hens could discriminate between two objects, and even between photographs of these objects, when they were presented at a distance of 5–25 cm, but this ability diminished when hens were required to choose from 120 cm. Similarly, hens were able to recognize a familiar flockmate if they could approach within

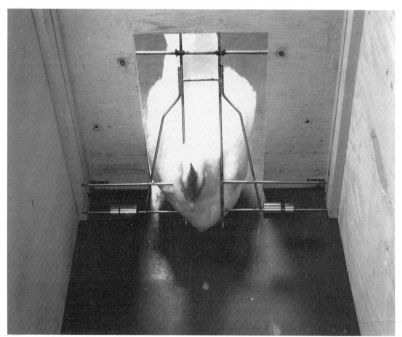

Fig. 14.1.   By indirectly measuring an animal's preferences, scientists gain insight into how animals think. In this experiment a hen is pushing though a weighted door to reach a social partner. By increasing the amount of weight, scientists can 'ask' her about the importance of social contact.

30 cm to inspect the head and comb area, and they were able to recognize familiar flockmates from life-size photographs, often pecking at the comb on the photograph (Dawkins, 1995, 1996) (see Section 7.2.3).

Animals may be trained to perform discrimination tasks, but this does not tell us whether recognition is important outside the laboratory and, if so, what mechanisms might be used to accomplish it. Although Kendrick *et al.* (1996) found that ewes could be trained to recognize their lambs using visual cues in a Y-maze, they took several weeks to accomplish this task. This seems surprising when one considers the strength of the maternal bond in sheep (see Section 8.1.6). However, sheep are a 'follower' species, in which lambs maintain close proximity to their dams, and these conditions may be more conducive to other modalities for recognition, such as olfaction (Fig. 14.2). Indeed, sheep can discriminate between individuals based on odours from saliva, wool, droppings and inter-digital secretions (Baldwin and

Fig. 14.2. Although it is likely that multiple sensory modalities are utilized in natural conditions, ewes are able to recognize their own lambs using only visual cues, olfactory or auditory cues.

S.T. Millman and I.J.H. Duncan

Meese, 1977). Recognition of lambs by maternal ewes is probably based on olfactory cues, emitted from the wool and skin of the lamb (Alexander and Stevens, 1981), as anosmic ewes do not reject nursing attempts of alien lambs (Baldwin and Shillito, 1974). Vocal cues appear to be important in maternal recognition by offspring, since young calves spent more time next to speakers emitting playbacks of vocalizations by their own dams than by other cows (Barfield *et al.*, 1994). Although sonograms indicate individual differences in maternal calls by cows (Barfield *et al.*, 1994) and by broody hens (Kent, 1989), there is no clear evidence that these differences are important for maternal recognition by their offspring. Kent (1993) found that chicks preferred maternal clucks of higher frequencies, suggesting that chicks may be responding to information about arousal or context, rather than identity.

Given that animals are able to discriminate between individuals, do they form a mental image of that animal? 'Recall', defined as the ability to form a mental image of an object in its absence, is one measure of complex cognitive abilities. Of course, recall is not directly accessible to scientific investigation and evidence is usually indirect, such as being able to describe or draw the image. This indirect evidence is thus limited to human beings. However, there is some indication that other species may possess this ability in response to social stimuli. When cocks were presented with photographic slides of other cocks, they were able to recognize individuals, even from views of them never seen before (Ryan, 1982). Similarly, sheep that were rewarded for responding to an image of the frontal view of a familiar individual also responded to a profile view of the same individual without further training (Kendrick, Leigh, Hinton, Pierce and da Costa, unpublished data, cited in Kendrick, 1998). However, this is not conclusive evidence that these animals have recall. Furthermore, there is no evidence that cattle or sheep recognize themselves when presented with a mirror image (Franklin and Hutson, 1982; Piller *et al.*, 1999).

Although it would appear that farm animals are capable of recognizing and categorizing stimuli when presented to them, as yet there are no techniques to determine if mental images of social companions are formed in their absence. Such information could be important to understand suffering that arises from deprivation of contact with closely bonded social companions (see Chapter 11). However, suffering could also occur without recall. At this time there is no evidence to support the notion that farm animals think about their companions, but they may think about, and hence anticipate, social interactions in a general sense. Van Kampen (1997) suggests that male jungle fowl anticipate social interactions with females, since males performed food calls in the absence of females at locations where the male had encountered females previously. As researchers begin to tackle difficult subjective issues, such as the existence of boredom in farm animals (Wemelsfelder, 1997), techniques may be devised to answer the nature of feelings that animals possess for each other.

14.1.3   Classifying individuals

If animals are capable of recognizing individuals and placing them in particular categories, what do these categories look like? First, one would expect farm animals to categorize individuals into broad categories of species, gender and kin, because of their effects on reproductive fitness. Lill (1968c) found that hens in mixed groups of Brown and White Leghorn strains tended to avoid birds of the other strain, and directed agonistic behaviour towards individuals of their own strain. Hens and cockerels also preferred to direct their sexual behaviour towards individuals of their own strain (Lill, 1968a, b). Similarly, Evans and Marler (1992) found that hen appearance markedly affected duration of courtship displays by cockerels, with the strongest response evoked by hens of the cockerel's own strain. These classifications seem to be context dependent, since hen appearance did not affect alarm calling by cockerels, in response to an aerial predator.

Categorizing individuals according to familiarity and social status is important for social cohesion and to decrease aggression within a group. Individuals alter their behaviour according to whether social companions are familiar or unfamiliar (Grigor *et al.*, 1995; Boissy and Le Neindre, 1997). They also discriminate individuals according to social status (Hogue *et al.*, 1996; Nicol and Pope, 1998). Some species also categorize individuals as kin and non-kin. For example, golden hamsters display less flank and vaginal marking in response to flank odours from their brothers than they do to odours from their non-brothers, even when siblings were cross-fostered and hence unfamiliar (Heth *et al.*, 1998). However, a recent study found that pigs behaved as aggressively towards kin as towards non-kin when groups were mixed (Stookey and Gonyou, 1998). However, whether farm animals alter their behaviour towards individuals from different categories depends on a variety of motivational factors, based on their physiological states and the novelty or complexity of the environment. For example, ewes presented with images of familiar male or female sheep in a Y-maze preferred to approach the male when they were in oestrus, but preferred to approach the female when in anoestrus (Kendrick *et al.*, 1995, 1996). Similarly, ewes are responsive to the odour of a newborn lamb for only a few hours, following stimulation of the vagina and cervix, which occurs at parturition (Kendrick *et al.*, 1991). Some of these factors will be discussed further below in Sections 14.2 and 14.3.

14.2   Social Learning

The social environment is known to have an impact on learning in
animals (for review see Nicol, 1995). The presence of companions may
aid trial-and-error learning in an individual by decreasing fear or
arousal. For example, Boissy and Le Neindre (1990) found that Aubrac
heifers learned an operant task faster when tested with other heifers
than when tested alone. Aubrac heifers were shown to exhibit greater
exploratory behaviour when grouped than when isolated (Veissier
and Le Neindre, 1992). In addition, the behaviour of companions
may increase the likelihood that the same behaviour will be performed
by an animal in that environment as a result of social facilitation
or contagious behaviour. The presence of penmates that were feeding
stimulated feeding behaviour in chickens (Keeling and Hurnik, 1993,
1996a) and pigs (Keeling and Hurnik, 1996b). However, from the stand-
point of social cognition, it is more interesting to examine situations in
which a naive animal is provided with a knowledgeable demonstrator
and adjusts its behaviour accordingly. The demonstrator may direct the
performance of behaviour by making a particular stimulus or location
more conspicuous within the animal's environment through processes
of stimulus enhancement or location enhancement. In a more complex
form of social learning, the exact motor patterns of a demonstrator
may be copied, through the process of imitation. These forms of social
learning suggest fairly high levels of cognitive ability in the observing
animal. They indicate that the observing animal is forming a concept
about the behaviour of the demonstrator and extrapolating from this to
the observer's own behaviour.

It is not possible to know precisely what information an individual
obtains as a result of observing or interacting with social companions.
As a result, research in the area of social learning relies on indirect
measurements. However, inferences may be drawn by designing
experiments in which individuals act on such information. Although
social learning may have an important impact on management and
welfare of farm animals, this is a new area of research for applied
ethologists. Hence, we will draw from studies using other species to
explain the mechanisms of social learning, where information on farm
animals is lacking.

14.2.1   Developing new skills

Social learning is an important mechanism through which an animal
can expand its repertoire of behaviour, developing new skills and
motor patterns to solve problems. Chesler (1969) found that, when
kittens were provided with trained demonstrators, they were able to

acquire skills necessary for an operant task, whereas kittens without a demonstrator were unable to learn the task. Similarly, Slabbert and Rasa (1997) found that working dogs' pups that observed their mothers locate and retrieve narcotics performed the task significantly better during training at 6 months of age, even though they had received no reinforcement during the observation period. This type of observational learning may have application when farm animals have to learn the use of new technologies, such as automatic milkers and feeders.

The effect of a demonstrator on learning of a task has been examined using farm animal species. Nicol and Pope (1993) found that trained demonstrators facilitated learning of a key-peck response by laying hens. Since hens were provided with untrained demonstrators in the control treatment, the improvement in their performance can be attributed to transmission of information, and not simply to the presence of a social companion. Conversely, Veissier (1993) found that heifers provided with a trained demonstrator did not learn an operant task faster. However, heifers spent more time near the training box when they were provided with a demonstrator. Similarly, Nicol and Pope (1994a) found that provision of a demonstrator did not improve the ability of young pigs to learn an operant task, but time spent facing the operant panels and the number of unrewarded nose presses were significantly greater when pigs had a demonstrator. It is not clear if these results reflect genuine species differences in social effects and cognitive abilities or differences in experimental procedures.

Social transmission of undesirable behaviour patterns such as cribbing is a particular concern of horse owners (McGreevy *et al.*, 1995), but confounding environmental influences make research in the long-term development of stereotypies difficult. However, there is some evidence that observational learning may be involved in the development of stereotypies. Cooper and Nicol (1994) found that voles developed stereotypies earlier when housed next to stereotyping voles, and voles housed adjacent to somersaulting demonstrators were more likely to develop that variation of stereotypy, suggesting that imitation may have been involved.

14.2.2   Refining search skills

Social learning may also affect the way in which an animal gains access to resources. For example, naive rats learned that food had become available at a familiar feeding site after they interacted with a colony member who had recently eaten there (Galef and White, 1997). The rats were also able to determine the nature of the food available at the site, since individuals that were conditioned to avoid that food would not travel to the feeding site after interacting with the demonstrators.

Similarly, adolescent jungle fowl showed a significant preference for a location within an enclosure and for a type of dish after they observed successful foraging by demonstrators (McQuoid and Galef, 1992, 1993). Veissier and Stefanova (1993) found that groups of lambs that were reared with an older experienced lamb learned to suck from a teat bucket faster than groups without a demonstrator. However, demonstrators did not improve the performance of horses learning a food location task (Baer *et al.*, 1983; Baker and Crawford, 1986), a stimulus discrimination task (Clarke *et al.*, 1996) or an operant task for a food reward (Lindberg *et al.*, 1998).

14.2.3   Influencing preferences

Social learning may have the greatest importance in managing farm animal species with respect to the influence of a demonstrator on the decisions an animal makes. Animals may learn what foods to avoid through their interactions with social companions. Johnston *et al.* (1998) found that day-old chicks avoided pecking a bead coated with a bitter-tasting fluid after observing strong aversive reactions by another chick. Chicks observing another chick pecking at a water-coated bead did not show this aversion. Similarly, lambs have been shown to avoid a palatable food after observing the aversion of their mothers (Mirza and Provenza, 1990; Thorhallsdottir *et al.*, 1990a).

Social companions may also influence the decisions an animal makes by demonstrating choices available. Animals tend to be neophobic, particularly with respect to food items. Although one characteristic of domestication is a willingness to consume novel foods, this may be a result of management. Animals may be more apt to consume a novel food item after they have observed a social companion consuming it. Chapple *et al.* (1987) found that weaned Merino lambs did not learn to eat wheat during 5 days of exposure, whereas lambs provided with experienced wheat-eating demonstrators rapidly learned to consume wheat. Similarly, lambs provided with experienced barley-eating social models consumed more barley than did controls (Thorhallsdottir *et al.*, 1990a).

In addition to shaping responses towards novel foods, interactions with social companions may strengthen or reverse associations that farm animals have formed. Thorhallsdottir *et al.* (1987) found that lambs with ewes learned to avoid a novel food following poisoning with lithium chloride, whereas pairs of orphan lambs did not. Ewes displayed a stronger aversion towards the novel food than did other lambs, and responses by their dams may have increased the persistence of aversion by lambs tested with ewes. However, social effects on conditioned responses may differ between species and between types of stimuli. For example, Galef and White (1997) found that rats conditioned to avoid a food would not travel to a feeding site after interacting with a colony member who had recently eaten the food, while non-averted rats did. However, another study (Galef *et al.*, 1997) found that rats averted to a flavoured fluid increased their intake of this fluid after interacting with a demonstrator that had drunk it. Rats also showed a greater motivation to seek out this flavour in a T-maze. Conditioned aversions have been extinguished following exposure to non-averted social companions in cattle (Ralphs and Olsen, 1990; Ralphs, 1996) and sheep (Thorhallsdottir *et al.*, 1990c; Provenza and Burritt, 1991), which may be important where naive individuals are introduced to grazing herds or flocks (Fig. 14.3).

In addition to effects of feeding behaviour, observational learning may affect preferences such as mate choice. Galef and White (1998) found that mate choice in quail could be reversed in females after they observed another female mating with an unpreferred male. Copying of mate choice has not been studied in other farm animal species, although mate selection based on observational learning could have an impact on production systems in which large groups are maintained and random mating assumed.

Fig. 14.3. Animals may learn to consume some foods and to avoid others by observing and interacting with their social companions. Animals particularly attend to information from social companions when they are exposed to unfamiliar foods or environments.

14.2.4   Factors involved in social learning

From principles of behavioural ecology (see Chapter 1), one would expect social learning to be a particularly important mechanism for acquiring information and skills by species that interact socially. However, these presumptions may be misleading, and may reflect differences in opportunity for social learning, rather than differences in cognitive abilities. For example, Zenaida doves are territorial and have a solitary lifestyle. However, these doves were equally adept at acquiring information from a demonstrator to find food, compared with feral pigeons that live in flocks and are gregarious (Lefebvre et al., 1998). In fact, animals can acquire information, not only from conspecifics, but also from individuals from other species, when there is opportunity to do so. For example, turkey chicks sometimes fail to begin feeding post-hatch and eventually die from 'starve-out'. Savory (1982) found that adding some broiler chicks to groups of turkey chicks stimulated turkey chicks to consume more food than groups of turkeys chicks on their own. It is not clear if turkey chicks acquired information by observing the broiler chicks, or whether they were stimulated to explore by the feeding behaviour of broiler chicks. However, this information may be important where different stocks are reared together, and where certain behaviour patterns may be adaptive, such as response to predators or shelter-seeking behaviour in range cattle and sheep.

Individuals may obtain information about the environment from social companions, but they must also balance this with information obtained from personal experience. For example, information from social companions has not been found to override sensory information. When lambs were prevented from receiving olfactory, visual and auditory sensory information about a novel food, they would not consume it, even when they could observe their social companions doing so (Chapple et al., 1987). However, there may be thresholds involved, since lambs did consume the novel food in the presence of other lambs eating it, when only one sensory system was blocked.

Information from social companions probably takes on greater significance when animals are processing information about unfamiliar environments. Galef (1993) found that demonstrator rats influenced the food preferences of observers more when eating unfamiliar than familiar foods. Similarly, chicks were found to drink from a familiar coloured tube when available and preferred a tube of the same colour as a demonstrator was drinking from only when provided with two unfamiliar drinking tubes (Franchina et al., 1986). Ralphs and Olsen (1990) found that, although aversion to larkspur was extinguished in conditioned heifers when grazing with non-averted heifers on larkspur-infested rangeland, the aversion was renewed when the heifers were returned to the situation where aversion was created. Similarly, Scott *et al.* (1996) found that, when subgroups of lambs were familiar with a pasture, they consumed different foods according to prior conditioned preferences. However, when subgroups naive or familiar to the pasture were grazed together, familiar lambs retained their conditioned food preference while naive lambs acquired preferences for both foods. Because of effects of familiarity with the environment, experiments set up in a laboratory may more successfully demonstrate the existence of social learning, and run the risk of overemphasizing the importance of social learning in natural or commercial environments.

Social influences may have a greater effect during critical periods, most notably in the young animal. Persistence of food aversion was greater in lambs conditioned at 6 weeks of age compared with 12 weeks (Mirza and Provenza, 1990). Thorhallsdottir *et al.* (1990b) found food aversion persisted for 3 months post-weaning when lambs were conditioned at 8 weeks of age, whereas lambs in a similar experiment conditioned at 11 weeks of age did not show aversion post-weaning. It is not known if these differences reflect differences in memory retention or reflect a difference in the influence of social companions at different times.

Food deprivation might be expected to increase attentiveness of observers to a demonstrator's performance. However, Nicol and Pope (1993) found that food-deprived laying hens were less discriminatory in a key-peck response when provided with a trained demonstrator. It may be that success of a demonstrator actually distracts hungry observing individuals and inhibits learning. Interestingly, learning may be facilitated most when animals are able to observe the mistakes of a demonstrator. Johnston *et al.* (1998) found that chicks only learned to avoid a bitter-tasting fluid after observing another chick during training when mistakes occurred and the demonstrator showed aversion, as opposed to during testing. Naive starlings provided with a demonstrator making only incorrect responses performed significantly better at an operant task than those observing demonstrators making only correct responses (Templeton, 1998). When demonstrators made only incorrect responses observers were more likely to pick the opposite (correct) response than when both correct and incorrect responses were observed. Templeton suggests that learning to avoid may be primary to learning to copy a response.

Information from social companions may be filtered according to the identity of the demonstrator. Maternal influence appears to have the greatest effect on the social transmission of information in many species, and may be mediated through familiarity, attentiveness or discrimination based on expectation of information quality. Flavour preferences in the lamb have been shown to be affected by postnatal exposure to flavoured milk (Nolte and Provenza, 1991), and also by prenatal exposure to flavours from their dam's feed, which may be transmitted through the amniotic fluid (Schaal *et al.*, 1995). Chesler (1969) found that kittens observing their mothers as demonstrators in an operant task acquired and discriminated the response sooner than those with a strange female cat demonstrator. Similarly, while lambs learned to avoid pellets treated with lithium chloride after observing their mothers eat and then avoid them, lambs did not develop this aversion when provided with a non-lactating ewe as a social model (Thorhallsdottir *et al.*, 1990a). Strong maternal influences on behaviour may be due to greater familiarity. However, Boissy and Le Neindre (1990) found that degree of familiarity between heifers did not affect social learning of an operant task. Tutoring may be involved in some relationships, since domestic hens increase their rate of ground pecking and scratching when observing chicks eating unpalatable food (Nicol and Pope, 1996). As an individual matures, it is likely that maternal influence decreases relative to that of social peers (for review see Veissier *et al.*, 1998).

Animals may discriminate between demonstrators, based on social status. Galef and Whiskin (1998) found that age of demonstrator did not have an effect on social influence of food choices of young rats. However, Nicol and Pope (1994b) found that laying hens learned a key-peck response to obtain food faster when a dominant hen was used as a demonstrator, even though dominant and subordinate demonstrators did not differ in their key-pecking rate or accuracy. This could be due to an increased attentiveness to a dominant hen, but some information filtering is probably occurring, since hens observing a cockerel demonstrator performed very few operant or general pecks (Nicol and Pope, 1998). Cockerels are dominant to females (Guhl, 1949). However, cockerels rarely show aggression to females (Kruijt, 1964), and it is possible that hens would be less attentive to them despite their dominant status. In addition, cockerels use food-calling during courtship of females which may affect both the reliability of information transmitted by the demonstrator (Marler *et al.*, 1986a) and the motivation of the observing hen (Van Kampen, 1994).

14.3   Communication

Exploring the complexities of communication (see Section 2.2) has the potential to provide important information for our understanding of social cognition. In some ways, communication is similar to the process of recognition, since an animal does not broadcast, or attend to, all of the information available to it. Whereas social learning occurs when animals use other individuals as a resource for information, communication occurs when information is shared between individuals. However, the mechanisms of communication may be similar to those for social learning, since an animal balances information provided by another individual with its own experiences.

Animals gather information using all of their sensory modalities and hence it is likely that modes of communication exploit all these senses. However, the majority of research into animal communication has focused on signals involving visual displays and vocalizations, since these signals are easily perceived by and manipulated by researchers. By examining the variety and context of signals that farm animals transmit and the responses of animals receiving these signals, researchers are attempting to tease apart the complexity of animal communication. Studies of animal communication are relevant to helping our understanding of how farm animals relate to each other, and they may also provide tools by which internal states, such as suffering and contentment, may be comprehended.

14.3.1   Maintaining group cohesion

One function of communication is to maintain group cohesion, which is necessary for individuals to coordinate their activities. There is some evidence of communication between an incubating hen and her developing embryos and between the developing embryos themselves, which helps to synchronize the hatching of these embryos. When all the sounds generated by incubating hens from the 15th day of incubation and all the sounds generated by chick embryos from the 18th day of incubation were played into an incubator during artificial incubation, hatching was significantly more synchronized compared with eggs incubated without sounds (Greenlees, 1993). Interestingly, Greenlees (1993) found that, after hatching, chicks exposed to maternal sounds during incubation approached maternal feeding calls more quickly than did control chicks not exposed to those sounds during incubation. The two treatment groups did not differ in their responses to white noise or alarm calls. Panning (1998) described another example of communication maintaining group cohesion. She showed that the presence of a mother turkey hen, surprisingly, had no effect on stimulating or synchronizing turkey chicks to feed or drink. However, her presence had a major effect on synchronizing rest in the chicks, and this led to a rhythm with all chicks in the brood being active or quiescent together.

14.3.2   Communicating internal states

Due to concerns for animal welfare, there has been interest in under-standing how farm animals communicate their internal states, such as frustration, fear and distress, which may act as indicators of suffering (Fig. 14.4). Vocalizations have been proposed as a method to assess the suffering and pain of cattle in slaughter plants (Grandin, 1998) and during hot-iron branding (Schwartzkopf-Genswein *et al.*, 1998). Similarly, vocalizations by swine have been used as a measure of excitability (von Borell and Ladewig, 1992) and of pain at castration (Weary *et al.*, 1998).

Although vocalizations may increase as a correlation with a painful procedure, are these calls involuntary reactions to pain, or do they convey information to an audience? Recent studies indicate that distress vocalizations do indicate honestly the level of need, at least with respect to young animals. Weary and Fraser (1995) found that, when piglets were isolated, those which were of low body weight, slow weight gain or had missed a milk ejection called for longer and at a higher frequency than did heavy piglets with good weight gains and fed piglets. Similarly, piglets isolated at 14°C called for longer duration and at higher frequencies than did piglets isolated at 30°C (Weary *et al.*, 1997). Pitch and frequency of calls may reflect distress due to pain, since significantly more high-frequency calls were given by piglets during castration than by piglets that were only restrained and sham castrated (Weary *et al.*, 1998).

There is some suggestion that distress calls may be intended for a particular audience, since piglets were found to increase their rate of calling twofold when provided with a playback of sow calls (Weary

Fig. 14.4. Social communication may provide information on the suffering and pleasure experienced by farm animals. For example, analysis of vocalizations during elective surgeries may give valuable information on the amount of pain that animals are experiencing.

et al., 1997). If piglets are accurately transmitting information about their level of need through their vocalizations and direct their calls towards a specific audience, then one would expect a difference in the responses of sows. Weary *et al.* (1996) found that sows showed a stronger response to playback calls of needy piglets that called at a higher frequency, higher rate and longer duration. Sows also showed a greater response to calls from their own piglets. Hormonal status also appears to influence the responsiveness of sows to distress calls by piglets, since responsiveness is strongest immediately following parturition (Hutson *et al.*, 1992).

In addition to expressing distress, animals may express other emotional states, and may influence the behaviour of others by doing so. For example, the gakel-call of domestic fowl may reflect frustration (Zimmerman and Koene, 1998). It seems likely that these signals are intended as information directed at other individuals, since domestic fowl frustrated individually do not give the gakel-call (Duncan, 1970).

14.3.3 Communicating information about the environment

Investigation of alarm calling by domestic fowl indicates that at least some farm animals have the capacity of communicating specific information to their social companions, and suggests the rudiments of language. Cockerels produce alarm calls to indicate the presence of predators, with a cackle-type alarm call produced in response to ground predators and a scream-type alarm call produced in response to aerial predators (Gyger *et al.*, 1987). Cockerels also respond with the appropriate alarm calls and non-vocal responses of crouching and visual fixation from video images (Evans *et al.*, 1993a) and computer-generated images (Evans *et al.*, 1993b) of potential predators. In addition to sharing information about the type of predator detected, cockerels appear to calibrate the level of threat that these potential predators pose, since faster moving images evoke a significant increase in calling.

It appears that males call for an intended audience, since few alarm calls are produced when males are alone or with another male (Karakashian *et al.*, 1988). However, the audience towards which cocks produce alarm calls is very general, since alarm calling by cockerels is not affected by the familiarity (Karakashian *et al.*, 1988) or the appearance (Evans and Marler, 1992) of hens. However, fewer alarm calls are produced for an audience of female quail, indicating a species effect.

It also appears that specific information is conveyed with these alarm calls, since hens respond to playbacks of aerial-type alarm calls by running towards an area of cover, and they adopt an erect and vigilant posture in response to playbacks of ground-type alarm calls (Evans *et al.*, 1993a). Although specific information appears to be transmitted by them, alarm calls can represent only the most basic elements of language. For example, there is no evidence that cocks invent new calls to communicate information about novel predators or situations.

14.3.4 Influencing the behaviour of others

Another example of animals conveying specific information about the environment is the production of food calls by domestic fowl. Both cockerels and hens produce repetitious single-note calls when they find food items, and the rate of calling increases with food quality (Marler *et al.*, 1986a). Moffatt and Hogan (1992) found that jungle fowl chicks approached a goal box faster in response to playbacks of a maternal food call that indicated high-quality food. Chicks seemed to possess an innate responsiveness to maternal food calls, but expectations were reversed when calls indicating low food quality were paired with preferred food items. Since hens change their rate of ground pecking and ground scratching when they observe chicks feeding on low-quality food (Nicol and Pope, 1996), it appears that hens communicate to an intended audience and are deliberately attempting to alter the behaviour of their chicks. If animals produce food calls to alter the behaviour of another individual, this suggests that they possess complex cognitive capabilities. Animals would need to form an abstract idea of the desired behaviour, and puzzle through different ways to elicit this behaviour in another individual.

Cockerels also seem to produce food calls for an intended audience, since they eat the food items and do not call when they are alone or in the presence of other cockerels (Marler *et al.*, 1986b). Cockerels also vary their calls according to food quality, and hens are more likely to approach a cockerel that is producing calls indicative of high, than low, food quality (Marler *et al.*, 1986a). However, cockerels also produce food calls for non-edible items, raising the possibility that cockerels deliberately attempt to deceive hens. Gyger *et al.* (1988) found that, when cockerels called for preferred food items, females approached 86% of the time and usually ate the food item. Conversely, hens only approached in 29% of the cases where food was not visible to them. Cockerels were also more likely to produce food calls in the absence of food if hens were far away and when cockerels were placed with unfamiliar hens (Marler *et al.*, 1986b). It is possible that cockerels produce food calls with the intention of establishing contact with hens.

Although food calls may provide information about food quality, it is not clear if this information is important to hens. Food calling also occurs in the context of courtship, and it is difficult to interpret why hens approach calling males. There is some evidence that hens expect food when they hear food calls by cockerels, since they scratched and pecked at the ground in front of cockerels when they responded to food calls (Gyger *et al.*, 1988). Furthermore, hens responded to playbacks of food calls by looking downward, rather than looking or moving towards the speaker, and they did not respond in this manner to playbacks of alarm or contact calls by cockerels (Evans and Evans, 1999). However, Nicol and Pope (1998) found that cockerels were not effective demonstrators for hens that were learning an operant task using food rewards, which suggests hens do not expect cockerels to provide reliable information about food. If cockerels are deliberately deceiving hens about food items to elicit approach, hens must form an expectation of food based on these calls. Van Kampen (1994) suggests that sexual or exploratory motivational systems cause hens to approach food-calling cockerels, since he found that hunger did not affect hen responses to food calls. Moreover, the hens that approached a food-calling cockerel were the ones more likely to produce a sexual crouch. Hence, while it appears likely that food-calling cockerels intend to influence the behaviour of hens, it is difficult to establish whether deception occurs.

14.4   Implications for Animal Production

Applied ethologists are only beginning to explore the cognitive abilities of farm animals. Although there is evidence that suggests complex cognitive abilities in some species, there has been a tendency for topics to be explored in only particular species. For example, most of the exploration of maternal–offspring recognition has involved sheep (e.g. Shillito-Walser, 1980). Similarly, vocal communication has focused on chickens (e.g. Marler *et al.*, 1986a,b; Evans and Marler, 1991, 1992, 1994; Evans *et al.*, 1993a,b) and, more recently, piglets (Weary and Fraser, 1995; Weary *et al.*, 1996, 1997, 1998). Although this approach has yielded a good understanding of cognitive mechanisms within these species, it may not be appropriate to extrapolate these findings to other species, and there is a need for information on all domestic species if we are to understand the implications for their well-being.

Understanding of social cognition may be useful for animal production in the following ways. First, recognition and processing of social information may be important to understand and manage social groups effectively, as well as to set limits on group sizes. Second, it may be possible to put social learning to use in situations in which animals are expected to feed from automated feeders, enter automated milking systems and use passageways to gain access to weighing-scales and other facilities. The great animal welfare benefit of these systems is that the animals can have control over when to visit the various facilities. If the animal can also learn how to use them from flock- or herd-mates without being driven by human caretakers, then husbandry systems will become kinder and gentler. Third, a more sophisticated understanding of animal communication may provide more detailed information about negative and positive emotional states that our farm animals experience. We will then be better able to assess their welfare.

The social environment has the potential to both improve and reduce animal welfare. Social interactions may alleviate boredom, decrease fear and provide pleasure through play or mutual grooming (Fig. 14.5). However, if groups are not managed properly, aggression and hysteria can cause injuries and anxiety in particular individuals. A number of researchers have used social stimuli to reduce fear in novel situations, such as handling and transportation. For example, since cattle and sheep react to a mirror image as if it were an unfamiliar individual, mirrors incorporated into handling restraint chutes appear to decrease fear that arises from social isolation (Piller *et al.*, 1999).

Individual recognition may occur, but how important is it in large, homogeneous flocks in commercial conditions? Hughes *et al.* (1997) found little evidence of individual recognition in large flocks and hypothesized that birds generalize their social status. Chickens roamed throughout the barn, and did not form territories in a large flock. However, pair bonds between particular individuals could still occur in these situations, and it would be interesting to track the movement of such individuals relative to each other in large flocks. If pair bonding occurs commonly in commercial conditions, then perhaps we should be making use of it to reduce stressful situations. For example, perhaps during transportation every effort should be made to transport bonded animals together. When sows farrow, perhaps bonded sows should

Fig. 14.5. Within a group, animals often form special associations with certain other individuals and spend more time close to them than they do to the rest of the group. To date, there has been little research into the effects of these associations.

be given neighbouring farrowing crates and eventually returned to the social group together. Similarly, perhaps closely bonded dairy replacement heifers should be introduced to the milking herd together and then dried off together at the end of their lactations. When we know more about friendly associations between animals, all these manipulations might be possible and beneficial.

Certain husbandry procedures might also benefit from knowledge of animals' social cognition. For example, animal scientists have already explored the stimuli necessary to elicit sexual behaviour in male farm animals, and have utilized this information in the design of mounting dummies for semen collection. However, there may be other steps that could be taken to facilitate the whole of the artificial insemination process. For example, semen collection from turkey toms and domestic cocks and insemination of the females still involve restraint with its accompanying stress. If we really understood what recognition processes occur between the male and female during natural sexual activity, then we might be able to arrange for a much less stressful artificial process.

It has long been recognized that social learning occurs in farm animals. In flocks of hill sheep and herds of range cattle, with fairly stable memberships, there is transmission to new recruits of information such as location of water sources and shelters. However, apart from the use of a 'Judas sheep' or 'Judas goat' to lead naive animals at the abattoir, there has been little application of social learning. With the ever-increasing use of technology in animal production, the potential would seem to be great. For example, as previously mentioned, there would seem to be lots of opportunities to use demonstrators to teach naive animals how to use automated feeders, milkers and other facilities. It might even be possible to use cross-species demonstrators to overcome particular problems, such as using broiler chicks to stimulate feeding by turkey chicks, as discussed previously.

It might also be possible to use the vocalizations of animals to make more accurate assessment of their welfare. This is currently being done to assess painful procedures, such as castration (Weary *et al.*, 1998). However, there may also be vocalizations indicating pleasure or contentment. Might it be possible to calm down animals during potentially stressful situations, such as shearing time for sheep, by broadcasting such contentment vocalizations? Could we encourage synchronized rest in poultry chicks by broadcasting vocalizations normally given by the broody hen to elicit a bout of brooding?

Animal science has made huge advances in the last 50 years. We now have comprehensive knowledge about the health, nutritional, physiological, environmental and behavioural needs of our farm animals. The final frontier will be in understanding their cognitive processes and, because they are social species, particularly their social cognition. As has been said many times, we will never know exactly what animals are thinking, but indirect probes are being developed by which we can make informed guesses about their cognitive processes. We are just scratching the surface at the moment. Only by building up a substantial body of knowledge in the areas of social cognition can we have a complete picture of farm animal biology. Only by having a complete picture can we give our farm animals the care and management that they require and deserve.

References

Alexander, G. and Stevens, D. (1981) Recognition of washed lambs by merino ewes. *Applied Animal Ethology* 7, 77–86.

Baker, A.E.M. and Crawford, B.H. (1986) Observational learning in horses. *Applied Animal Behaviour Science* 15, 7–13.

Baldwin, B.A. and Meese, G.B. (1977) The ability of sheep to distinguish between conspecifics by means of olfaction. *Physiology and Behavior* 18, 803–808.

Baldwin, B.A. and Shillito, E.E. (1974) The effects of ablation of the olfactory bulbs on parturition and maternal behaviour in Soay sheep. *Animal Behaviour* 22, 220–223.

Barfield, C.H., Tang-Martinez, Z. and Trainer, J.M. (1994) Domestic calves (*Bos taurus*) recognize their own mothers by auditory cues. *Ethology* 97, 27–264.

Bindra, D. (1976) *A Theory of Intelligent Behavior*. John Wiley & Sons, New York.

Boissy, A. and Le Neindre, P. (1990) Social influences on the reactivity of heifers: implications for learning abilities in operant conditioning. *Applied Animal Behaviour Science* 25, 149–165.

Boissy, A. and Le Neindre, P. (1997) Behavioral, cardiac and cortisol responses to brief separation and reunion in cattle. *Physiology and Behavior* 61, 693–699.

Bradshaw, R.H. (1991) Discrimination of group members by laying hens. *Behavioural Processes* 24, 143–151.

Chapple, R.S., Wodzicka-Tomaszewska, M. and Lynch, J.J. (1987) The learning behaviour of sheep when introduced to wheat. II. Social transmission of wheat feeding and the role of the senses. *Applied Animal Behaviour Science* 18, 163–172.

Chesler, P. (1969) Maternal influence in learning by observation in kittens. *Science* 166, 901–903.

Clarke, J.V., Nicol, C.J., Jones, R. and McGreevy, P.D. (1996) Effects of observational learning on food selection in horses. *Applied Animal Behaviour Science* 50, 177–184.

Cooper, J.J. and Nicol, C.J. (1994) Neighbour effects on the development of locomotor stereotypies in bank voles, *Clethrionomys glareolus*. *Animal Behaviour* 47, 214–216.

Dawkins, M.S. (1995) How do hens view other hens? The use of lateral and binocular visual fields in social recognition. *Behaviour* 132, 591–606.

Dawkins, M.S. (1996) Distance and social recognition in hens: implications for the use of photographs as social stimuli. *Behaviour* 133, 663–680.

Dawkins, M.S. and Woodington, A. (1997) Distance and the presentation of visual stimuli to birds. *Animal Behaviour* 54, 1019–1025.

Duncan, I.J.H. (1970) Frustration in the fowl. In: Freeman, B.M. and Gordon, R.F. (eds) *Aspects of Poultry Behaviour.* Oliver & Boyd, Edinburgh, pp. 15–31.

Evans, C.S. and Evans, L. (1999) Chicken food calls are functionally referential. *Animal Behaviour* 58, 307–319.

Evans, C.S. and Marler, P. (1992) Female appearance as a factor in the responsiveness of male chickens during anti-predator behaviour and courtship. *Animal Behaviour* 43, 137–145.

Evans, C.S., Evans, L. and Marler, P. (1993a) On the meaning of alarm calls: functional reference in an avian vocal system. *Animal Behaviour* 46, 23–38.

Evans, C.S., Macedonia, J.M. and Marler, P. (1993b) Effects of apparent size and speed on the response of chickens, *Gallus gallus*, to computer-generated simulations of aerial predators. *Animal Behaviour* 46, 1–11.

Franchina, J.J., Dyer, A.B., Zaccaro, S.J. and Schulman, A.H. (1986) Socially facilitated drinking behavior in chicks (*Gallus domesticus*): relative effects of drive and stimulus mechanisms. *Animal Learning and Behavior* 14, 218–222.

Franklin, J.R. and Hutson, G.D. (1982) Experiments on attracting sheep to move along a laneway: III. Visual stimuli. *Applied Animal Ethology* 8, 457–478.

Galef, B.G. Jr (1993) Functions of social learning about food: a causal analysis of effects of diet novelty on preference transmission. *Animal Behaviour* 46, 257–265.

Galef, B.G. Jr and Whiskin, E.E. (1998) Determinants of the longevity of socially learned food preferences of Norway rats. *Animal Behaviour* 55, 967–975.

Galef, B.G. Jr and White, D.J. (1997) Socially acquired information reduces Norway rats' latencies to find food. *Animal Behaviour* 54, 705–714.

Galef, B.G. Jr and White, D.J. (1998) Mate-choice copying in Japanese quail, *Coturnix coturnix japonica. Animal Behaviour* 55, 545–552.

Galef, B.G. Jr, Whiskin, E.E. and Bielavska, E. (1997) Interaction with demonstrator rats changes observer rats' affective responses to flavors. *Journal of Comparative Psychology* 111, 393–398.

Grandin, T. (1998) The feasibility of using vocalization scoring as an indicator of poor welfare during cattle slaughter. *Applied Animal Behaviour Science* 56, 121–128.

Greenlees, B. (1993) Effects of enriching the acoustical environment during incubation on hatching and post-hatch chick responses. Unpublished Master's thesis, University of Guelph, Canada.

Grigor, P.N., Hughes, B.O. and Appleby, M.C. (1995) Social inhibition of movement in domestic hens. *Animal Behaviour* 49, 1381–1388.

Guhl, A.M. (1949) Heterosexual dominance and mating behaviour in chickens. *Behaviour* 2, 106–120.

Gyger, M., Marler, P. and Pickert, R. (1987) Semantics of an avian alarm call system: the male domestic fowl, *Gallus gallus domesticus*. *Behaviour* 102, 13–40.

Gyger, M., Karakashian, S.J., Duffy, A.M. Jr and Marler, P. (1988) Alarm signals in birds; the role of testosterone. *Hormones and Behavior* 23, 305–314.

Heth, G., Todrank, J. and Johnston, R.E. (1998) Kin recognition in golden hamsters: evidence for phenotype matching. *Animal Behaviour* 56, 409–417.

Hogue, M.E., Beaugrand, J.P. and Lague, P.C. (1996) Coherent use of information by hens observing their former dominant defeating or being defeated by a stranger. *Behavioural Processes* 38, 241–252.

Hughes, B.O., Carmichael, N.L., Walker, A.W. and Grigor, P.N. (1997) Low incidence of aggression in large flocks of laying hens. *Applied Animal Behaviour Science* 54, 215–234.

Hutson, G.D., Argent, M.F., Dickenson, L.G. and Luxford, B.G. (1992) Influence of parity and time since parturition on responsiveness of sows to a piglet distress call. *Applied Animal Behaviour Science* 34, 303–313.

Johnston, A.N.B., Burne, T.H.J. and Rose, S.P.R. (1998) Observational learning in day-old chicks using a one-trial passive avoidance learning paradigm. *Animal Behaviour* 56, 1347–1353.

Karakashian, S.J., Gyger, M. and Marler, P. (1988) Audience effects on alarm calling in chickens (*Gallus gallus*). *Journal of Comparative Psychology* 102, 129–135.

Keeling, L.J. and Hurnik, J.F. (1996a) Social facilitation acts more on the appetitive than the consummatory phase of feeding behaviour in domestic fowl. *Animal Behaviour* 52, 11–15.

Keeling, L.J. and Hurnik, J.F. (1996b) Social facilitation and synchronization of eating between familiar and unfamiliar newly weaned piglets. *Acta Agriculturae Scandinavica, Section A, Animal Science* 46, 54–60.

Kendrick, K.M. (1998) Intelligent perception. *Applied Animal Behaviour Science* 57, 213–231.

Kendrick, K.M., Atkins, K., Hinton, M.R., Heavens, P. and Keverne, E.B. (1996) Are faces special for sheep? Evidence from facial and symbolic discrimination learning tests showing effects of inversion and social familiarity. *Behavioural Processes* 38, 19–36.

Kent, J.P. (1989) On the acoustic basis of recognition of the mother hen by the chick in the domestic fowl (*Gallus gallus*). *Behaviour* 108, 1–9.

Kent, J.P. (1993) The chick's preference for certain features of the maternal cluck vocalization in the domestic fowl (*Gallus gallus*). *Behaviour* 125, 177–187.

Kruijt, J.P. (1964) Ontogeny of social behaviour in Burmese Red Junglefowl (*Gallus gallus spadiceus*). *Behaviour Supplement* XII, 201 pp.

Lefebvre, L., Palameta, B. and Hatch, K.K. (1998) Is group living associated with social learning? A comparative test of a gregarious and a territorial columbid. *Behaviour* 133, 241–261.

Lill, A. (1968a) An analysis of sexual isolation in the domestic fowl: I. The basis of homgamy in males. *Behaviour* 30, 107–126.

Lill, A (1968b) An analysis of sexual isolation in the domestic fowl: II. The basis of homogamy in females. *Behaviour* 30, 127–145.

Lill, A. (1968c) Some observations of the isolating potential of aggressive behaviour in the domestic fowl. *Behaviour* 31, 127–143.

Lindberg, A.C. and Nicol, C.J. (1996) Effects of social and environmental familiarity on group preferences and spacing behaviour in laying hens. *Applied Animal Behaviour Science* 49, 109–123.

Lindberg, A.C., Kelland, A. and Nicol, C.J. (1998) Effects of observational learning on acquisition of an operant response in horses. *Applied Animal Behaviour Science* 61, 187–199.

Marler, P., Dufty, A. and Pickert, R. (1986a) Vocal communication in the domestic chicken: I. Does a sender communicate information about the quality of a food referent to a receiver? *Animal Behaviour* 34, 188–193.

Marler, P., Dufty, A. and Pickert, R. (1986b) Vocal communication in the domestic chicken: II. Is a sender sensitive to the presence and nature of a receiver? *Animal Behaviour* 34, 194–198.

McGreevy, P.D., French, N.P. and Nicol, C.J. (1995) The prevalence of abnormal behaviours in dressage, eventing and endurance horses in relation to stabling. *Veterinary Record* 137, 36–37.

McQuoid, L.M. and Galef, B.G. Jr (1993) Social stimuli influencing feeding behaviour of Burmese fowl: a video analysis. *Animal Behaviour* 46, 13–22.

Mirza, S.N. and Provenza, F.D. (1990) Preference of the mother affects selection and avoidance of foods by lambs differing in age. *Applied Animal Behaviour Science* 28, 255–263.

Moffatt, C.A. and Hogan, J.A. (1992) Ontogeny of chick responses to maternal food calls in the Burmese Red Junglefowl (*Gallus gallus spadiceus*). *Journal of Comparative Psychology* 106, 92–96.

Nicol, C.J. (1995) The social transmission of information and behaviour. *Applied Animal Behaviour Science* 44, 79–98.

Nicol, C.J. and Pope, S.J. (1993) Food deprivation during observation reduces social learning in hens. *Animal Behaviour* 45, 193–196.

Nicol, C.J. and Pope, S.J. (1994a) Social learning in sibling pigs. *Applied Animal Behaviour Science* 40, 31–43.

Nicol, C.J. and Pope, S.J. (1994b) Social learning in small flocks of laying hens. *Animal Behaviour* 47, 1289–1296.

Nicol, C.J. and Pope, S.J. (1996) The maternal feeding display of domestic hens is sensitive to perceived chick error. *Animal Behaviour* 52, 767–774

Nicol, C.J. and Pope, S.J. (1998) The effects of demonstrator social status and prior foraging success on social learning in laying hens. *Animal Behaviour* 57, 163–171.

Nolte, D.L. and Provenza, F.D. (1991) Food preferences in lambs after exposure to flavors in milk. *Applied Animal Behaviour Science* 32, 381–389.

Panning, L.M. (1998) The effects of maternal presence on early feeding and resting behavior in turkey poults with special reference to starve-out. Unpublished Master's thesis, University of Guelph, Canada.

Piggins, D. and Phillips, C.J.C. (1998) Awareness in domesticated animals – concepts and definitions. *Applied Animal Behaviour Science* 57, 181–200.

Piller, C.A.K., Stookey, J.M. and Watts, J.M. (1999) Effects of mirror-image exposure on heart rate and movement of isolated heifers. *Applied Animal Behaviour Science* 63, 93–102.

Provenza, F.D. and Burritt, E.A. (1991) Socially induced diet preferences ameliorate conditioned food aversion in lambs. *Applied Animal Behaviour Science* 31, 229–236.

Ralphs, M.H. and Olsen, J.D. (1990) Adverse influence of social facilitation and learning context in training cattle to avoid eating larkspur. *Journal of Animal Science* 68, 1944–1952.

Ryan, C.M.E. (1982) Concept formation and individual recognition in the domestic chicken. *Behaviour Analysis Letters* 2, 213–220.

Savory, C.J. (1982) Effects of broiler companions on early turkey performance. *British Poultry Science* 23, 81–88.

Schaal, B., Orgeur, P. and Arnould, C. (1995) Olfactory preferences in newborn lambs: possible influence of prenatal experience. *Behaviour* 132, 351–365.

Schwartzkopf-Genswein, K.S., Stookey, J.M., Crowe, T.J. and Genswein, B.M.A. (1998) Comparison of image analysis, exertion force, and behavior measurements for use in the assessment of beef cattle responses to hot-iron and freeze branding. *Journal of Animal Science* 76, 972.

Scott, C.B., Banner, R.E. and Provenza, F.D. (1996) Observations of sheep foraging in familiar and unfamiliar environments: familiarity with the environment influences diet selection. *Applied Animal Behaviour Science* 49, 165–171.

Shillito-Walser, E. (1980) Maternal recognition and breed identity in lambs living in a mixed flock of Jacob, Clun Forest and Dalesbred sheep. *Applied Animal Ethology* 6, 221–231.

Slabbert, J.M. and Rasa, O.A.E. (1997) Observational learning of an acquired maternal behaviour pattern by working dog pups: an alternative training method? *Applied Animal Behaviour Science* 53, 309–316.

Stolba, A. and Wood-Gush, D.G.M. (1989) The behaviour of pigs in a semi-natural environment. *Animal Production* 48, 419–425.

Stookey, J.M. and Gonyou, H.W. (1998) Recognition in swine: recognition through familiarity or genetic relatedness? *Applied Animal Behaviour Science* 55, 291–305.

Templeton, J.J. (1998) Learning from others' mistakes: a paradox revisited. *Animal Behaviour* 55, 79–85.

Thorhallsdottir, A.G., Provenza, F.D. and Balph, D.F. (1987) Food aversion learning in lambs with or without a mother: discrimination, novelty and persistence. *Applied Animal Behaviour Science* 18, 327–340.

Thorhallsdottir, A.G., Provenza, F.D. and Balph, D.F. (1990a) Ability of lambs to learn about novel foods while observing or participating with social models. *Applied Animal Behaviour Science* 25, 25–33.

Thorhallsdottir, A.G., Provenza, F.D. and Balph, D.F. (1990b) The role of the mother in the intake of harmful foods by lambs. Applied *Animal Behaviour Science* 25, 35–44.

Van Kampen, H.S. (1994) Courtship food-calling in Burmese Red Junglefowl: I. The causation of female approach. *Behaviour* 131, 261–275.

Van Kampen, H.S. (1997) Courtship food-calling in Burmese red junglefowl: II. Sexual conditioning and the role of the female. *Behaviour* 134, 775–787.

Veissier, I. (1993) Observational learning in cattle. *Applied Animal Behaviour Science* 35, 235–243.

Veissier, I. and Le Neindre, P. (1992) Reactivity of Aubrac heifers exposed to a novel environment alone or in groups of four. *Applied Animal Behaviour Science* 33, 11–15.

Veissier, I. and Stefanova, I. (1993) Learning to suckle from an artificial teat within groups of lambs: influence of a knowledgeable partner. *Behavioural Processes* 30, 75–82.

Veissier, I., Boissy, A., Nowak, R., Orgeur, P. and Poindron, P. (1998) Ontogeny of social awareness in domestic herbivores. *Applied Animal Behaviour Science* 57, 233–245.

von Borell, E. and Ladewig, J. (1992) Relationship between behaviour and adrenocortical response pattern in domestic pigs. *Applied Animal Behaviour Science* 34, 195–206.

Weary, D.M. and Fraser, D. (1995) Calling by domestic piglets: reliable signals of need? *Animal Behaviour* 50, 1047–1055.

Weary, D.M., Lawson, G. and Fraser, D. (1996) Sows show stronger responses to isolation calls of piglets associated with greater levels of piglet need. *Animal Behaviour* 52, 1247–1253.

Weary, D.M., Ross, S. and Fraser, D. (1997) Vocalizations by isolated piglets: a reliable indicator of piglets need directed towards the sow. *Applied Animal Behaviour Science* 53, 249–257.

Weary, D.M., Braithwaite, L.A. and Fraser, D. (1998) Vocal response to pain in piglets. *Applied Animal Behaviour Science* 56, 161–172.

Wemelsfelder, F. (1997) The scientific validity of subjective concepts in models of animal welfare. *Applied Animal Behaviour Science* 53, 75–88.

Zimmerman, P.H. and Koene, P. (1998) The effect of frustrative nonreward on vocalisations and behaviour in the laying hen, *Gallus gallus domesticus*. *Behavioural Processes* 44, 73–79.

Index

LIVERPOOL
JOHN MOORES UNIVERSITY
AVRIL ROBARTS LRC
TEL. 0151 231 4022